OZONATION IN ORGANIC CHEMISTRY

Volume I

Olefinic Compounds

This is Volume 39 of
ORGANIC CHEMISTRY
A series of monographs
Editors: ALFRED T. BLOMQUIST and HARRY H. WASSERMAN

OZONATION IN ORGANIC CHEMISTRY

Volume I

Olefinic Compounds

Philip S. Bailey

Department of Chemistry
The University of Texas at Austin
Austin, Texas

Editorial Advisor:
WALTER TRAHANOVSKY
Iowa State University of Science and Technology
Ames, Iowa

ACADEMIC PRESS New York San Francisco London 1978
A Subsidiary of Harcourt Brace Jovanovich, Publishers

ACADEMIC PRESS, INC.
111 Fifth Avenue, New York, New York 10003

United Kingdom Edition published by
ACADEMIC PRESS, INC. (LONDON) LTD.
24/28 Oval Road, London NW1 7DX

Library of Congress Cataloging in Publication Data

Bailey, Philip S.
Ozonation in Organic Chemistry

(Olefinic Compounds ; v. 1) (Organic Chemistry ;)
Includes bibliographical references and idex.
1. Olefins. 2. Ozonization. 3. Ozone.
I. Title. II. Series: Organic chemistry, a series of monographs ; .
QD281.095B35 vol. 1 [QD305.H7] 546'.721'1s
ISBN 0-12-073101-0

PRINTED IN THE UNITED STATES OF AMERICA

This book is dedicated to the memory of the author's esteemed friend, colleague, and mentor, the late Professor Rudolf Criegee of the University of Karlsruhe, West Germany.

His dedication to chemistry, especially ozone chemistry, was so intense that even while weakened with emphysema (perhaps caused by breathing ozone fumes in his makeshift laboratory after World War II) he worked against time in his laboratory until the work was complete which helped remove all remaining doubt concerning the validity of the carbonyl oxide mechanism of ozonolysis, known as the Criegee mechanism. Less than a year later he died.

Professor Criegee will always live in the memories of those who knew him as the epitome of a true gentleman, a dedicated scientist and teacher, and a genuine friend.

Contents

CHAPTER IV

Ozonolysis of Olefins: Initial Ozone Attack and Adduct

CHAPTER V

Ozonolysis of Olefins: The Peroxidic Products

CHAPTER VI

Ozonolysis of Olefins: Routes to Peroxidic Products

CHAPTER VII

Ozonolysis of Olefins: Competitions in Peroxidic Product Formation

CHAPTER VIII

Ozonolysis of Olefins: Routes from Peroxidic to Nonperoxidic Products

CHAPTER IX

"Anomalous" Ozonolysis of Olefins

CHAPTER X

"Special" Liquid-Phase Ozonolyses

CHAPTER XI

Electrophilic Ozone Attack on Olefins. Epoxides and Other "Partial Cleavage" Products

CHAPTER XII

Gas-Phase Ozonation of Olefins

EPILOGUE

Overview and New Developments 243

Preface

A revival of interest in ozone-organic chemistry began approximately 25 years ago when the genius of the late Professor R. Criegee of the University of Karlsruhe, West Germany, transformed the classical ozonolysis reaction from an art into a science and inspired not only studies in regard to the ozonolysis of olefins, but also researches which have established the fact that ozone is a very versatile oxidizing agent which will react, with varying ease, with most organic species.

More recently, the interest in ozone-organic chemistry has centered around its role in air pollution, in water and wastewater purification, and in the potential danger that air pollutants are decreasing the ozone layer in the stratosphere and thereby endangering life on earth. Although much of the utility of ozone is being investigated from the engineering viewpoint, it is my strong belief that a thorough understanding of the theories, or mechanisms, of ozone reactions is essential for the development of efficient and economical processes.

Also, ozone is being used more and more in the chemical and pharmaceutical industries as an oxidizing agent in syntheses of useful products.

It is the purpose, therefore, of this two-volume treatise to make available in one place present-day knowledge concerning the reactions of ozone with organic substances and the mechanisms of these reactions. In this way all investigators will have easy access to this information and can bring themselves up to date in the shortest possible time. It is further hoped that these monographs will excite interest in exploring further the frontiers of ozone-organic chemistry.

This volume (Vol. I) deals with ozone itself, a historical background of its reactions with organic substances, a detailed description of the step-by-step mechanism of the classical ozonolysis reaction of olefins and how it evolved, and a thorough discussion and review of the reactions which compete with ozonolysis, i.e., epoxidation and/or other "partial cleavage" reactions. Both liquid and gas phase reactions are included. The literature is reviewed through mid-1976 and more than 700 references are cited.

A future volume (Vol. II) will review the literature concerning the reactions of ozone with acetylenic compounds, aromatics of various types, heterocyclic compounds, various nucleophiles such as amines, sulfides, sulfoxides, selenides, phosphines, and phosphites, carbon–nitrogen double bonds, carbon–hydrogen bonds, silicon–hydrogen bonds, silicon–silicon bonds, and organometallics. It will also bring the treatise up to date in regard to interim developments in the olefin area.

PHILIP S. BAILEY

Acknowledgments

It gives me pleasure to thank my colleague and friendly competitor in ozone-organic chemistry, Professor Robert W. Murray, University of Missouri, St. Louis, for reading most of the manuscript of this volume and providing many helpful suggestions and constructive criticisms. I am also grateful to Professor Walter S. Trahanovsky, Iowa State University, for many useful suggestions. In addition, I wish to thank various past students and post-doctoral fellows, especially those who were with me at the inception of this treatise some fourteen years ago and who helped find almost every mention of ozonation of organic compounds in the literature up to that time. Especially helpful in this respect was Professor Per Kolsaker of the University of Oslo, Norway. Finally, I am appreciative of my wife, Jean, for her patience, encouragement, and understanding during the many hours I spent in my study during the final two years of writing this monograph.

Tentative Contents of Volume II

ORGANIC CHEMISTRY

A SERIES OF MONOGRAPHS

EDITOR

HARRY H. WASSERMAN
Department of Chemistry
Yale University
New Haven, Connecticut

1. Wolfgang Kirmse. CARBENE CHEMISTRY, 1964; 2nd Edition, 1971
2. Brandes H. Smith. BRIDGED AROMATIC COMPOUNDS, 1964
3. Michael Hanack. CONFORMATION THEORY, 1965
4. Donald J. Cram. FUNDAMENTALS OF CARBANION CHEMISTRY, 1965
5. Kenneth B. Wiberg (Editor). OXIDATION IN ORGANIC CHEMISTRY, PART A, 1965; Walter S. Trahanovsky (Editor). OXIDATION IN ORGANIC CHEMISTRY, PART B, 1973; PART C, 1978
6. R. F. Hudson. STRUCTURE AND MECHANISM IN ORGANO-PHOSPHORUS CHEMISTRY, 1965
7. A. William Johnson. YLID CHEMISTRY, 1966
8. Jan Hamer (Editor). 1,4-CYCLOADDITION REACTIONS, 1967
9. Henri Ulrich. CYCLOADDITION REACTIONS OF HETEROCUMULENES, 1967
10. M. P. Cava and M. J. Mitchell. CYCLOBUTADIENE AND RELATED COMPOUNDS, 1967
11. Reinhard W. Hoffmann. DEHYDROBENZENE AND CYCLOALKYNES, 1967
12. Stanley R. Sandler and Wolf Karo. ORGANIC FUNCTIONAL GROUP PREPARATIONS, VOLUME I, 1968; VOLUME II, 1971; VOLUME III, 1972
13. Robert J. Cotter and Markus Matzner. RING-FORMING POLYMERIZATIONS, PART A, 1969; PART B, 1; B, 2, 1972
14. R. H. DeWolfe, CARBOXYLIC ORTHO ACID DERIVATIVES, 1970
15. R. Foster. ORGANIC CHARGE-TRANSFER COMPLEXES, 1969
16. James P. Snyder (Editor). NONBENZENOID AROMATICS, VOLUME I, 1969; VOLUME II, 1971
17. C. H. Rochester. ACIDITY FUNCTIONS, 1970
18. Richard J. Sundberg. THE CHEMISTRY OF INDOLES, 1970
19. A. R. Katritzky and J. M. Lagowski. CHEMISTRY OF THE HETEROCYCLIC *N*-OXIDES, 1970
20. Ivar Ugi (Editor). ISONITRILE CHEMISTRY, 1971

ORGANIC CHEMISTRY

21. G. Chiurdoglu (Editor). Conformational Analysis, 1971

22. Gottfried Schill. Catenanes, Rotaxanes, and Knots, 1971

23. M. Liler. Reaciton Mechanisms in Sulphuric Acid and Other Strong Acid Solutions, 1971

24. J. B. Stothers. Carbon-13 NMR Spectroscopy, 1972

25. Maurice Shamma. The Isoquinoline Alkaloids: Chemistry and Pharmacology, 1972

26. Samuel P. McManus (Editor). Organic Reactive Intermediates, 1973

27. H. C. Van der Plas. Ring Ttransformations of Heterocycles, Volumes 1 and 2, 1973

28. Paul N. Rylander. Organic Syntheses with Noble Catalysts, 1973

29. Stanley R. Sandler and Wolf Karo. Polymer Syntheses, Volume I, 1974; Volume II, 1977

30. Robert T. Blickenstaff, Anil C. Ghosh, and Gordon C. Wolf. Total Synthesis of Steroids, 1974

31. Barry M. Trost and Lawrence S. Melvin, Jr. Sulfur Ylides: Emerging Synthetic Intermediates, 1975

32. Sidney D. Ross, Manuel Finkelstein, and Eric J. Rudd. Anodic Oxidation, 1975

33. Howard Alper (Editor). Transition Metal Organometallics in Organic Synthesis, Volume I, 1976; Volume II, 1978

34. R. A. Jones and G. P. Bean. The Chemistry of Pyrroles, 1976

35. Alan P. Marchand and Roland E. Lehr (Editors). Pericyclic Reactions, Volume I, 1977; Volume II, 1977

36. Pierre Crabbé (Editor). Prostaglandin Research, 1977

37. Eric Block. Reactions of Organosulfur Compounds, 1978

38. Arthur Greenburg and Joel Liebman, Strained Organic Molecules, 1978

39. Philip S. Bailey. Ozonation in Organic Chemistry, Vol. I, 1978

40. Arthur Greenburg and Joel F. Liebman. Strained Organic Molecules, 1978

41. Harry H. Wasserman and Robert W. Murray (Editors). Singlet Oxygen, 1978

CHAPTER I

Introduction

I. Importance of Ozone Chemistry

Beginning with the time that the Earth was first able to support life, ozone has existed. The odor of ozone in the air during a thunder and lightning storm has been observed since ancient times and was described as "sulfurous" by Homer in both the Odyssey and the Iliad (*1*). Homer also alluded to a "freshness" after a storm. Man has since copied nature and uses electrical discharge as his principal method of generating ozone. The first commercial ozone-generating machine (ozonator) was manufactured by von Siemens in 1857 (*2*).

Ozone is a highly reactive allotrope of ordinary atmospheric oxygen in which the molecule is composed of three, rather than two, atoms of oxygen. Like so much else in the world ozone is both good and bad, or beneficial and harmful. On the positive side, it is generated photochemically in the stratosphere and is believed to protect life on Earth from overexposure to ultraviolet and other high energy radiation. A present-day concern is that chlorofluoromethanes (from aerosols) may greatly decrease or destroy the ozone layer, thereby seriously affecting climatic conditions and life itself on Earth (*3*). The same fear has been expressed in regard to the exhaust from supersonic airplanes (*4*), products from nuclear explosions (*5*), and other pollutants (*6*). Second, ozone has proved to be a highly useful and powerful oxidizing agent in the chemical industry, the cost of generating it being the only deterrent to its wider use. At the present time the largest user of ozone in this manner is still Emery Industries in the manufacture of azelaic and pelargonic acids by the ozonolysis of oleic acid (*7*). More important, however, is the growing use of ozone in water and wastewater purification. Ozone has been used in this way in Europe for many years, but only now is the United States beginning to realize the potential of ozone in disinfection, in sewage and industrial waste treatment, and in odor, color, and taste control (*2,8–11*).*

* It is ironic that as early as 1916 an authority on ozone wrote: "The purification of drinking water has ceased to be a problem. The solution as found in the treatment with ozone is an

On the minus side is the fact that ozone is highly toxic in concentrations greater than 0.1 ppm by volume (*2,13,14*). Being highly reactive toward all kinds of organic substances, it can, of course, react with protein and other biomolecules of plant and animal tissues. Of great concern today is the smog problem. Among the complex reactions involved is the generation of ozone by the photochemical decomposition of nitrogen dioxide; ozone then reacts with various hydrocarbons and other pollutants—from automobile exhaust, industrial plant emissions, and incineration—to give materials harmful to plant and animal life (*15,16*).

For other properties and points of interest concerning ozone, the reader is referred to other reviews (*2,13,14*).

II. Historical

The history of ozone–organic chemistry divides itself naturally into three eras, each of which is dominated by the genius of a particular investigator.

A. DISCOVERY: THE SCHÖNBEIN PERIOD, 1840–1902

Schönbein* (*17*), in 1840, is generally credited with being the discoverer of ozone as a chemical substance. He gave it the name "ozone," derived from the Greek word *ozein*, meaning "to smell" (*18,19*). Schönbein (*20*) also appears to be the first to study the reactions of ozone with organic, and also inorganic, materials, although in most cases products were not identified. An exception is his experiment with ethylene (*21*), which appears to be the first ozonolysis ever performed, although the usual reference (*22*) to the first ozonolysis, cited by previous reviewers (*23–27*), has nothing to do with the subject. Other noteworthy ozonation studies during this period, in which some product identification was made, were by von Gorup-Besanez (*28*), Houzeau (*29*), Berthelot (*30*), Mailfert (*31*), Legler (*32*), Besson (*33*), Otto (*34*), Swarts (*35*), and Trillat (*36*). Additional details concerning the discoveries of this period can be found in other reviews (*18,19,23–26,37–44*).

---

accomplished fact, and the general application is only a matter of time—probably a long time, because it is not the difficulty of the matter itself that has to be solved, but the ignorance and prejudice of the public . . ." [Vosmaer (*12*), quotation reprinted by permission of D. Van Nostrand Co., copyright holder].

* Christian Friedrich Schönbein was born October 18, 1799 in Metzingen. He studied at Tübingen, Erlangen, and Paris, but received an honorary doctorate at Basle, after accepting a Professorship there. He is noted not only for his ozone studies, but also for the discovery of guncotton. He died August 29, 1868. For further details concerning his life and work, see Oesper (*27a*).

B. DEVELOPMENT: THE HARRIES PERIOD, 1903–1948

This important period in the history of ozone chemistry began with the first paper of Harries* (*45*) in 1903. Harries published nearly 100 papers on ozone chemistry during a period of 15 years (*46–47*). He not only established ozonolysis as one of the classical reactions of organic chemistry (see Chapter III, Section I, A) but also pioneered most of the known reactions of ozone with organic compounds and gave us our first insight into how these reactions occur. It is interesting to note that although Harries' interest in ozone chemistry apparently began with his interest in rubber, it reached fulfillment with his marriage in 1899 to the daughter of the founder of the von Siemens firm, which manufactured the first commercial ozonators (*48,48a*). His publications essentially ceased, however, when he left the University of Kiel in 1916 to join the von Siemens firm (*48,48a*).

Concurrently with Harries, Molinari (*49*) independently began a study of the ozonolysis of unsaturated fatty acids and other substances which showed promise of paralleling the work of Harries in importance. However, after a polemic with Harries, Molinari published very little more in the field (*50*).

Of great importance also during this period was the mechanism proposed by Staudinger (*51*) for the ozonolysis reaction (Chapter III, Section I, B). Others who made important contributions were Rieche (*52–54*), who established the structure of an ozonide as a 1,2,4-trioxolane, Pummerer (*55*), and Briner (*56,57*). During a 42-year period, Briner published approximately 150 papers on ozone chemistry (including nearly 70 on ozonolysis of olefins), making him the most prolific publisher in the field. Although he certainly made important contributions, many of the papers published before Criegee's mechanistic papers lost much of their significance because they assumed any peroxidic ozonolysis product, most of which were not purified, to be an ozonide. For further discussion of the contributions of this period, see earlier sources and reviews (*18,24–26,43,46,48,48a,53,54,58,59*).

C. UNDERSTANDING AND UTILIZATION: THE CRIEGEE PERIOD, 1949 TO PRESENT

The late Professor R. Criegee,† University of Karlsruhe, can be considered to be the father of modern ozone–organic chemistry. His work, beginning

* Carl Dietrich Harries was born August 5, 1866 in Luckenwald. He studied at Jena, Munich, and Berlin, receiving his doctorate at Berlin in 1892 under Hoffmann. After several years as an assistant to Emil Fischer and, later, as an associate professor at Berlin, he was called in 1904 to the University of Kiel as a full professor and director of the Organic Chemistry Institute. He is noted both for his pioneering ozone studies and for his work with rubber. He died November 3, 1923. For additional details see (*48,48a*).

† Rudolf Criegee was born May 23, 1902 in Düsseldorf. He studied at Tübingen, Greifswald, and Würzburg, receiving his doctorate at the latter University under Otto Dimroth in 1925.

in 1949 (*60*) and culminating with his first major paper on the mechanism of ozonolysis in 1953 (*61*), ushered in an era of great interest in ozone–organic chemistry in which most of our understanding of the mechanisms of ozone attack has developed and is developing. Accompanying this has been a corresponding increase in practical applications. The Criegee mechanism of ozonolysis is outlined in Chapter III, Section I, C).

III. Scope and Terminology

This treatise will deal largely with the Criegee era of ozone–organic chemistry and will build upon the two preceding general reviews in English (*26,62*). There are also other, smaller, reviews on some of the topics covered by this review, which will be mentioned as needed. This treatise, however, will be comprehensive, making an effort not only to bring everything up to date in one place, but also to show the development of present-day thought.

There are a number of similar-sounding terms in ozone chemistry, often used imprecisely and with variable meanings. In this book the following usage will be employed. An *ozonator* is an apparatus for generating ozone. The terms *ozonization* and *ozonation* are synonymous and refer to the action of ozone on any substance in a broad sense. The term *ozonolysis* is a much more specific and narrow term, used solely to designate the cleavage of bonds composed of sp^2 or sp carbon atoms.

The term *ozonide* has often been very carelessly used. To inorganic chemists it means an ionic compound $M^+O_3^-$ in which M^+ is a metal cation and O_3^- is an anion-radical having one more electron than ozone itself. The term has also been used for ozone-phosphite, etc., adducts (to be discussed in Vol. Two). In this treatise such substances wiil be called *ozonates*. To most organic chemists, ozonides are certain specific peroxidic products in the ozonolysis reaction. Too often, however, the term has been used for any peroxidic ozonolysis product. In this treatise, when the word *ozonide*

He was an Assistant at Würzburg and Marburg before going to Karlsruhe as an Associate Professor in 1937. The period 1939–1943 was spent in the German army, part of the time as a private on the Russian front. He returned to teaching and research in a bombed and burned-out chemistry building and, in spite of these difficulties, was promoted to Professor and Director of the Institute for Organic Chemistry in 1947. From 1955 to 1971 he was Editor in charge of content for *Chemisches Berichte*. He became emeritus professor of chemistry in 1968. He died November 7, 1975. He is considered to have been one of the most distinguished organic chemists of our time. Although only his classical studies in ozone–organic chemistry are discussed in this book, Professor Criegee also made outstanding contributions to the fields of organic peroxides, osmium tetroxide and lead tetraacetate oxidations, α-glycol cleavage, and small ring compounds. For additional information see Maier (*59a*).

is used alone, it will refer to a 1,2,4-trioxacyclopentane (or 1,2,4-trioxolane), which is a stable peroxidic ozonolysis product. More specifically, ethylene ozonide would be 1,2,4-trioxacyclopentane and propylene ozonide would be 3-methyl-1,2.4-trioxacyclopentane. A "*primary ozonide*" is the initial adduct formed between ozone and a carbon–carbon double bond; it is a 1,2,3-trioxacyclopentane (or 1,2,3-trioxolane) (Chapter IV, Section I). The term "initial ozonide" also has been used for this intermediate. Various papers in the literature erroneously refer to the primary ozonide as a "molozonide." In actuality, a "molozonide" is the hypothetical, now disproved, four-membered ring adduct proposed by Staudinger (Chapter III, Section I, B). In this treatise, only the term "primary ozonide" will be used for the initial adduct. The term "isozonide" is sometimes found, especially in the older literature. This is the term Staudinger used for the isolable ozonide, a 1,2,4-trioxacyclopentane, to distinguish it from his "molozonide" (Chapter III, Section I, B). The terms "normal," or "parent," ozonide and "cross" ozonide will be defined in Chapter VI.

Unless specifically stated to the contrary, the ozonation procedure referred to in this book involves attack of gaseous ozone in an oxygen stream (2–6% ozone by volume) upon an organic substance in the liquid phase, pure or in solution. Gas-phase and solid-gas-phase ozonolyses will be discussed in Chapter XII.

CHAPTER **II**

The Ozone Molecule

Essential to an understanding of the reactions of ozone with organic compounds is a knowledge of the structure of the ozone molecule itself. The development of a satisfactory structural concept took a very long time, general agreement concerning bond angles and lengths coming only in 1952, more than 110 years after the discovery of ozone. Detailed discussions of the early structural ideas can be found elsewhere (*18,19,24,26,38,39,43,63,64*).

I. Resonance and Molecular Orbital Descriptions

The work of Hughes (*65*) and of Gordy and co-workers (*66*), in the period 1952–1956, with the microwave spectrum of ozone has established the facts that the ozone molecule in the ground state has an obtuse angle of 116°45 ± 35′ and an oxygen–oxygen bond length of 1.278 ± 0.003Å. The molecule has no appreciable paramagnetism (*66*), and its dipole moment is low, approximately 0.53 debye (*66*). Additional references to the physical properties of ozone can be found elsewhere (*13,67–70*).

On the basis of the properties just given, the structure of the ozone molecule can be described as a resonance hybrid of the four canonical forms represented by **1** (*66*). A simplified molecular orbital description is illustrated by structure **2** (*71*). Each oxygen atom is sp^2-hybridized, with two of the three sp^2 orbitals of the apex atom forming localized sigma molecular orbitals by combining with bonding sp^2 orbitals of the terminal oxygen atoms. All nonbonded sp^2 orbitals contain two electrons, and the entire molecule is engrossed by a π cloud containing four electrons, which arise from the "p" atomic orbital distribution shown in **2**. The explanation for the low dipole moment is that the large dipole resulting from the unshared electron pair in the third sp^2 orbital of the apex atom opposes the large moment predicted by the canonical forms **1** (*66*). More detailed molecular orbital, as well as other, discussions concerning the structures of both ground and excited state ozone molecules can be found elsewhere (*72–78*).

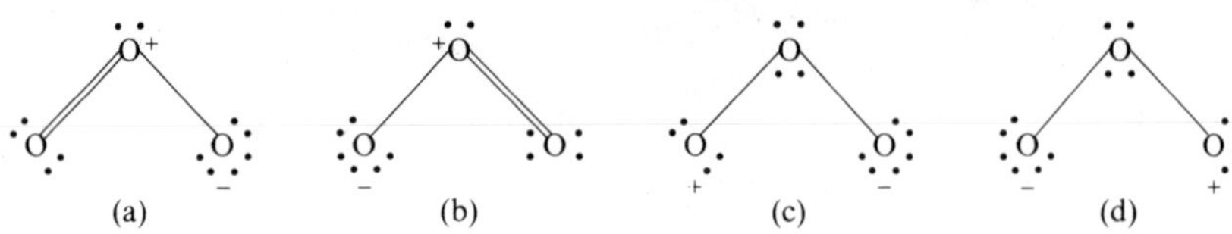

1

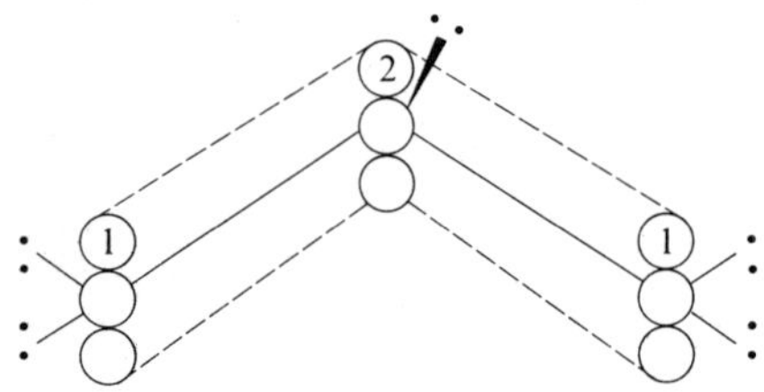

2

II. The Singlet Diradical Description

Goddard and co-workers (*76,78*) have made extensive *ab initio* calculations regarding the ozone molecule, using generalized valence-bond and configuration interaction methods. From these calculations they reached the conclusion that the ozone molecule is a singlet diradical in its ground state. Although it is difficult to see how such a structure can explain the electrophilic character of ozone (see Chapter IV, Section II, A; Chapter XI, Section III; Vol. Two), it is important to realize that a singlet diradical can have a continuous scale of dipolar character all the way up to a full dipolar ion because its electrons remain paired (*79*).

III. Types of Ozone Attack

On the basis of the resonance description of the ozone molecule given in Section I of this chapter, one could predict that ozone should be able to function as a 1,3-dipole (*80*), an electrophile (*81*), or a nucleophile. The reaction of ozone as a 1,3-dipole is discussed in Chapter IV, Section II; an additional example will be taken up in Vol. Two, under ozonation of carbon–hydrogen bonds. Purely electrophilic ozone attack is discussed in Chapter XI; other examples will be brought forth in Vol. Two, under ozonation of nucleophiles. The few examples of nucleophilic attack will be described in Vol. Two. It is also possible, in view of the finding of Goddard *et al.* (*76,78*), that ozone can react as a diradical; this will be discussed in Vol. Two under reactions of ozone with saturated molecules (ozonation of carbon–hydrogen bonds).

CHAPTER III

Ozonolysis of Olefins: Introduction

I. Historical Perspective

This section is concerned with the development of present-day theory regarding the best known of all reactions of ozone, the important ozonolysis reaction, used extensively both for structure proof (and other analytical purposes) and for synthesis. As indicated in Chapter I, Section II, the mechanistic theories of both Harries and of Staudinger played very important roles in the development of present-day theory and are reflected in the Criegee mechanism of ozonolysis. It is important, therefore, to summarize briefly the contributions of these workers to present-day thought.

A. CONTRIBUTIONS OF HARRIES

At the very outset of his research with ozone, Harries observed that certain olefinic compounds reacted with ozone in the absence of ionic solvents to give peroxidic oils (*45,82*), which he later called ozonides and to which he assigned structure **3** (*23,83*). Later he changed the structural assignment to **4** (*84–86*). Structure **4** will be recognized as the structure presently given to the highly unstable initial ozone–olefin adduct called a "primary ozonide" (see Chapter IV, Section I).

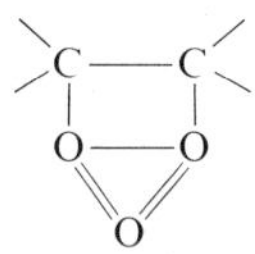

3

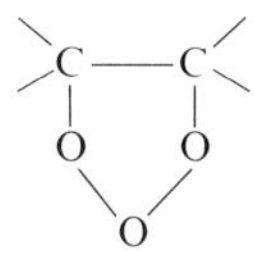

4

Harries proposed that ozonides decompose in two ways, one to give aldehydes and/or ketones and hydrogen peroxide [Eq. (1)] and the other to give an aldehyde or ketone and a "peroxide," **5** [Eq. (2)]. Modernized, this "peroxide" becomes the Criegee carbonyl oxide zwitterion (Section I, C of this chapter). Harries suggested that the "peroxide" could either rearrange to a carboxylic acid (if a hydrogen atom is attached to the carbon atom of the peroxy group) [Eq. (3)] or, in some cases, dimerize (*23,85,87,88*). In addition, he even explained the "abnormal" ozonolysis of camphene on the basis of rearrangement of the "peroxide" [Eq. (4)] (*85,89*). Harries also recognized that from an "ozonide" of an unsymmetrical olefinic compound the decomposition represented by Eq. (2) could occur in two ways, as indicated by Eq. (5) (*85,88*). The modern version of this is discussed in Chapters VI and VII.

$$\mathbf{4} \xrightarrow{H_2O} \;>C{=}O + H_2O_2 + O{=}C< \qquad (1)$$

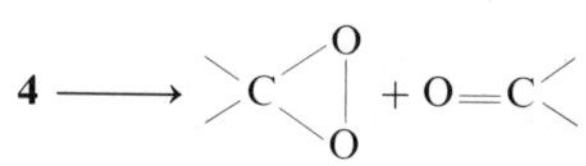

(2)

5

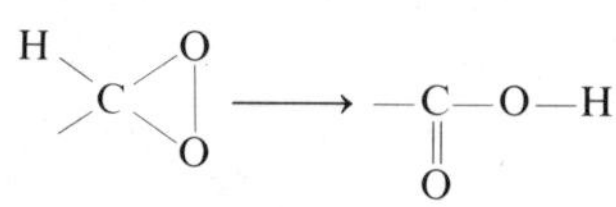

(3)

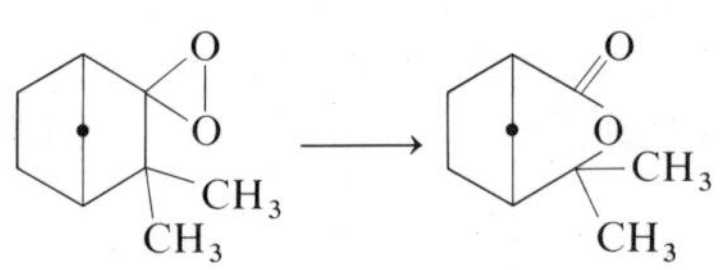

(4)

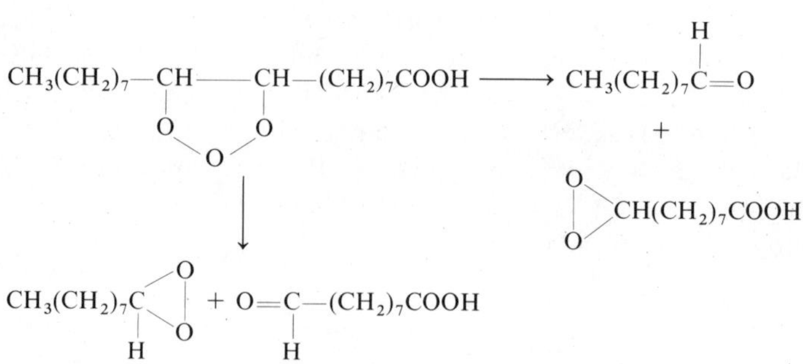

(5)

Even though some other theories of Harries turned out to have no basis in fact, too much cannot be said concerning Harries' astuteness in suggesting the above mechanisms, which are amazingly close to some of our present-day concepts.

B. THE STAUDINGER MECHANISM

In 1925, Staudinger (*51*) suggested a more intricate mechanism of ozonolysis than that of Harries, but one that represented a further step toward present-day concepts. The useful parts are outlined in Scheme 1; as with the theories of Harries, some parts are no longer meaningful and are omitted.

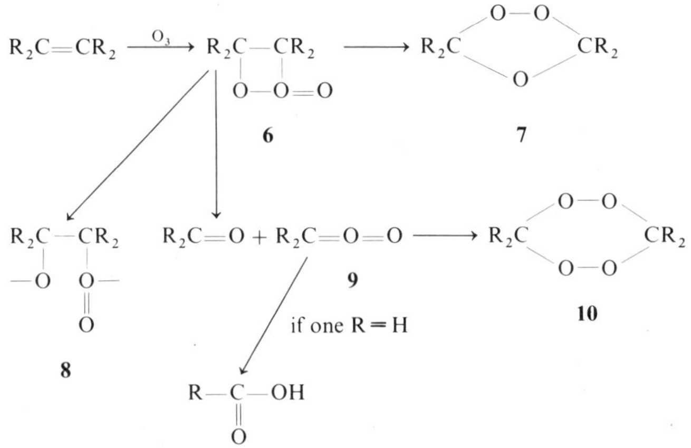

SCHEME 1. Staudinger mechanism.

Staudinger's proposal was that ozone reacts with olefins to form an unstable initial adduct with a four-membered ring (**6**), which he called a "mol-ozonide" by analogy to the "mol-oxide" thought then to be formed in the first step of autoxidation (*90*). He attributed the explosiveness of certain ozonolysis reaction mixtures, in contrast to pure ozonides (*84*), to the molozonide. Staudinger suggested several fates for the molozonide, the most important of which was rearrangement to a true ozonide (**7**), which he called an "isozonide." In support of structure **7** he pointed out that reduction of ozonides invariably gave aldehydes and ketones, etc., rather than glycols, which would be expected from Harries' ozonide structure **4**. Structure **7** was later established by Rieche (*52–54*). In addition, Staudinger (*51*) suggested that the molozonide could polymerize (**6** → **8**) and decompose to an aldehyde or ketone and **9**. Structure **9** (which is a step closer to the Criegee zwitterion) could then rearrange to a carboxylic acid (if one R is hydrogen) or dimerize (to **10**), as also proposed by Harries.

Between the proposals of Staudinger and the publication of the Criegee mechanism, several mechanisms for the rearrangement of the Staudinger molozonide were suggested (*91–95*). If such rearrangements do play a minor role in the mechanism of ozonolysis, they involve a 1,2,3-trioxolane rather than a molozonide (Chapter IV, Section I). It is interesting that the Staudinger molozonide (**6**) was recently resurrected in a proposal described as "A New and Unifying Concept" (*96–98*). This proposal, however, has since been shown to have no basis in fact (*99–102*) (see Chapter VI, Section V.)

C. THE CRIEGEE MECHANISM

An intensive and extensive study of the active oxygen–containing products of a wide variety of unsaturated substances in different types of solvents resulted in the development of the Criegee mechanism of ozonolysis during the period 1949–1957 (*60–62,103–110*). The mechanism is outlined in Scheme 2 (*62*).

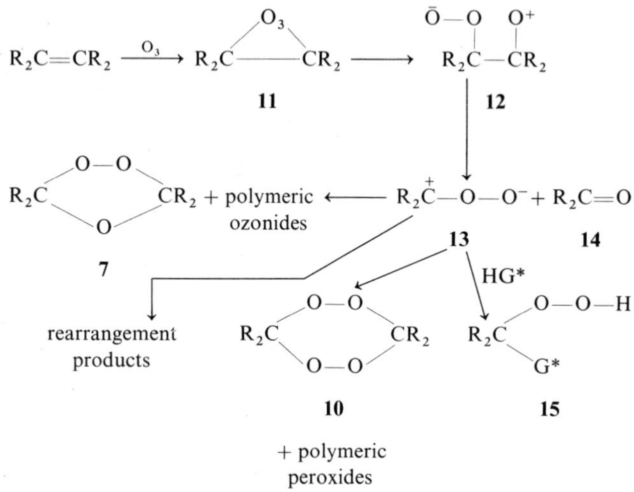

*G = OH, OR, O—C(=O)—R, etc.

SCHEME 2. Criegee mechanism.

The first intermediate, **11**, an ozone–olefin adduct the structure of which Criegee left in doubt, was assumed to be very unstable and to cleave to a zwitterion, **13**, and an aldehyde or ketone, **14**, presumably through **12**. Zwitterion (or carbonyl oxide) **13**, the key intermediate in the Criegee mechanism, is a modernization of minor intermediates proposed earlier by Harries and Staudinger (Section I, A and I, B of this chapter). The fate of **13**, according to the Criegee mechanism, depends partly on its source and partly on its environment. There were suggested at least four ways in which it can stabilize itself: (a) reaction with an aldehyde or ketone (**14**), usually an aldehyde, to give an ozonide (**7**) or polymeric ozonides (aprotic solvents); (b) dimerization and polymerization to give **10** and polymeric peroxides (favored by aprotic solvents and by **14** being a ketone); (c) reaction with a protic, nucleophilic solvent molecule to give **15**; (d) rearrangement to give "abnormal" ozonolysis products. Further discussion of the original Criegee mechanism and of the competitive fates of the key zwitterion intermediate can be found in other reviews (*62,110–116*).

II. Chronological Mechanistic Description

The Criegee mechanism set the proper course and perspective for thinking and investigation concerning the mechanism of ozonolysis. However, it left several questions unanswered. A complete understanding of the ozonolysis reaction must include the following: (a) the nature of the initial ozone attack; (b) the structure of the initial ozone–olefin adduct, if any; (c) the mechanistic routes from the initial adduct (or attack) to the peroxidic ozonolysis products; (d) the structures and properties of the peroxidic ozonolysis products and special conditions favoring the formation of each; (e) the mechanistic routes from peroxidic to nonperoxidic ozonolysis products. Each of these will be discussed in the following chapters—not necessarily in the chronological order just mentioned, however.

CHAPTER IV

Ozonolysis of Olefins: Initial Ozone Attack and Adduct

Although the Criegee mechanism for ozonolysis (Chapter III, Section I, C) proposes an initial adduct, as did the Staudinger mechanism (Chapter III, Section I, B), it gives no clue concerning its structure or how it is formed mechanistically. In attacking these problems it is best to start off with the questions: Is there an initial ozone adduct? If so, what is its structure?

I. The Primary Ozonide

There has been much discussion in the literature concerning the primary ozonide, and several structures have been proposed for it, including **4**, **6** (and its modernization **16**), **17**, **18**, and **19** (*26,55,60–62,91,94,108,110–112, 117–132*). Structures **17** and **18** are π and σ complexes, respectively, whereas **19** is a peroxy epoxide.

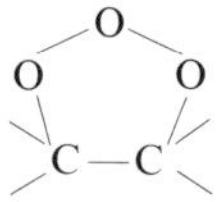

4

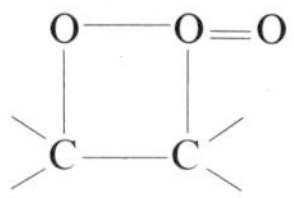

6

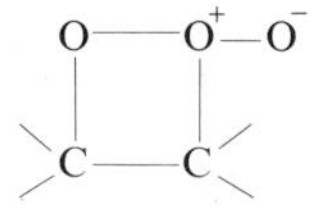

16

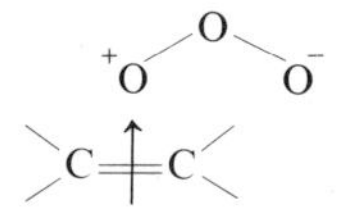

17

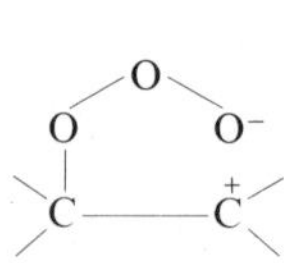

18

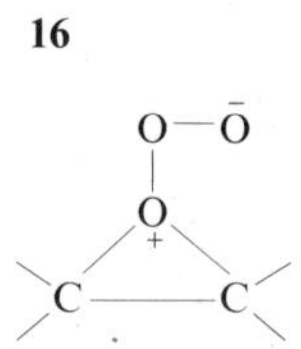

19

It was not until 1959 that the actual existence of a primary ozonide was finally established, when Criegee and Schröder (*133,134*) observed the precipitation of a crystalline material during ozonolysis of *trans*-1,2-di-*tert*-butylethylene (**20**) in pentane at $-75°C$. The substance decomposed with the evolution of 37–39 kcal/mole of heat when the temperature of the reaction mixture was allowed to rise to −60°C. Work-up of the reaction mixture gave only the normal ozonide, **24** (mostly trans, see Chapter VI, Section II), and polymeric peroxides. That the material was indeed the initial ozone adduct (**21**), with the σ bond of the double bond of **20** still intact, was established in two ways (*134*). First, when a cold ether solution of the substance was treated with methanol, the product was the methoxy hydroperoxide (**23**), obtained also by ozonolysis of **20** in methanol (Criegee mechanism, Chapter III, Section I, C, Scheme 2). If, however, the ether solution of the initial adduct was allowed to come to room temperature before the methanol was added, no methoxy hydroperoxide (**23**) was obtained, showing that the primary ozonide had already decomposed and that the Criegee zwitterion had reacted in other ways before the addition of methanol. Second, reduction of the cold ether solution of the primary ozonide with isopropyl magnesium bromide gave the racemic glycol **22**. Under these same conditions neither the warm ozonation mixture nor trimethylacetaldehyde gave the glycol. These reactions are outlined in Scheme 3. Similar studies carried out with *cis*-1,2-di-*tert*-butylethylene failed to give any evidence for a primary ozonide (*134*), showing either that the cis isomer does not yield a primary ozonide, or that the adduct is too unstable to be observed under the conditions employed.

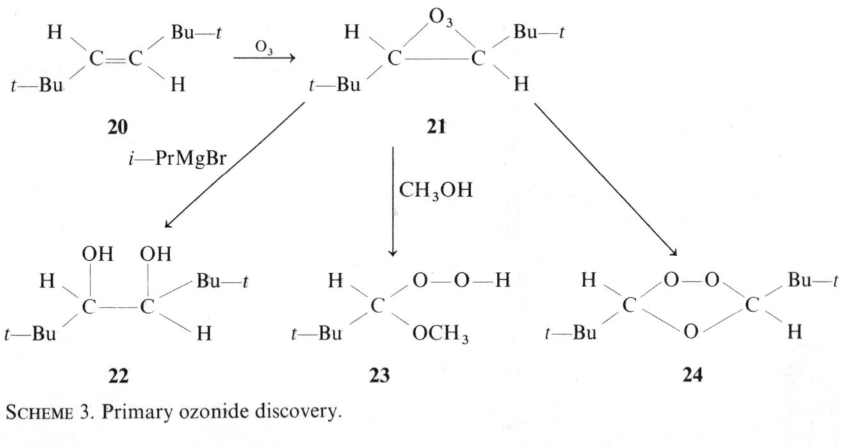

SCHEME 3. Primary ozonide discovery.

Criegee (*112*) also reported that a primary ozonide was observed for *trans*-1,2-diisopropylethylene. Following the pioneering work of Criegee and Schröder (*134*), Greenwood, by similar techniques, established the existence of primary ozonides of simpler alkenes: *trans*-3-hexene (*135,136*), *trans*-2-pentene (*136*), 1-pentene (*136*), *trans*-2-butene (*137*), 1-butene (*137*), and, presumably, *trans*-3-octene (*135,136*).

Although the investigations of Criegee and of Greenwood and co-workers confirmed the existence of the an initial ozone–olefin adduct, at least for certain *trans*-alkenes and 1-alkenes, they did not settle the question of ring size of the primary ozonide: i.e., a four-membered ring structure (**16**), as originally proposed by Staudinger (Chapter III, Section I, B), or the five-membered ring structure (**4**) suggested by Harries for normal ozonides (Chapter III, Section I, A). This question was settled in 1966 by means of low-temperature nuclear magnetic resonance (NMR) studies of the ozonolysis reaction mixtures of *trans*-1,2-di-*tert*-butylethylene in Freon 11 ($-110°C$) and acetone-d_6 ($-95°C$) (*99,138*). In each case the NMR spectrum consisted only of two singlet peaks, characteristic of the *tert*-butyl and the methine protons in the expected ratio of 9:1. At temperatures above $-60°C$ these peaks disappeared and peaks characteristic of the decomposition products appeared. These spectra prove that the primary ozonide is symmetrical in structure, thus eliminating structure **16**. Although the NMR data alone do not eliminate a symmetrical complex such as **17** or a peroxy epoxide (**19**), these data coupled with the reduction of the initial ozonides to glycols (*134–137*) establishes the five-membered ring (**4**) beyond reasonable doubt. Since this work, π complexes (**17**) have been observed and shown to be highly unstable (*102,139–141*).

Following the NMR study of the primary ozonide of *trans*-1,2-di-*tert*-butylethylene, Durham and Greenwood (*142,143*) similarly established the 1,2,3-trioxolane structure for primary ozonides of *trans*-2-butene, *trans*-3-hexene, and *trans*-2-pentene. Greenwood (*135,136,143*) strongly favors the term "molozonide" for the initial ozone–olefin adduct, because this was the term first used by Staudinger (Chapter III, Section I, B) for such an adduct. However, as discussed in Chapter III, Section I, B, Staudinger thought that the initial adduct had a four-membered ring, **4** (or **16**), by analogy to the "mol-oxide" (dioxetane) thought then to be the intermediate in autoxidation of olefins (*51,90*). The use of the term "molozonide," therefore, should be limited to the Staudinger structure and not be used for the 1,2,3-trioxolane structure, which has been established for the initial adduct. A better term for the latter is "primary ozonide" (*55,60,134*).

Recently, Alcock and Mile (*144*) observed, via infrared (IR) spectra, the primary ozonide of *trans*-1,2-diisopropylethylene in carbon disulfide solution at $-105°C$. By adding propionaldehyde to similar reaction mixtures at $-78°C$ and determining the yields of the normal and "cross" ozonides (Chapter VI, Section II) obtained, they were able to show that the primary ozonide persisted for as long as 2 hours and had a half-life of approximately

1 hour. Earlier, Greenwood and Durham (*145*) had calculated the half-life of the primary ozonide of *trans*-3-hexene to be about 55 minutes in Freon 12 at −100°C.

Low-temperature (−110°C) NMR studies with *cis*-1,2-di-*tert*-butylethylene failed to give any evidence for the presence of a primary ozonide (*138*). Similarly, the first papers of Greenwood (*135,136*) reported failure to establish the existence of a cis primary ozonide via the reductive method of Criegee and Schröder (*134*), and the suggestion was made that *cis*-olefins go directly to peroxidic ozonolysis products without the formation and decomposition of a primary ozonide. Later, however, Greenwood (*137,146*) reported the formation of explosive precipitates upon ozonation of *cis*-2-butene, *cis*-2-pentene, *cis*-3-hexene, and ethylene in pentane at −110°C, all of which decomposed to give normal nonexplosive peroxide ozonolysis products. That these actually were primary ozonides, even though Greenwood failed in attempts to reduce them to glycols, was established in the cases of *cis*-2-butene, *cis*-2-pentene, and *cis*-3-hexene by means of their NMR spectra at −130°C (*143,147*).

The NMR spectra of the cis primary ozonides do not rigorously distinguish between the 1,2,3-trioxolane structure (**4**) and the 3-membered peroxy epoxide structure (**19**) suggested for certain cis primary ozonides by Story and Murray (*127,128*) and Freedman *et al.* (*130*). However, the absorptions of the ring hydrogens occur where one would expect from the 1,2,3-trioxolane structures for both the cis and the trans isomers. The cis ring hydrogens in each case are slightly downfield from the trans ring hydrogens by about the same amount as observed for other ring systems (*143*), indicating that the ring sizes are the same for the cis and trans primary ozonides. Moreover, the greater complexity of the ring proton absorptions in the NMR spectra of the cis primary ozonides in comparison to those of the trans primary ozonides tends to favor the 1,2,3-trioxolane structures, which Bauld, Bailey *et al.* (*148*) have described as having a carbon–carbon half-chair conformation. With such, the ring hydrogens should be identical (either diaxial or diequatorial) for *trans*-1,2,3-trioxolanes, but different (one axial and one equatorial) for *cis*-1,2,3-trioxolanes. The spectra of Durham and Greenwood (*143*) fit this picture. The ring proton absorptions were the expected quartet and triplet for the primary ozonides of *trans*-2-butene and *trans*-3-hexene, respectively, but were unresolvable broad peaks for the corresponding cis primary ozonides.

The work of Greenwood and Durham (*142,143,145*) also established the fact that the formation of the primary ozonide is not just incidental, but that the 1,2,3-trioxolane is an actual intermediate in ozonolysis. They monitored its decomposition to peroxidic ozonolysis products by means of NMR spectra, at times using an oscilloscope.

It is obvious from the above discussion that cis primary ozonides are much less stable than trans primary ozonides, and the importance of steric factors in this regard has been observed (*136,137,148*). Bauld, Bailey *et al.* (*148*) have pointed out that the factors that destabilize a *cis*-1,2,3-trioxolane should stabilize a *trans*-1,2,3-trioxolane. In the trans case, concerted disrotatory cleavage of the carbon–carbon bond of the trioxolane ring should be increasingly hindered as the substituents increase in size, whether these substituents are diaxial or diequatorial, since one of the substituents must rotate inward, thereby encountering severe repulsions; therefore, the trioxolane should increase in stability as the substituents increase in bulk. In contrast, the *cis*-1,2,3-trioxolane can avoid this by rotating both substituents outward in disrotatory cleavage; moreover, the distortion created by the cis bulky groups causes an instability that sets the cleavage operation into motion.

It is interesting to note in this connection, however, that, although all the primary ozonides observed and characterized during the above-described period were those of 1,2-disubstituted ethylenes, two other types have recently been observed (*149*). These are the primary ozonides of methyleneadamantane (**25a**), a 1,1-disubstituted ethylene, and of isopropylideneadamantane (**25b**), a tetrasubstituted ethylene. The primary ozonides were reduced by isopropylmagnesium chloride to the corresponding 1,2-diols (*149*).

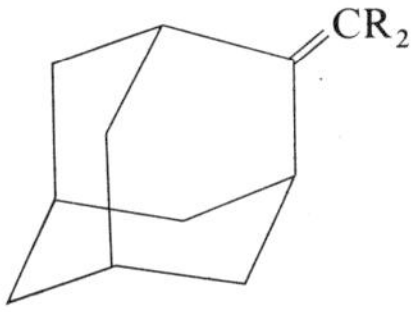

25a: R=H
25b: R=CH_3

II. 1,3-Dipolar Cycloaddition

An earlier review (*62*) discusses the disagreement that existed before about 1960 concerning whether the initial ozone attack on an olefinic double bond or an aromatic bond was a one-step, four-center, concerted process, or was a two-step addition with the first step involving ozone as an electrophile. The evidence for each concept dealt largely with aromatic compounds and consisted of: (a) the position of attack on a polycyclic aromatic system, whether it occurred at the bond of lowest bond-localization energy or at atoms of lowest atom-localization energy; and (b) rate studies showing that factors that increase the rate of electrophilic substitution also increase the rate of ozonolysis of benzene and substituted benzenes (*62*).

During the 1960s, these conflicts were reconciled by Huisgen's classification of ozone as a typical 1,3-dipole and of the addition of ozone to a multiple bond as a 1,3-dipolar cycloaddition (*80,112,150*). The 1,3-dipolar cycloaddition of ozone to an olefinic double bond is illustrated in Eq. (6).

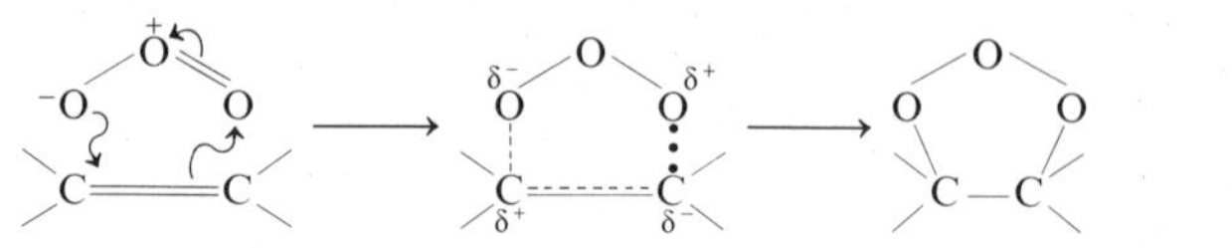

26 4 (6)

A. ELECTROPHILIC NATURE OF OZONE IN 1,3-DIPOLAR CYCLOADDITION

Kinetic studies with ozonolyses of various olefins show clearly that substituents on the carbons of a double bond increase the rate of ozone attack if they are electron donating and decrease the rate of ozone attack if they are electron withdrawing, just as found earlier for substituents on a benzene ring (*62,151*). This has been shown to be true both for vapor-phase (*152–155*) and liquid-phase (*156–164*) ozonolyses. For example, alkyl (*156,158,160,163*) and alkoxy (*165*) groups activate the double bond, whereas chloro (*120,157, 166–169,169a*) (see also Table I), bromo (*163,169–171*), nitro (*163,172*), carbonyl (*163,166,173*), and cyano (*163*) groups deactivate the double bond for ozone attack. Also, substituents on the phenyl group of substituted styrenes either activate or deactivate the olefinic double bond for ozone attack, depending on whether they are electron repelling or attracting (*159, 161,162*). In most cases the kinetics for the above ozonolyses showed the reactions to be second order (first order in regard to each reactant) (*153–157, 159,160,162,163,174*).

Although these data clearly show the electrophilic character of ozone attack on a double bond, they are also consistent with the 1,3-dipolar cycloaddition mechanism and a semipolar transition state, **26**, in which the electrophilic oxygen of ozone is more strongly bonded than the nucleophilic oxygen. Thus, electron-donating and -withdrawing groups can stabilize or destabilize the transition state through decreasing or increasing the positive charge on carbon in **26**. Huisgen (*150*) emphasizes that "concerted addition" should not be taken to mean that the two new σ bonds form in the transition state to the same degree. Fliszár *et al.* (*162*) have shown that in certain cases rate constants increase to some extent with an increase in the dielectric constants of the solvent, clearly indicating the semipolar character of the transition state (**26**).

The kinetic data involving the effects of substituents on a double bond fit much better the concerted addition, with a transition state such as that pictured in Eq. (6), than a purely electrophilic attack as the first step of a two-step attack involving a carbonium ion intermediate. This is illustrated in Table I. For a purely electrophilic attack the relative rates for ozonolyses of 1,1-disubstituted ethylenes should be approximately equal to those for 1,2-disubsituted ethylenes. Comparisons, in Table I, of isobutylene (k_{rel} = 97,000) with *cis*-2-butene (k_{rel} = 163,000) and of 1,1-dichloroethylene (k_{rel} = 22) with the 1,2-dichloroethylenes (trans k_{rel} = 591) show that this is not true. Also, the enhancement of rate in going from the dimethylethylenes to the tri- and tetramethylethylenes (k_{rel} = 200,000) is not as great as expected from an initial electrophilic ozone attack (Table I). A likely explanation for

TABLE I

SECOND-ORDER RELATIVE RATE CONSTANTS FOR OZONATIONS OF SELECTED OLEFINS IN CCl_4 AT ROOM TEMPERATURE[a]

Olefin	k^b	Olefin	k^b
$Cl_2C{=}CCl_2$	1.0	$H_2C{=}CHPr$	81,000
$ClCH{=}CCl_2$	3.6	$H_2C{=}CMe_2$	97,000
$H_2C{=}CCl_2$	22.1	*cis*-MeCH=CHMe	163,000
cis-ClCH=CHCl	35.7	$Me_2C{=}CHMe$	167,000
trans-ClCH=CHCl	591	$Me_2C{=}CMe_2$	200,000
$H_2C{=}CHCl$	1180	1,3-Butadiene	74,000
$H_2C{=}CH_2$	~25,000	Styrene	103,000

[a] Reprinted, in part, with permission, from Williamson and Cvetanović (*157*). Copyright by the American Chemical Society.

[b] k_{rel}, liters/mole second.

this is a steric retardation of the cycloaddition. Huisgen (*150*) pointed out that 1,1-disubstituted ethylenes react much more slowly than 1,2-disubstituted ethylenes in almost all 1,3-dipolar cycloadditions. Cvetanović (*152,157*) and Fliszár (*160*), however, suggested another explanation in place of, or in addition to, the steric effect. They suggested the reversible initial formation of a π complex (**27**), which can then enter into the 1,3-dipolar cycloaddition as illustrated in Eq. (7). In such a scheme the second step (**27** to **4**, through **28**), which would involve the nucleophilic end of the ozone molecule, would be retarded either by the inductive effects of alkyl groups or the resonance effects of chloro groups (R in **27** or **28**). Fliszár (*160*) calculated the ρ^* constants for simple *cis*- and *trans*-alkenes from the Taft equation and found them to be positive (5.37 for cis and 7.58 for trans). From this he concluded that the rate-determining step is nucleophilic (**27** → **4**) rather than electrophilic (alkene → **27** or **28**). This is to be contrasted with substituted styrenes where the Hammett ρ constant is negative (ca. -1) (*159–162*).

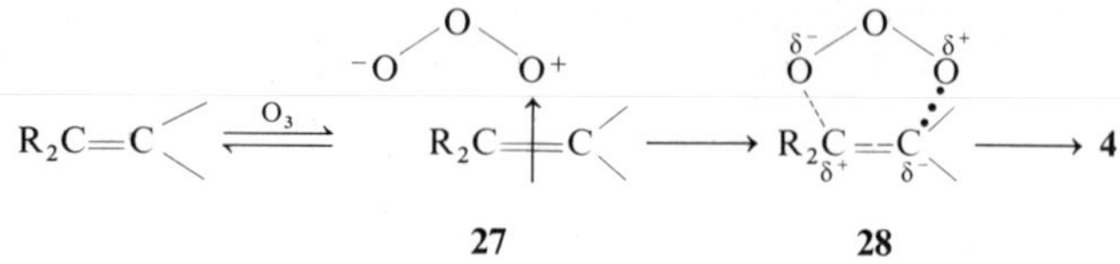

(7)

Most of the preceding kinetic studies involved the mixing together of solutions of ozone and of the olefin, or olefins, and second-order kinetics were observed. In contrast, for ozonolyses of olefins by the usual procedure of bubbling gaseous ozone-oxygen through a solution of the olefin (a heterogeneous procedure) the kinetics fit better a zero-order equation (*164,175*), and there were only slight differences in the rate constants of individual olefins (*164*). Competitive ozonolyses of pairs of olefins by this procedure, however, resulted in big differences in relative rates (*164*). In an attempt to explain these strange results, Altwicker and Basila (*164*) also invoked a reversible π complex.

Ozone π complexes have been shown to exist in the cases of homologs of benzene and of certain aryl-substituted ethylenes or ethanes (*141*). They have tentatively been observed in the case of simple olefins (*102,176*). Alcock and Mile (*176*) believe them to be precursors to primary ozonides, whereas Heicklen and co-workers (*102*) believe that they are merely coincidental. These π complexes are discussed in greater detail in Chapter VII, Section I, B, 6 and Chapter XI, Section IV.

B. OTHER EVIDENCE FOR 1,3-DIPOLAR CYCLOADDITION

Another characteristic of 1,3-dipolar cycloaddition is that the nature of the solvent has little effect upon the reaction. This appears to be true, in general, for ozonolysis (*160,174*). Razumovskii *et al.* (*174*) considered this to be evidence against a polar transition state. Their picture, however, of a polar transition state was, in essence, an actual carbonium ion, rather than as shown in Eq. (6).

Typical of 1,3-dipolar cycloadditions (*150*), steric factors influence the rate and, sometimes, even the course (Chapter XI) of the reaction between ozone and an olefin. This was discussed to some extent in the preceding subsection. More striking, however, is the faster rate of ozonolysis of a trans isomer than of a cis isomer (cf. *cis*- and *trans*-1,2-dichloroethylene in Table I), especially if large groups, such as carbonyl, chloro, or phenyl, are attached to the double bond (*152,153,157,160,164,166,177,178*). The opposite is generally true with the two-step addition of bromine to a double bond (*150*). Huisgen (*150*) explained this on the basis that during the concerted 1,3-dipolar cycloaddition the hybridization of the carbon atoms of the double bond changes from sp^2 to sp^3 in the transition state and, even though the carbon–carbon bond length increases somewhat, the accompanying decrease in the bond angle from 120° to 109° results in considerable compression of the van der Waals radii of the eclipsed cis substituents. This results in a higher activation energy for the cycloaddition of the cis isomer than of the trans isomer. Strangely enough, Razumovskii and Zaikov (*163*) reported maleic acid to be more reactive than fumaric acid, and Carles and Fliszár (*160*) reported *cis*-1, 2-di-*tert*-butylethylene to be more reactive than the trans isomer. Since these workers were using heterogenous procedures (passing ozone into solutions or suspensions), however, it is likely that the results were due to solubility differences in the *cis*-and *trans*-olefins. A later publication of Razumovskii (*179*) reported fumaric acid to be more reactive.

Steric effects also show up in other ways. Engel *et al.* (*180,181*) have shown that ozonolysis of the 9,11-double bond of a $\Delta^{9(11)}$-12-keto steroid with a trans fusion of rings A and B occurs in much better yield than ozonolysis of the 9,11-double bond of the corresponding A/B-cis-steroid.

Another striking phenomenon associated with 1,3-dipolar cycloadditions is the promoting effect that conjugation exerts on the dipolarophilic activity of a double bond (*150*). This is observed with ozonolysis provided that the conjugating group is not strongly electron attracting. For example, 1,3-butadiene is much more reactive (ca. 3:1) than ethylene (*157*) and styrene is more reactive than most simple 1-alkenes (*157,163*) (see Table I).

The strongest evidence for a concerted reaction, such as 1,3-dipolar cycloaddition, is stereochemical. It is obvious from the discussion of Section I of this chapter that cis and trans isomeric olefins give different primary ozonides. The reduction of the 1,2,3-trioxolanes of *trans*-alkenes to *dl*(or *threo*)-1,2-glycols (*134,136,137*) proves that the primary ozonide was the product of cis addition. Assuming that the primary ozonides of most *cis*-olefins also are 1,2,3-trioxolanes, they must have the cis configuration and, therefore, also must be products of a concerted cis addition.

Ozonolysis of olefins, like most reactions involving 1,3-dipolar cycloaddition (*150*), has a highly negative entropy of activation (*162*), indicating a high degree of order requirement in the transition state, as the preceding discussion clearly illustrates. However, 1,3-dipolar cycloadditions also have moderate to low enthalpy requirements (*150*). This is certainly true for ozonolysis of olefins, for which the activation energy in low, often close to zero (*162,174,182*). Benson (*183*) pointed out that only a concerted process can account for the low activation energy. The low activation energy also requires a transition state that resembles the reactants, as emphasized by Erickson and Clark (*184*). Transition state **26** might be pictured as an exaggerated envelope, as illustrated by **29**, with which overlap of the orbitals of the two adding systems can best occur. Ozone is a 4π electron system, and the addition is a symmetry-allowed $[_{\pi}4_s + {}_{\pi}2_s]$ process. Alternatively, the first ozone–olefin interaction might involve a π complex, as already discussed.

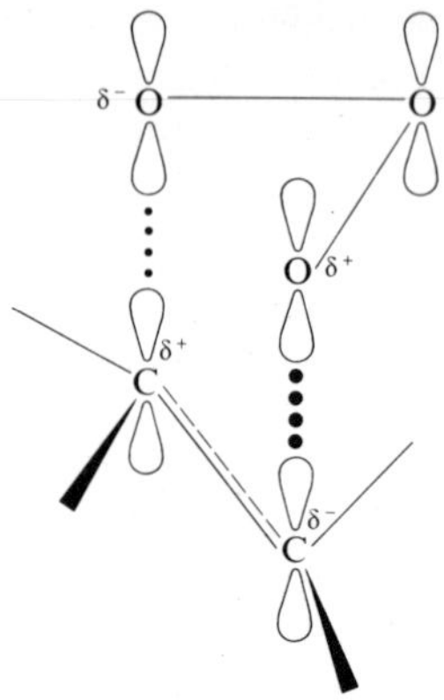

29

In summary, an accumulation of evidence indicates that the attack of ozone on an olefinic double bond is a stereospecific, one-step, concerted, 1,3-dipolar cycloaddition, in which, however, a reversible π complex might be formed initially. Whether or not this picture can be accommodated by the singlet diradical structure for ozone (*76,78*) remains to be seen. At present, it would appear that such a structure could explain all but the electronic effects discussed in Section II, A of this chapter. However, if, as suggested in Chapter II, Section II, the singlet diradical structural description of ozone is simply one extreme in a sliding scale of character involving the paired electrons, then ozone obviously can react as a zwitterion when the environment promotes such activity.

In this connection, it is interesting to note that Firestone (*185*) has presented a viewpoint quite different from Huisgen's (*150,186*) concerning the mechanism of 1,3-dipolar cycloaddition, in which a spin-paired diradical intermediate is suggested. The two different viewpoints probably will exist for a long time, since it is difficult to conceive of an experiment that will rigorously disprove one of the two mechanisms.

CHAPTER V

Ozonolysis of Olefins: The Peroxidic Products

Since the evidence for and against the simple Criegee mechanism has to do largely with modes of formation of ozonides and other peroxidic ozonolysis products, it is advantageous to discuss these before considering pathways leading to them.

I. Ozonides

A. FORMATION: PROPERTIES

The term "ozonide" should be used only for the 1,2,4-trioxolane structure originally proposed by Staudinger (*51*) and established first by Rieche (*52–54*) and later by Criegee and Lohaus (*104*). As emphasized in the earlier review (*62*) these substances are, in general, the major peroxidic ozonolysis products only when: (a) the solvent is aprotic (nonparticipating), and (b) the olefin is ethylene or a mono-, di-, or trisubstituted ethylene. This is understandable in terms of the Criegee mechanism, which involves ozonide formation by interaction of carbonyloxide **13** with a carbonyl moiety (**14**) (Chapter III, Scheme 2). Since this interaction is a 1,3-dipolar cycloaddition and is influenced by steric and electronic factors (*150*), ozonides are not generally produced if **14** is a ketone (**62**).

When pure, ozonides are liquids or solids with sharp boiling or melting points. They are much more stable than the older literature leads one to believe. Beginning in 1962 with the work of Schröder (*187*), it has become common practice to purify and separate ozonides from each other (even cis and trans isomeric ozonides) by means of gas chromatography (*128, 148,188*). This does not mean that one should not be cautious. Low molecular weight and highly strained ozonides can be explosive. However, most of the explosive "ozonides" of the early literature were not ozonides at all, but polymeric peroxides or ozonides contaminated with such. Criegee *et al.* (*109*) listed the physical properties of a number of pure ozonides of open-chain and cyclic olefins. A selection of others (mostly crystalline

TABLE II

CRYSTALLINE MONOMERIC OZONIDES OF OPEN-CHAIN OR EXOCYCLIC OLEFINIC COMPOUNDS

Example	Ozonide of	Yield (%)	M.p. (°C)	Reference
1	1,4-Dibromo-2,3-dimethyl-2-butene[a]	58	41–42	*189*
2	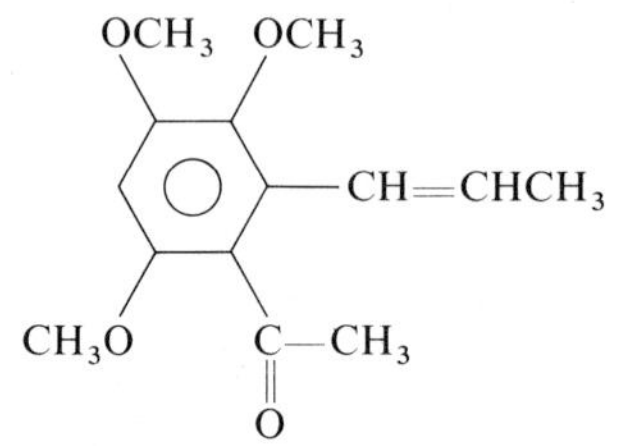	67	116–118	*190*
3	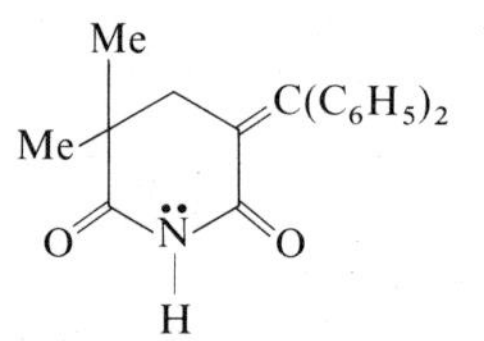	58	156 (dec.)	*191*
4	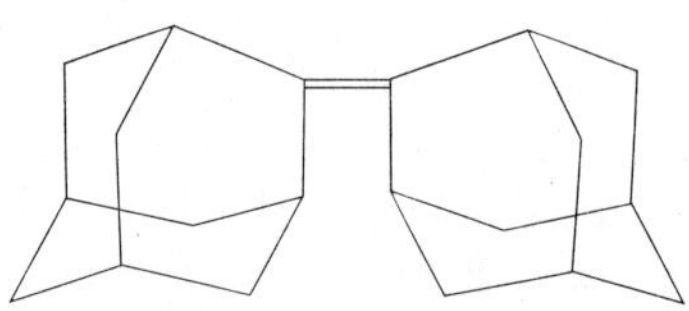 Bisadamantylidene	33	140 (dec.)	*192*
5	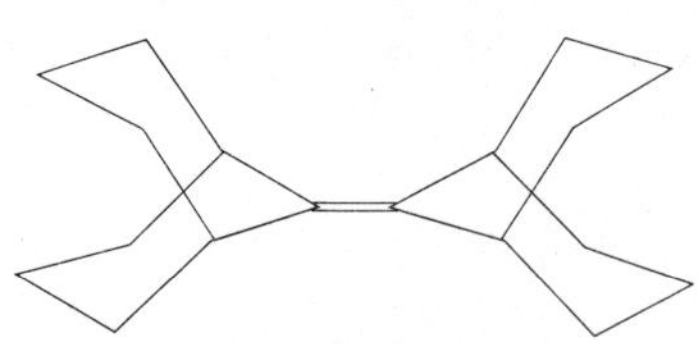 Bisbicyclo[3.3.1]non-9-ylidene	18	138 (dec.)	*192*
6	Methyl *p*-methoxycinnamate[b]	45–53 47–55	62–63 38–39	*193*

[a] These results were from the *trans*-olefin, and the crystalline ozonide was assigned the trans configuration. A small amount of the liquid *cis*-ozonide also was formed.

[b] Both the *cis*- and *trans*-olefins were ozonized, and both ozonides were isolated. The higher-melting ozonide was assigned the trans configuration and was obtained in 53% yield from the *trans*-olefin by vapor-phase chromatography. Isolated yields were 33% trans and 11% cis.

TABLE III

PURIFIED MONOMERIC OZONIDES OF CYCLIC OLEFINS

Example	Ozonide of	Yield (%)	M.p. (°C)	References
1	*cis*-1,2,3,4-Tetramethylcyclobutene	62	(liq.)	*204*
2	*trans*-1,2,3,4-Tetramethylcyclobutene	60	(liq.)	*204*
3	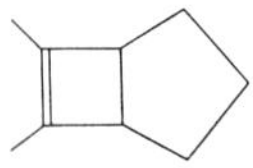	89	(liq.)	*205*
4	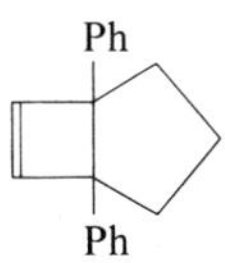	28	125–126	*206*
5	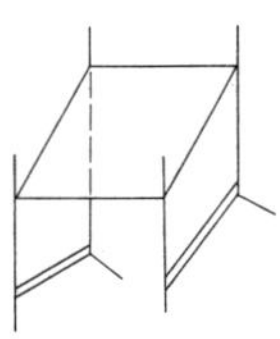	99	149–151	*207,208*
6	Diozonide of example 5	23	191–192	*207,208*
7	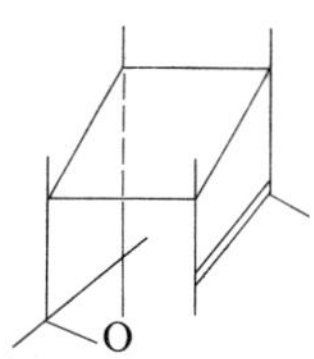	35	203	*207*
8	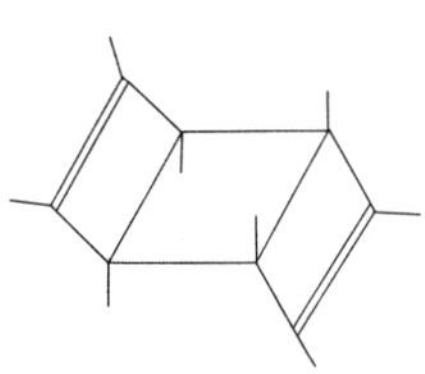	49	106	*207,208*
9	Diozonide of example 8	29	156–157	*207,208*
10	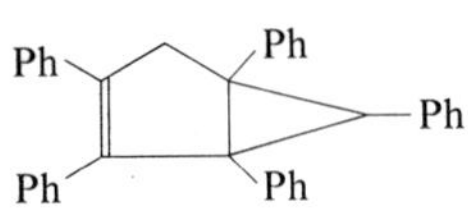	54	200–201	*209*
11	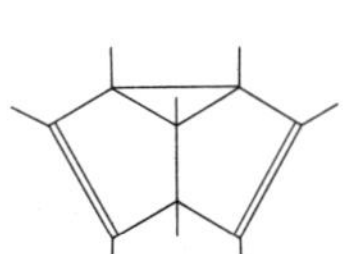	64 (diozonide)	133–135	*210,211*

TABLE III (*cont'd*)

Example	Ozonide of	Yield (%)	M.p. (°C)	References
12	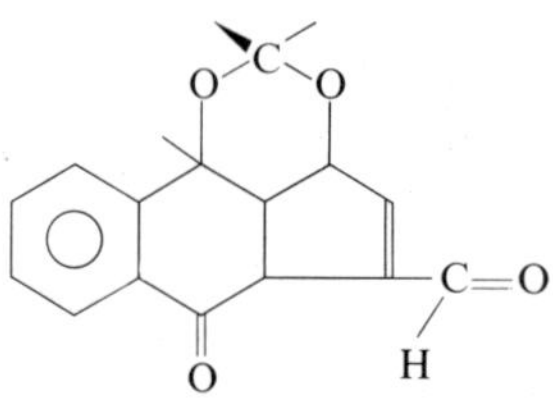	81	100 (dec.)	*212*
13	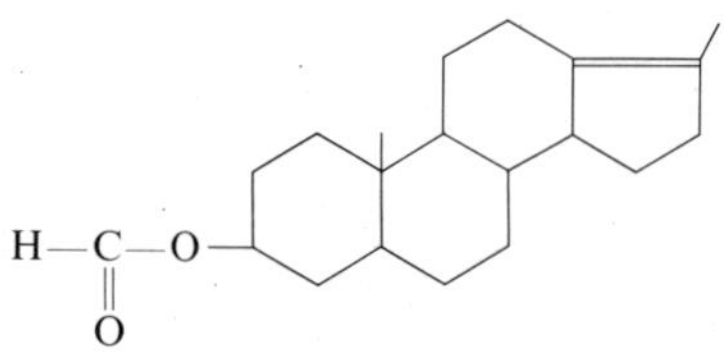	80	155–158	*213*
14	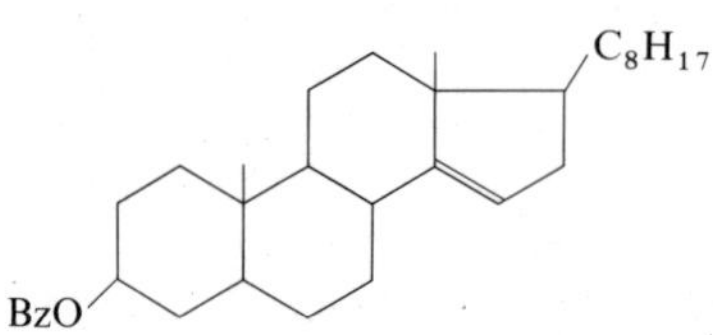	93	132–133	*214*
15	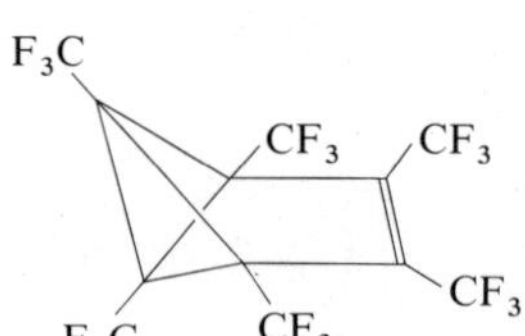	ca, 100	(liq.)	*215*
16	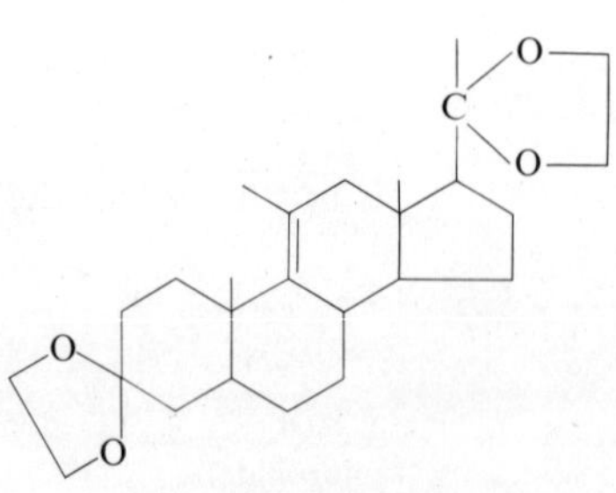	26	149–151	*216*

solids) are shown in Tables II (*189–193*) and III. A review by Murray (*14*) contains a longer list. It can be seen that many have high melting points.

There are at least three exceptions to the general rule (*62,109,194*) that tetrasubstituted ethylenes do not yield ozonides (or that carbonyl oxides do not add to ketones).

One exception is with certain cycloolefins with which the interaction between carbonyl oxide and carbonyl moieties is intramolecular. This was discussed in detail in the earlier review (*62*). At that time the only examples known involved 1,2-disubstituted cyclopentene types. Since then, ozonides of cyclobutene types have also been obtained in good yield. Several examples of ozonides of cycloolefins, most of them appearing since the earlier review (*62*), are shown in Table III. Additional examples of ozonides of the cyclobutene type have also been characterized by Criegee and co-workers (*195–203*).

The ozonides of cyclobutene and cyclopentene types are trioxabicycloheptanes and octanes with classical five-, six-, and seven-membered rings [Eqs. (8) and (9)]. The intramolecular driving force for formation of such rings would be great (*132*). Cyclohexenes give very low yields of ozonides (Table III, example 16), along with polymeric peroxides (*62*). Also, unsubstituted or 1-monosubstituted cyclopentenes give lower ozonide yields than do 1,2-disubstituted cyclopentenes, as discussed in the earlier review (*62*).

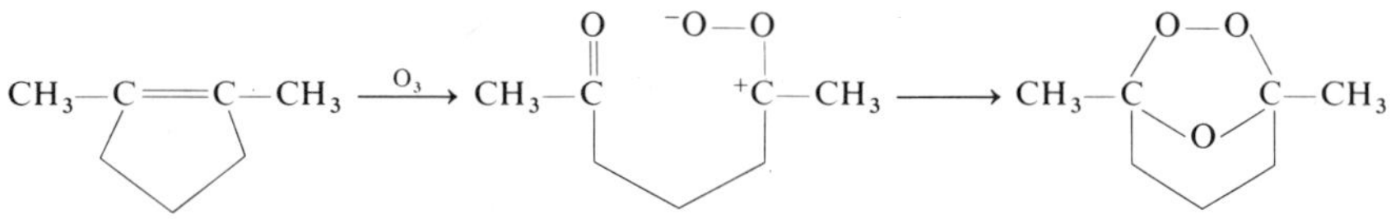

(8)

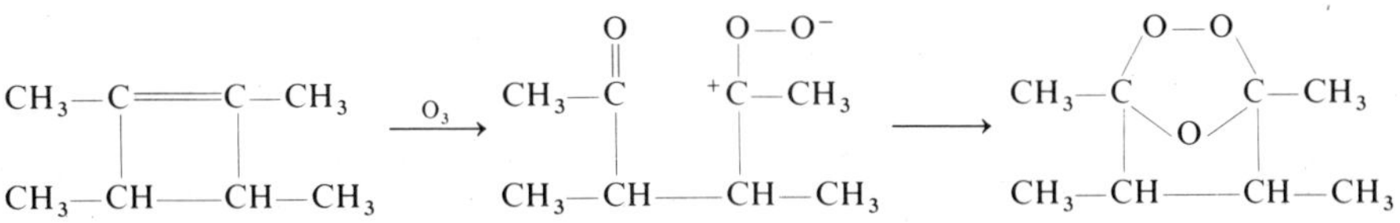

(9)

The second exception to the general rule that ozonides are not produced by addition of a carbonyl oxide (**13**) to a ketone (Chapter III, Scheme 2) concerns olefinic compounds that yield, along with the carbonyl oxide, a ketone that is activated by an electron-withdrawing group. This not only makes it a better electrophile, but also a better 1,3-dipolarophile. Two examples are shown in Scheme 4. Ozonides **33** and **34** were isolated in yields of 58–60% (*189*). Ozonide **33** also was obtained by ozonolysis of tetramethylethylene (**32**) in the presence of **31**, showing that it was produced from moieties **30** and **31** (*189*). Example 3 of Table II also is of this category.

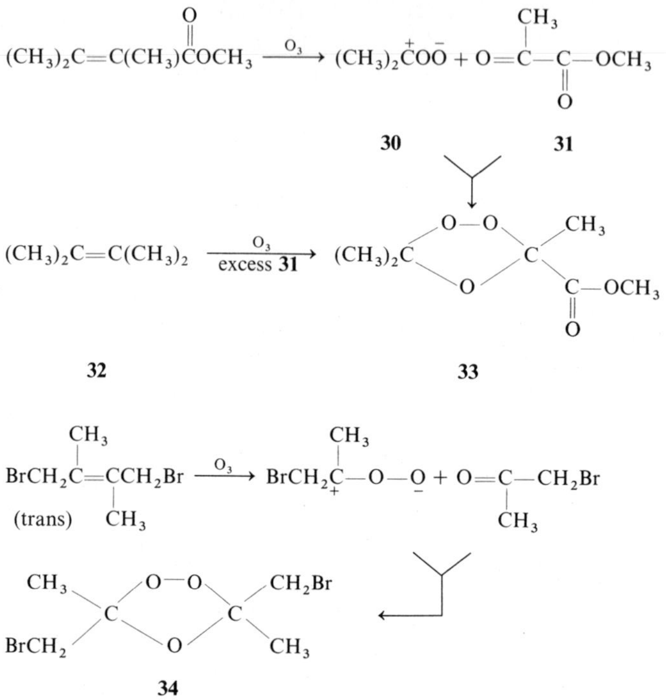

SCHEME 4. Ozonides of tetrasubstituted ethylenes.

A special case of this sort (or possibly it should be considered a fourth exception), is exemplified by the ozonolysis of bisadamantylidene (example 4, Table II) and of bisbicyclo[3.3.1]non-9-ylidene (example 5, Table II) to give ozonides, along with other products (see Chapter XI) (*192*). In fact, adamantanone proved to be a better 1,3-dipolarophile than benzaldehyde in that the carbonyl oxide from *cis*-stilbene reacted with adamantanone in preference to benzaldehyde to give an ozonide (*192*). The unusual dipolarophilic property of adamantanone and of bicyclo[3.3.1]nonan-9-one was explained on the basis of the driving force for the formation of the highly symmetrical and stable sp^3 bisadamantylidene and bisbicyclononylidene ozonide systems (which are similar to the parent adamantane and bicyclononane systems) from the unsymmetrical and strained sp^2 systems of the ketones (*132,192*).

The third exception to the rule that carbonyl oxides do not add to ketones to give ozonides involves ozonolysis of an olefin in the presence of excess ketone. Murray *et al.* (*217*) ozonized 2-pentene in the presence of excess acetone and obtained five different ozonides. Three were those obtained from 2-pentene alone, those of 2-pentene (normal ozonide) and 2-butene and 3-

hexene ("cross" ozonides, see Chapter VI, Sections I and II). The other two were **37** and **38**, produced by addition of carbonyl oxides **36** and **35** to acetone, respectively (Scheme 5). In these cases the carbonyl oxide was monosubstituted (an aldehyde oxide). These might be expected to react with

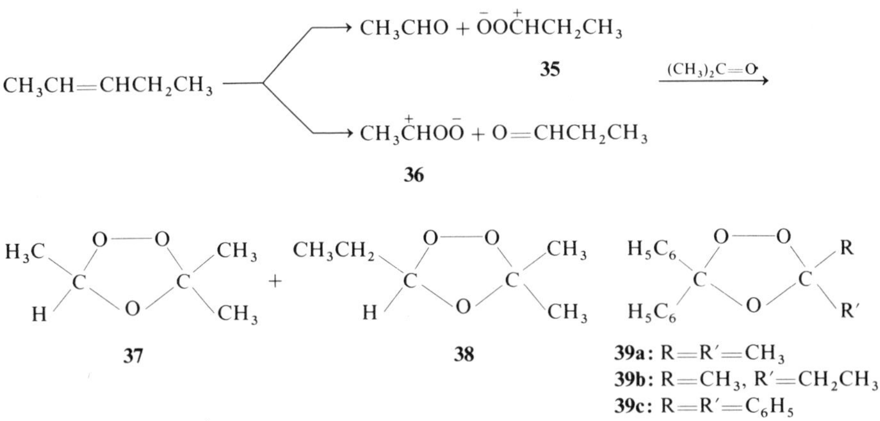

SCHEME 5. Ozonides from carbonyl oxides and ketones.

ketones more easily than would disubstituted carbonyl oxides (ketone oxides). Criegee and Korber (*218*), however, later obtained ozonides **39a** and **39b**, in good yield, and **39c** in 25% yield, from ozonolysis of tetraphenylethylene in the presence of excess acetone, 2-butanone, and benzophenone, respectively (*132*).

These reactions were possible because of the excess ketone. Otherwise, tetraphenylethylene produces a ketone diperoxide as the major product (see Section II). The ozonide of tetramethylethylene is as yet unknown. Criegee and Korber (*218*) believe that dimethylcarbonyl oxide is too unstable to react with a ketone. It either dimerizes (see Section II), polymerizes (see Section III), or rearranges to hydroxyacetone (*218–220*), in addition to reacting with an aldehyde when available (Chapter VI, Section I).

B. SPECTRAL PROPERTIES: IDENTIFICATION

There is much confusion in the literature prior to about 1955 concerning certain spectral and physical properties of ozonides because measurements were made on unpurified "ozonides," which contained starting material, decomposition products, and/or peroxidic ozonolysis products other than ozonides. For this reason, much of Briner's extensive work during that period concerning infrared (IR), Raman, and ultraviolet spectra, dielectric constants, dipole moments, and heats of formation of ozonides has lost its meaning; corrections have since been made (*221–224*).

Beginning with the Criegee era, IR and nuclear magnetic resonance (NMR) spectra have been used to characterize ozonides. Criegee *et al.* (*109*) found that thirteen out of eighteen pure ozonides absorb in the IR region of 9.4–9.6 μm (1060–1040 cm^{-1}). This has been verified by other workers (e.g., *149,187, 188,192,193,217,222,223,225–234*), whose results, however, indicated the range to be 8.9–9.85 μm (1113–1015 cm^{-1}). These bands are thought to be carbon–oxygen vibrational frequencies of the ether and/or peroxide linkages (*232–235*). Characteristic bands also have been reported for ozonides in the region 890–1000 cm^{-1} (*149,192,225,226,236*).

Infrared spectra have been used, at least tentatively, to distinguish between cis and trans isomeric ozonides (Chapter VI, Section II). The spectra of several pairs indicated that the trans isomer absorbs in the region 7.35–7.60 μm (1360–1320 cm^{-1}) and the cis isomer in the region 11.7–12.2 μm (855–820 cm^{-1}); in each case the absorptions were distinctive (*14,128,187, 193,227–229*).

Nuclear magnetic resonance spectra have been useful in confirming ozonide structures (*14,232,237*). Of special importance are the protons attached to the trioxolane ring, for which the δ values reported are in the 4.87–6.20 ppm range (4.87–5.38 ppm for ozonides of simple olefins with nonterminal double bonds). In most cases the cis ozonide methine protons absorbed at slightly lower field than did those of the trans ozonide (*14,128, 143,147,148,188,238,239*). Discussion of the use of spectra and other physical properties in the determination of cis:trans ozonide ratios can be found in Chapter VI, Section II.

Criegee *et al.* (*109*) have found that, for the several liquid ozonides they studied, good agreement could be obtained between determined and calculated molecular refraction values, using the value of 2.11 of Williams and Mosher (*240*) for the peroxidic oxygen atom. Razumovskii *et al.* (*234*) listed several other physical properties for 1-hexene ozonide.

II. Ketone and Aldehyde Diperoxides

As described by the Criegee mechanism (Chapter III, Scheme 2) ketone and aldehyde diperoxides (**10**) are products of ozonolysis of olefins in non-participating solvents, particularly of tetrasubstituted ethylenes which do not yield ozonides (*194*) (see Section I). Diperoxides were first described as ozonolysis products by Harries, who, however, thought them to be decomposition products of ozonides rather than independent peroxidic ozonolysis products (Chapter III, Section I, A).

These diperoxides have been known for a long time and have been synthesized from aldehydes and ketones in various unambiguous ways that establish their structures (*241–244*). The diperoxide ring has been shown to exist in a chair conformation (*245–247*). Most of the diperoxides are high melting and stable. Some examples are given in Table IV (*23,60,105,109,189, 194,224,241,248–254*).

III. Polymeric Ozonides and Peroxides: Oligomers

Ozonolyses in nonparticipating solvents yield as peroxidic ozonolysis products not only ozonides and ketone or aldehyde diperoxides (Section II), but also polymeric peroxides and/or polymeric ozonides (see Chapter VII, Section I for further discussion).

Polymeric ozonides were reported early by Harries (*86,255*) (See Chapter III, Section I, A), Staudinger (*51*) (see Chapter III, Section I, B), Rieche (*53,54,256,257*), Briner (*258*), and others. Rieche (*53,54*) suggested structure **40** for such substances. Criegee (*61,110*) proposed that polymeric peroxides, from tetrasubstituted ethylenes at least, were polymers of the carbonyl oxide zwitterion, i.e., $\left(\underset{\mathrm{R}}{\underset{|}{\mathrm{CH}}}\text{—O—O}\right)_{x}$. Fliszár *et al.* (*254*) presumed them to be open-chain polymers of carbonyl oxide.

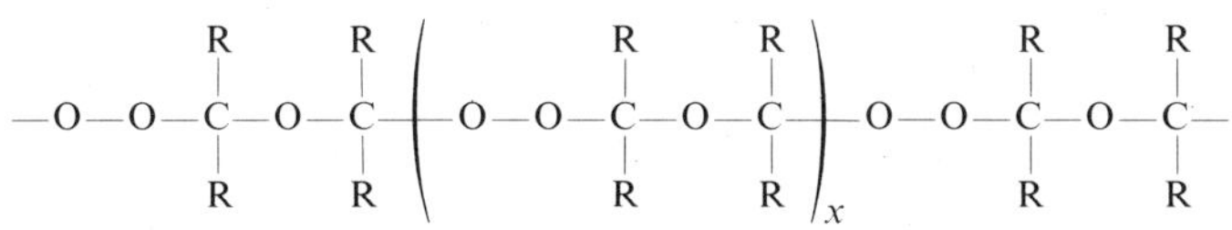

40

Some work has been done on the characterization of polyozonides of cycloolefins and of simple olefins. Studies with polyozonides of phenanthrene (*259*) (which has an almost olefinic double bond), norbornylene (*122*), 1,2,3,4,7,7-hexachlorobicyclo[2.2.1]hepta-2,5-diene (*260*), cyclohexene (*261*), and *cis*-Δ^4-tetrahydrophthalic acid (*262*) indicate that these have predominantly a normal ozonide structure, produced by interaction of the corresponding carbonyl oxide with itself, as illustrated in Eq. (10) for cyclohexene. The evidence in favor of such a structure came from IR spectra, which showed strong ozonide bands (Section I, B of this chapter) but only a weak aldehyde band (*259*), and NMR spectra, which showed a very low percentage of aldehyde proton (*260*). Molecular weight determinations indicated anywhere

TABLE IV

ALDEHYDE AND KETONE DIPEROXIDES FROM OZONOLYSIS

Example	Peroxide[a]	Olefin ozonized	Approx. yield (%)	M.p. (°C)	References
1	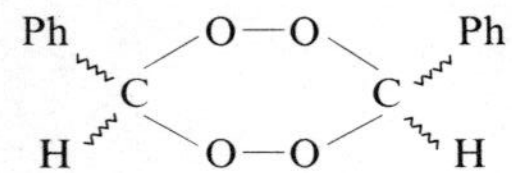	PhCH=CHPh	2–17	200	*109,241,248*
2	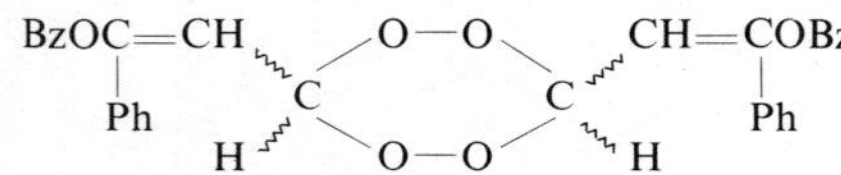	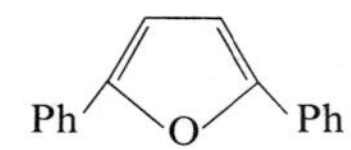	5–20	170	*249*
3	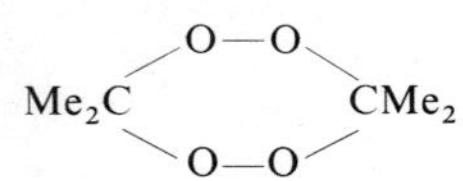	$Me_2C{=}CMe_2$ $Me_2C{=}CH_2$, etc.	15–20	131–132	*23,105,194,250,251*
4	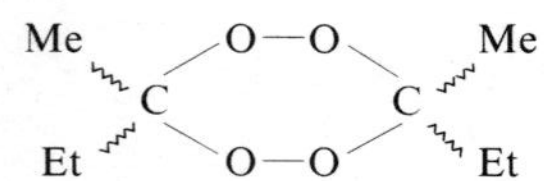	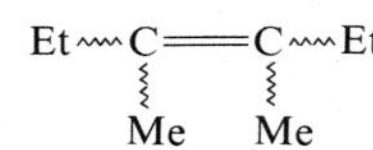	43	7–8	*251*
5	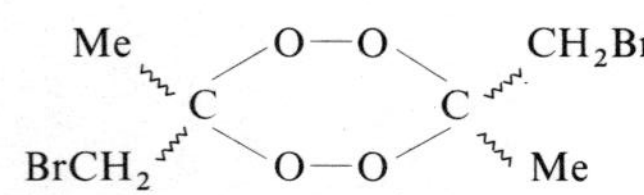	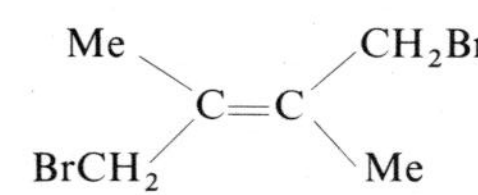	10	115–117	*189*
6	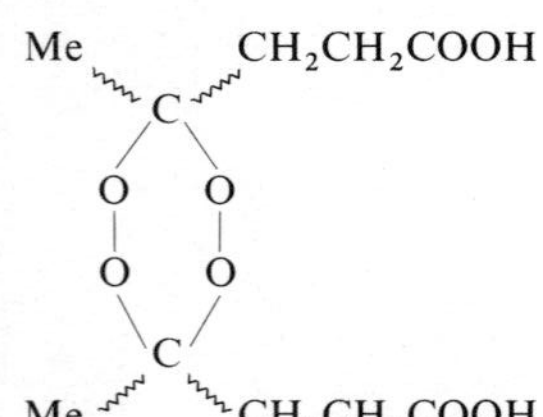	Rubber	—	194–196	*252*

7	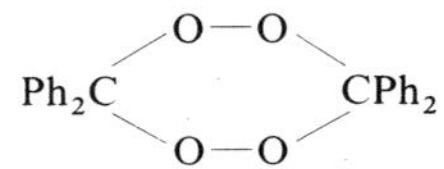	$Ph_2C{=}CPh_2$ $Ph_2C{=}CH_2$ $Ph_2C{=}CHR$ etc.	1–58	206–209 214–215	*224,241,253,254*
8	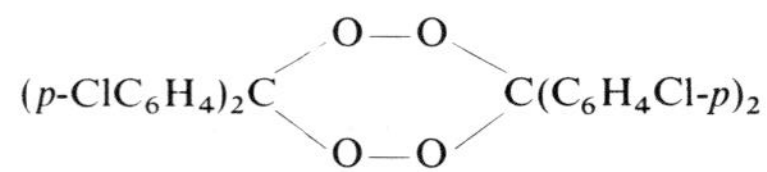	$(p\text{-}ClC_6H_4)_2C{=}CH_2$	5	218–219	*241*
9	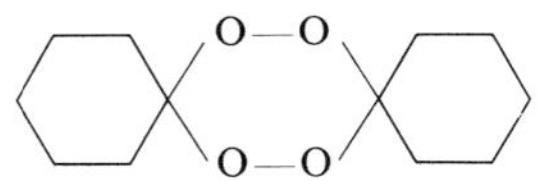	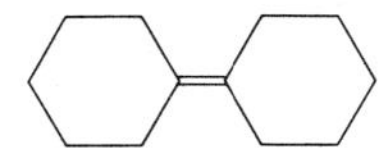	9	130	*194*
10	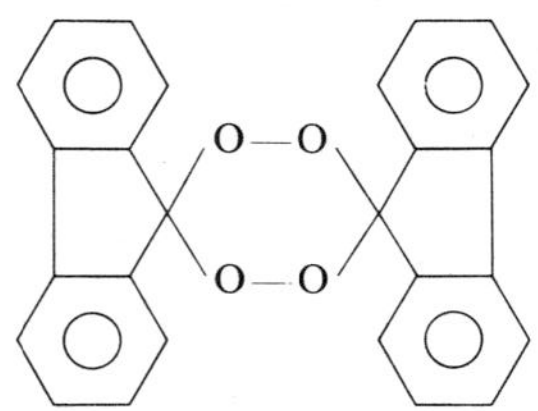	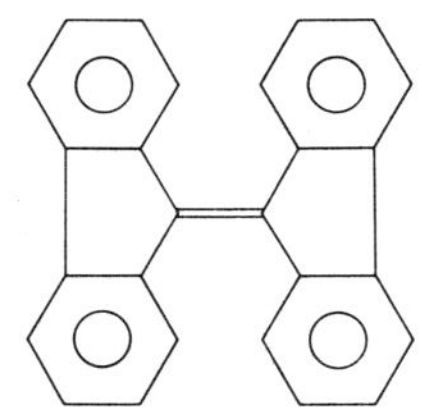	50	201–203	*194*
11	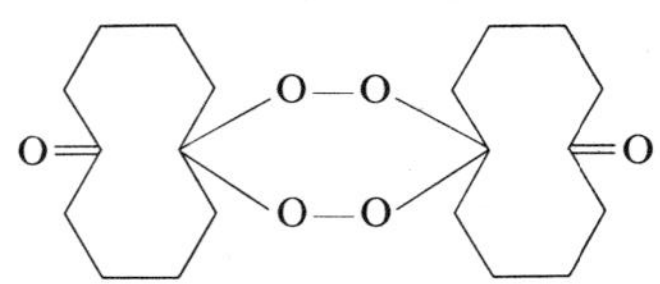	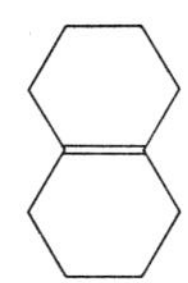	26	166	*60*

[a] Bz = benzoyl

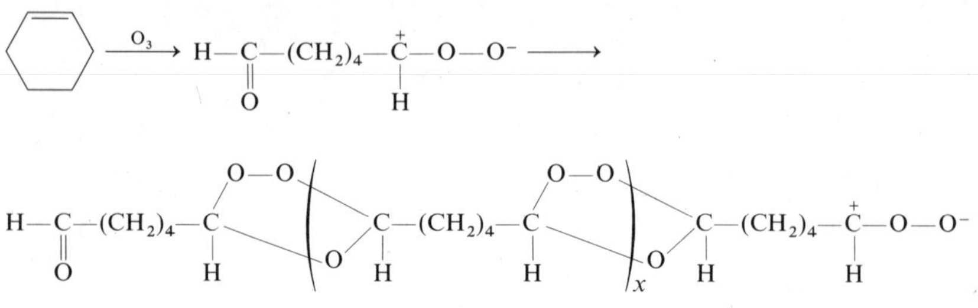

(10)

from two to six or more monomeric units in the polymer (*259–261*). These materials are reasonably stable, like monomeric ozonides.

Oligomers of simple olefins such as 1-hexene, 2-butene, 2-pentene, and 3-hexene have been studied by Razumovskii (*261*) and Greenwood (*263*) and co-workers. Razumovskii *et al.* (*234,261*), as well as other Russian investigators (*230*) suggested that they have largely cyclic structures of type **41**, where x is equal to at least 2 and no more than 12. They based this primarily on IR spectra that show strong ozonide bands (see Section I, B), but no

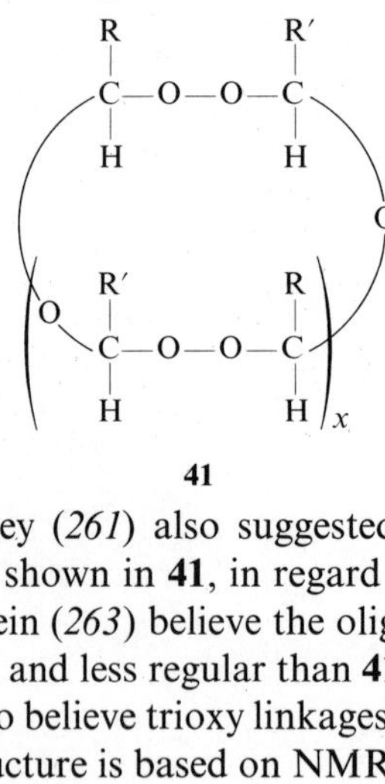

41

carbonyl absorption. They (*261*) also suggested that the structure is not always as regular as that shown in **41**, in regard to the R and R′ sequence. Greenwood and Rubinstein (*263*) believe the oligomer structure to be open chain, more complicated, and less regular than **41** in regard to the ether and peroxy linkages. They also believe trioxy linkages to be present occasionally, as shown in **42**. Their structure is based on NMR and IR spectra. The latter

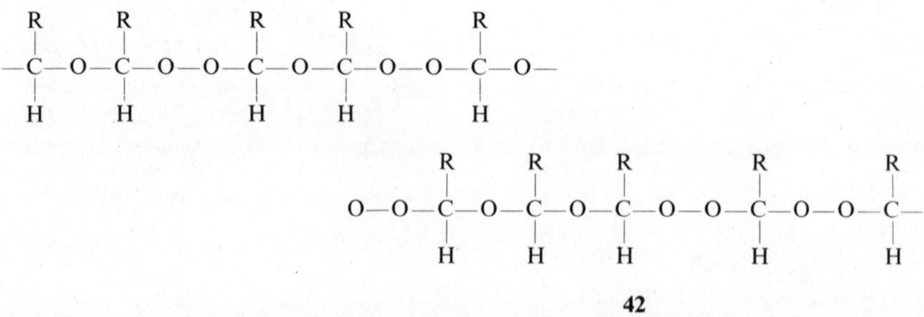

42

showed peroxy and ozonide peaks, but changed with time, giving rise to hydroxy and carbonyl peaks. They concluded that unstable linkages, such as trioxy, were present.

Tetrasubstituted ethylenes, upon ozonolysis in nonparticipating solvents, apparently give a wider variety of products, most of which are polymers of the carbonyl oxide, or of both the carbonyl oxide and ketone moieties, the former being predominant. Tetramethylethylene has been given the most study. Criegee and Lohaus (*194*) reported not only acetone diperoxide (Section II), but also the highly explosive acetone triperoxide **44** and a viscous explosive oil that appeared to be a higher polymer of carbonyl oxide **43**. Milas and Belic (*250*) reported, in addition to these, several other peroxides, which they believed to have structures **45**–**48**. In forming these, water and hydrogen peroxide also were reactants. It is important that further studies be carried out under completely anhydrous conditions.

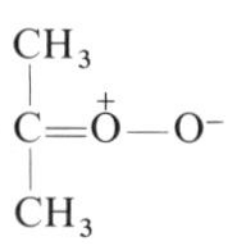

43

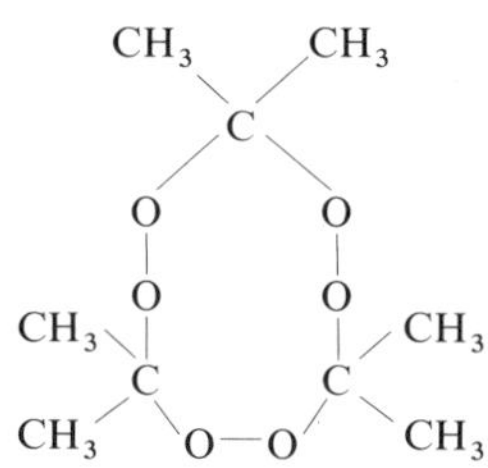

44

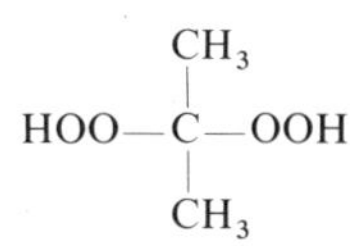

45

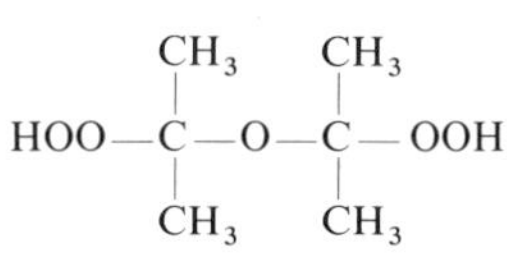

46

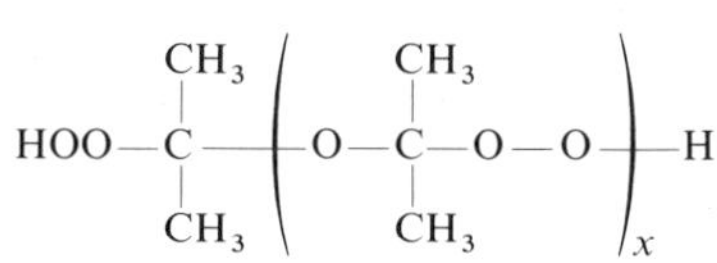

47 $x=3$

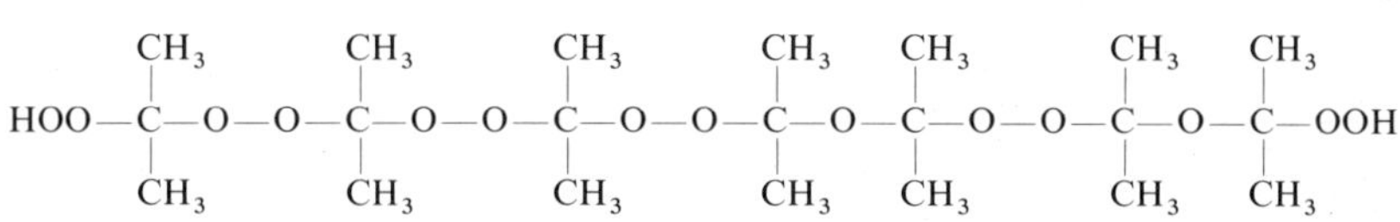

48

IV. α-Oxyalkyl Hydroperoxides

α-Alkoxy-, acyloxy-, hydroxy-, etc., alkyl hydroperoxides generally are the peroxidic ozonolysis products from olefin ozonations in participating (protic, nucleophilic) solvents (Chapter III, Section I, C and Chapter VII Section II). Select ones of these are listed in Table V. It is noteworthy that

TABLE V

α-Oxyalkyl Hydroperoxides Obtained by Ozonolysis

Example	Olefin	Solvent	Peroxide	Yield (%)	M.p. (°C)	References
1	$Me_2C{=}CMe_2$	MeOH	$Me_2C(OOH)(OMe)$	77	(liq.)	*194*
2	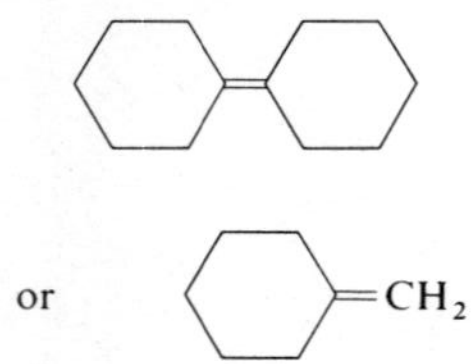	CCl_4 + MeOH	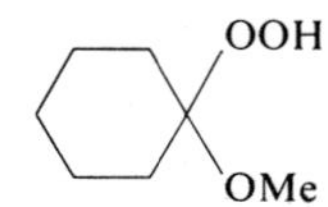	68 80	(liq.)	*194* *264*
3	$CH_3(CH_2)_5CH{=}CH_2$	EtOH	$CH_3(CH_2)_5CH(OOH)(OEt)$	92	(liq.)	*265*
4	$PhC(=O){-}CH{=}CH{-}C(=O)Ph$ or $PhC(=O){-}CH{=}C(OMe){-}C(=O)Ph$	CCl_4 + MeOH	$Ph{-}C(=O){-}CH(OOH)(OMe)$	60–84	82	*117* *120*
5	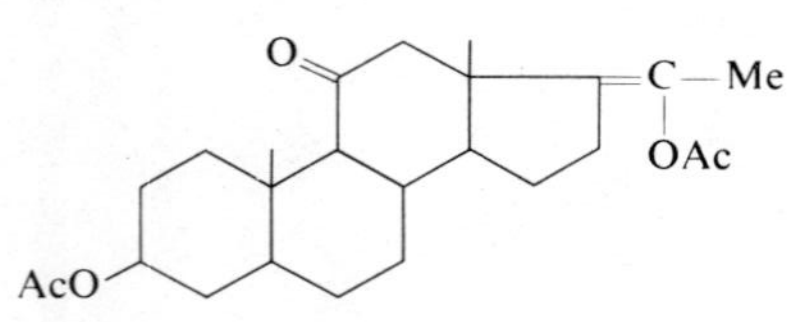		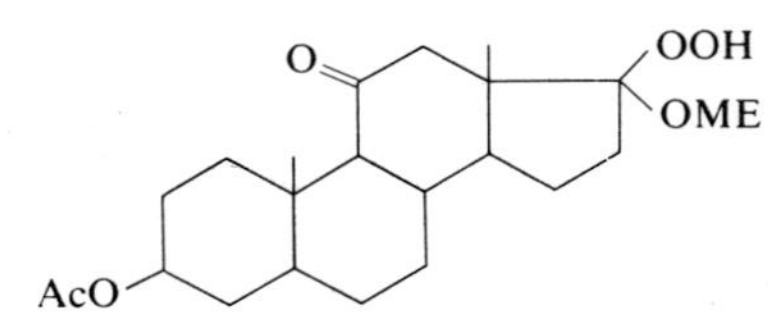	86	200–210 (dec.)	*266*

6	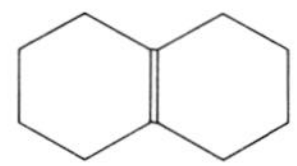	MeOH	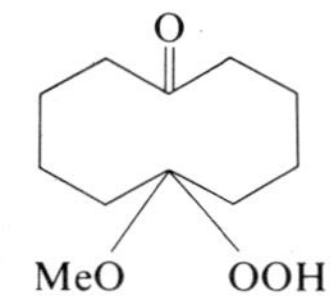	88	94–95	*60*
7	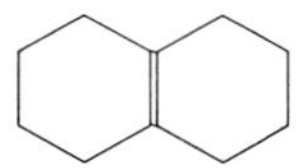	Acetone + $CH_3C(=O)$—OH	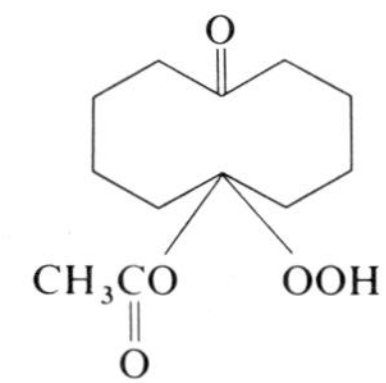	—	112–113 (perbenzoate derivative)	*60*
8	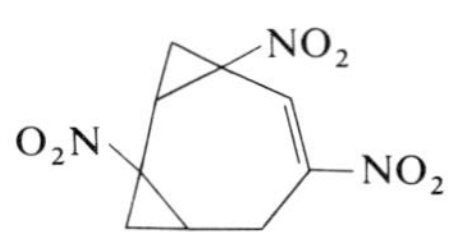	MeOH	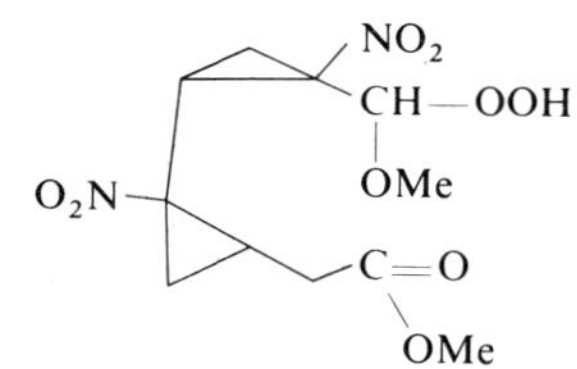	51	127–129	*267*
9	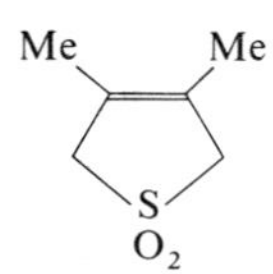	Acetone + HOH	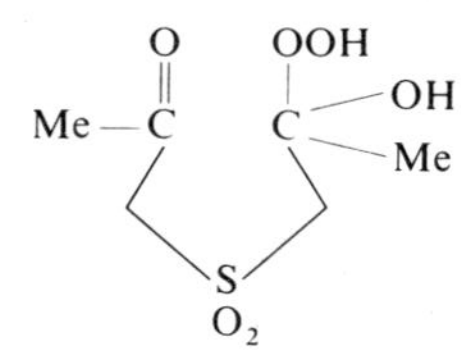	57	90	*268*

TABLE VI

CYCLIC TAUTOMERS OF α-OXYALKYL HYDROPEROXIDES FROM OZONOLYSES

Example	Olefin	Solvent	Peroxide	Yield (%)	M.p. (°C)	References
1	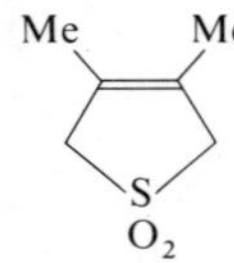	MeOH	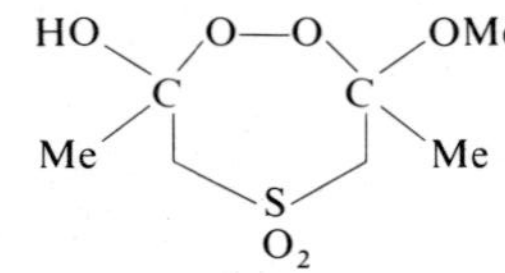	90	87–92	*268*
2	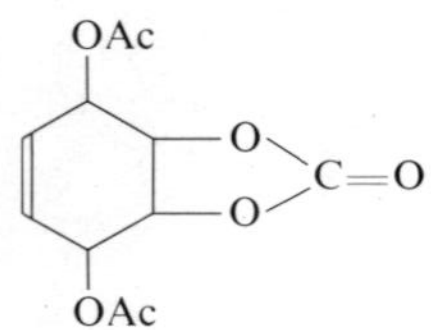	AcOH	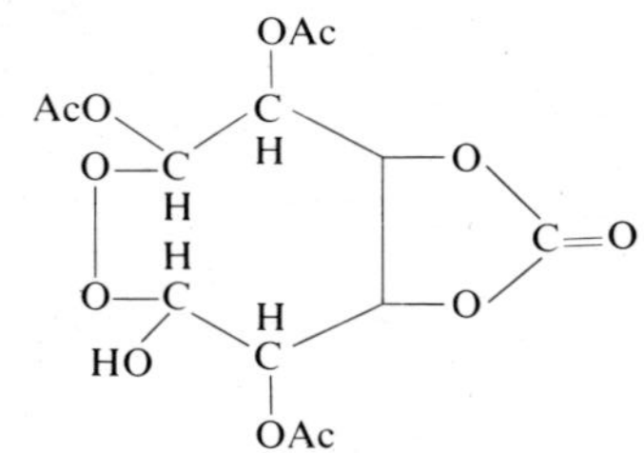	49	145–146	*280*
3	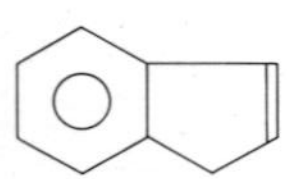	EtOH	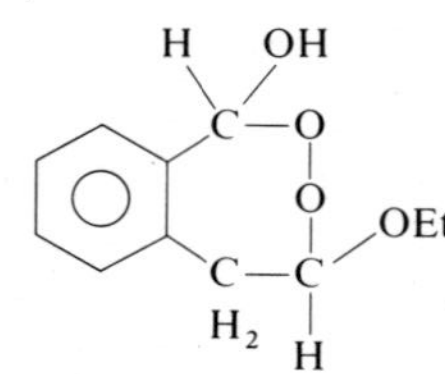 [a]	99	105–108	*281*
4	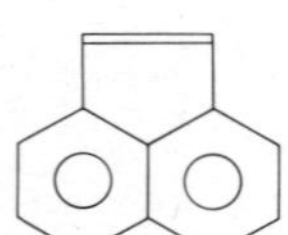	MeOH	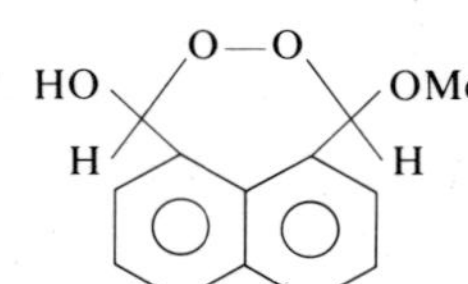	26	115–116	*282*

5	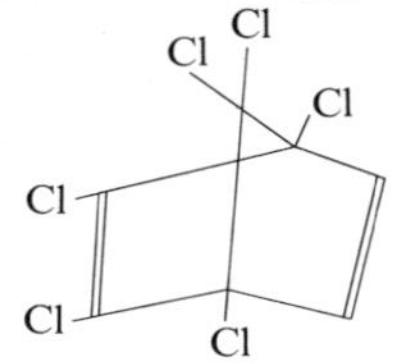	A: MeOH B: HOH	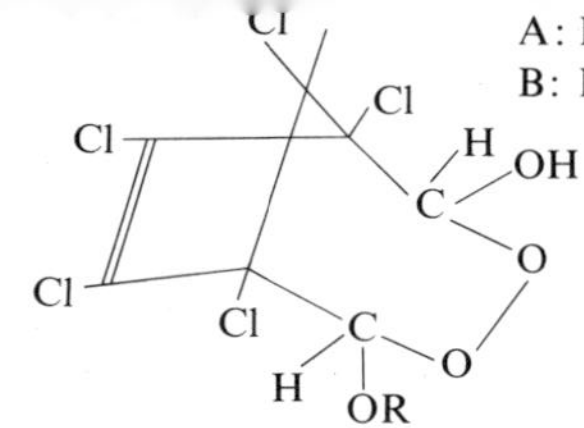 A: R=Me B: R=H	A: 68 B: 59	132–133 138–139	*260*
6	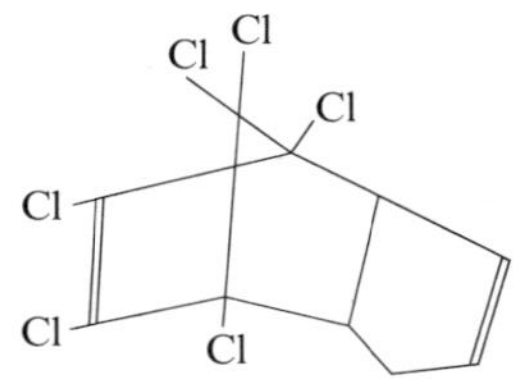	MeOH + CH_2Cl_2	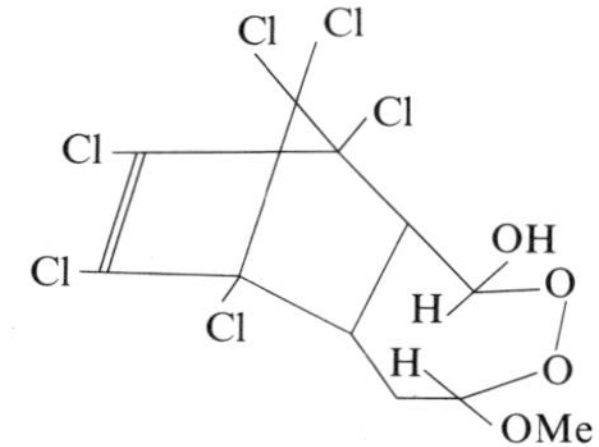	50	164 (dec.)	*262*
7	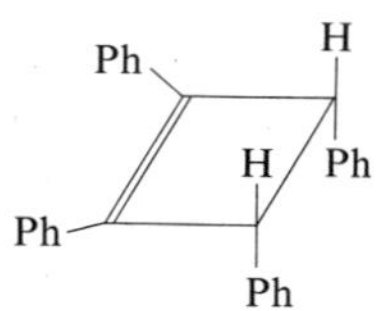	MeOH	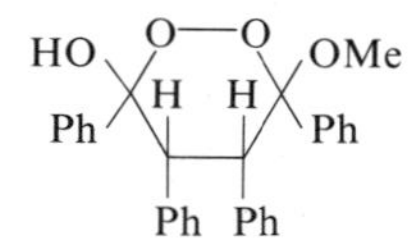	Quant	146	*130*
8	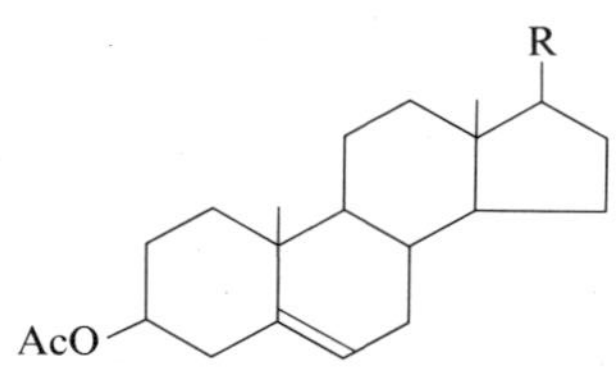 R=—CH(Me)$CH_2CH_2CH_2CHMe_2$	CCl_4 + R′OH	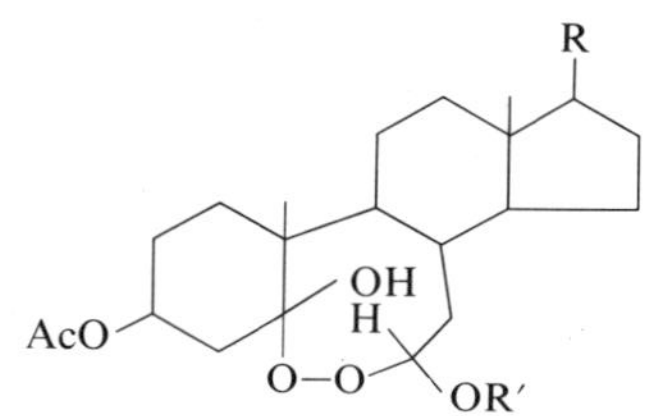 A: R′=Me, B: R′=Et, C: R′=*i*—amyl	A: 60 B: 60 C: —	145–146 140 131	*283* *284*

[a] The other possible isomer is also produced: see Chapter VII, Section II, A, structures **165** and **166**.

many of them have high melting points and are produced in high yield. Most of the oxyalkyl hydroperoxides in Table V are of the alkoxy type. A few, however, are acyloxy- (Table V, example 7), or hydroxy- (Table V, example 9) alkyl hydroperoxides. The fact that not as many of these have been characterized is probably due to their reduced stability, in comparison to the alkoxy type, rather than to a lesser tendency for the Criegee carbonyl oxide to react with water or carboxylic acids. Other examples of both the alkoxy and acyloxy types can be found in the literature (e.g., *94,134,149,264,266, 269–272*). Pospelov *et al.* (*273*) have synthesized a number of α-alkoxyalkyl hydroperoxides by ozonolyses of several olefins in various alcohols and, in several cases, have established their structure by NMR spectroscopy. The hydroperoxides in Table V were characterized in various ways, including elemental and peroxidic oxygen analyses, IR spectra that showed characteristic hydroxyl and peroxy bands, methoxy, etc. analyses, and the lead tetraacetate test for hydroperoxides (*274*).

α-Alkoxyalkyl hydroperoxides (or α-hydroperoxy ethers) are well known and have been prepared not only by ozonolysis, but also by autoxidation of ethers (*275,276*), addition of hydrogen peroxide to vinyl ethers (*277*), and interaction of acetals and hydrogen peroxide (*278*). In certain cases, α-alkoxyalkyl hydroperoxides have been made both by ozonolysis and by one of the other methods (*110,111,277–279*), as shown in Scheme 6.

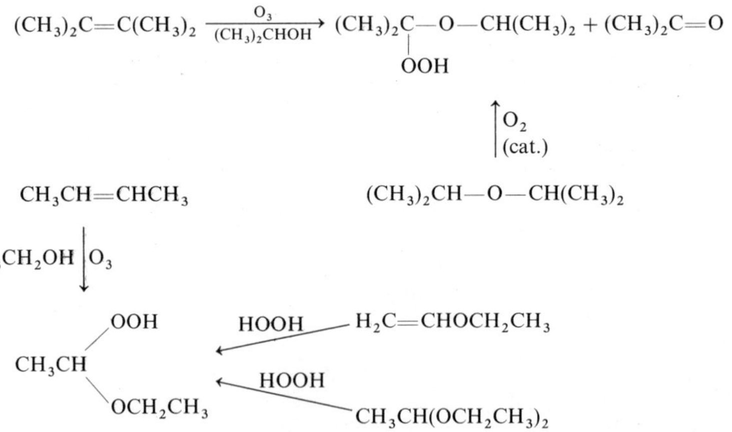

SCHEME 6. Syntheses of α-alkoxyalkyl hydroperoxides.

When certain cycloolefins are ozonized in an alcohol solvent, an intramolecular hemiperacetalization often occurs. Examples of such peroxides are shown in Table VI. There is indication that example 9 (Table V) is in equilibrium with such a structure. The examples in Table VI appear to exist entirely in the cyclic structure. In the cases examined, the IR spectrum showed no carbonyl group (unless one was present elsewhere in the molecule) and the peroxide gave a negative lead tetraacetate test (*274*). These peroxides are quite stable, often have high melting points, and usually are produced in good yields. Many other examples besides those in Table VI (*130,260, 262,268,280–284*) also have been characterized (e.g., *285–287*), including peroxides from ozonolyses of naphthalene and phenanthrene (to be discussed in Vol. II.)

CHAPTER VI

Ozonolysis of Olefins: Routes to Peroxidic Products

I. Evidence for Criegee Mechanism

The most important and unique concept of the Criegee mechanism (Chapter III, Section I, C) is that all peroxidic ozonolysis products result from various reactions of the key intermediate, the Criegee zwitterion (or carbonyl oxide). The arguments for this are strong. Perhaps the most obvious one is that ketone (or aldehyde) diperoxides and oligomers of the carbonyl oxide often are peroxidic ozonolysis products in nonparticipating solvents (Chapter V, Sections II and III), and that alkoxy- (etc.) alkyl hydroperoxides result from ozonolyses in participating solvents (Chapter V, Section IV). The simplest explanation for the formation of the latter is addition of a solvent molecule to the Criegee carbonyl oxide (Chapter III, Section I, C), although such hydroperoxides also could result by direct attack of the solvent molecule, a nucleophile, on the primary ozonide (*120*).

Evidence that ozonides are produced by recombination of the carbonyl oxide and aldehyde or ketone moieties (Scheme 2, Chapter III) is most convincing. Three different types of chemical evidence for this concept were presented early by Criegee and co-workers and were discussed in the 1958 review (*62*). The first of these dealt with ozonolyses of various olefins in the presence of an excess of a reactive aldehyde (*105,288*) to give a "*cross*" *ozonide*, presumably by interaction of the carbonyl oxide with the excess aldehyde rather than with the aldehyde moiety produced along with it (*61, 62,110*). Since these early examples, there have been many other similar examples involving cross ozonide formation with a foreign aldehyde or ketone (*96,148,188,192,217,218,228,231,237,289–298*). Some of these were mentioned and illustrated in Scheme 5 of Chapter V (*217,218*). Others have been used to argue against the Criegee mechanism (*96,237,289,292,293*). The most recent examples (*295–298*) have completely vindicated and confirmed the Criegee mechanism (see further discussion in Chapter VI, Section VI).

The second type of evidence presented early by Criegee and co-workers for the intermediacy of the carbonyl oxide in ozonide formation involved the production of 1,2-dimethylcyclopentene ozonide by ozonolysis of both 1,2-dimethylcyclopentene and *trans*-6,7-dimethyl-6-dodecene-2,11-dione, showing that the same carbonyl oxide was produced in both ozonolyses (*107*). This was illustrated in the earlier review (*62*). Since then, another similar example has been reported by Griesbaum (*299*), who showed that identical ozonides (**52**) were obtained from 1,2-dimethylcyclobutene (**51**) and *trans*-5,6-dimethyl-5-decene-2,9-dione (**49**), as illustrated in Scheme 7. The common carbonyl oxide for the two reactions is **50**. As pointed out by Griesbaum (*299*), although it is possible to explain the formation of **52** during ozonolysis of **49** by means of another mechanism (Chapter VI, Section III), it is not likely that **52** would be the sole ozonide from **49** if two different routes to ozonides were involved.

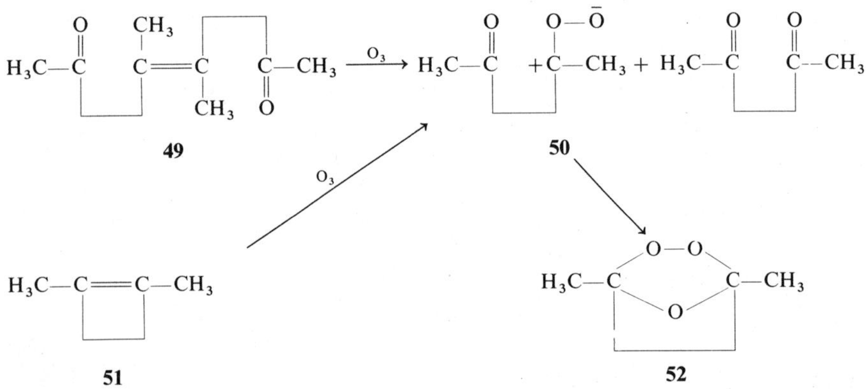

SCHEME 7. Intramolecular carbonyl oxide reactions; ozonide formation.

Similar intramolecular reactions have been carried out with compounds having terminal carboxy or hydroxy groups (*300–304*). The peroxidic ozonolysis products, however, were cyclic oxyalkyl hydroperoxides (**54**, **56**, and **59**) rather than ozonides, obtained, presumably, by interactions of the terminal zwitterion and hydroxy groups of **53**, **55**, and **58**, as illustrated in Scheme 8. It is possible, however, as pointed out specifically by Carman and Cowley (*304*) and, in general by the present author (*120*), that the primary ozonide (e.g., of **57**) rather than the zwitterion was the intermediate.

The third argument for the carbonyl oxide intermediate in ozonide formation, presented early by the Criegee school, involved the isolation of the same identical ozonide from two different olefinic compounds, as shown in Scheme 9 (*62,110,111,189*). The logical explanation is that the same carbonyl oxide and carbonyl moieties were intermediates in both ozonolyses (Scheme 9).

Since the early work of Criegee, other evidence has been forthcoming to support the intermediacy of the Criegee carbonyl oxide in ozonide forma-

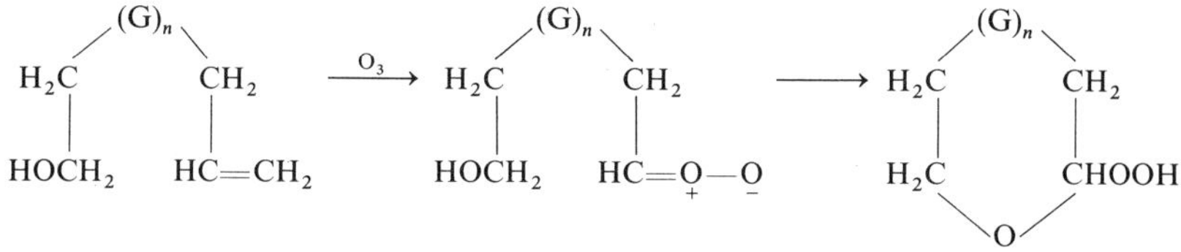

$G = O$ or CH_2
$n = O$ or 1

53 **54**

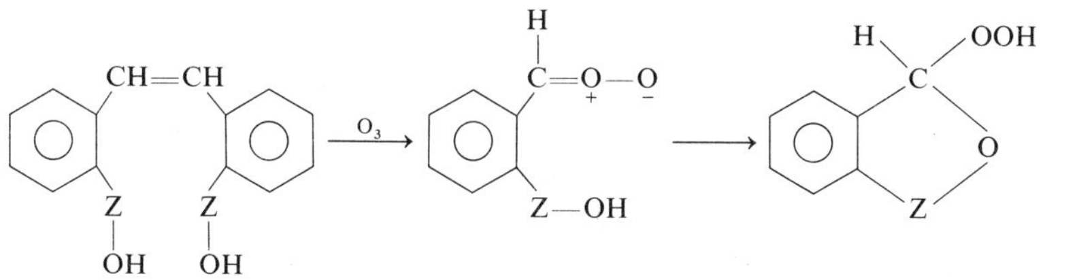

$Z = CH_2$ or $>C{=}O$

55 **56**

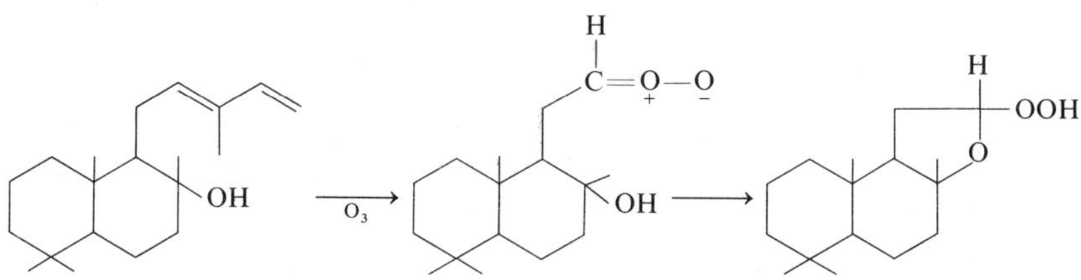

57 **58** **59**

SCHEME 8. Intramolecular carbonyl oxide reactions; α-oxyalkyl hydroperoxide formation.

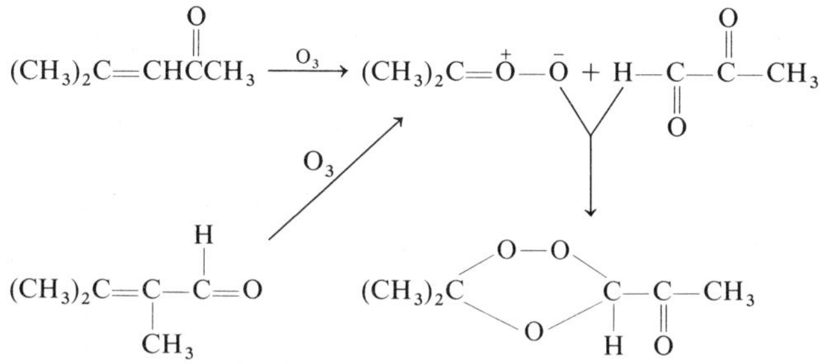

SCHEME 9. Same ozonide from two different olefins.

tion. Some of this is described in Section VI of this chapter. In addition, Murray and co-workers (*239,305–307*) have synthesized ozonides by photo-oxidation of diaryldiazomethanes in the presence of aldehydes and by oxidation of diaryl- or dialkyldiazomethanes with singlet oxygen, or tri-phenyl phosphite ozonate, in the presence of an aldehyde. In each case the oxidation of the diazo compound gave a carbonyl oxide, which in turn reacted with the aldehyde to form an ozonide. Earlier, Bartlett and Traylor (*308*) produced benzophenone diperoxide by dimerization of the similarly obtained carbonyl oxide. These reactions are outlined in Scheme 10.

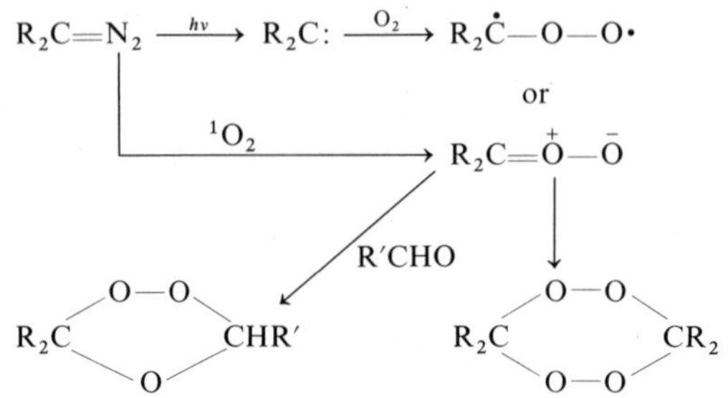

SCHEME 10. Synthesis of carbonyl oxides.

Wadt and Goddard (*78*) have stated that both ozone (see Chapter II, Section II) and the Criegee carbonyl oxide are singlet diradicals rather than zwitterions. According to them, it is both unnecessary and incorrect to show the zwitterion form as Murray and co-workers (*239,307*) did (Scheme 10), since their calculations show the zwitterion to be more energetic than the singlet diradical. These authors do not appear to consider the concept discussed in Chapter II, Section II that a singlet diradical can have a whole spectrum of character ranging from a pure diradical to a zwitterion (*79*) and, thus, react in the form best suited for the experimental conditions.

Criegee sought to obtain further evidence for a carbonyl oxide intermediate in ozonide formation by isolation of the ozonides of 3-hexene, 3-heptene, and 4-octene from ozonolysis of 3-heptene alone or of a mixture of 3-hexene and 4-octene (*110*). It was anticipated, as shown in Scheme 11, that the unsymmetrical olefin 3-heptene (**62**) should produce two pairs of moieties during ozonolysis—**63** and **64**, on the one hand, and **65** and **66** on the other—since the two alkyl groups attached to the double bond should not be different enough to cause any appreciable preferential cleavage of the primary ozonide (see Chapter VII, Section III). Recombination of these moieties should yield not only 3-heptene ozonide (**67**) from **63** plus **64** or **65** plus **66**, but also the "*cross*" *ozonides*, 3-hexene ozonide, (**68**) from **63** and **65**, and 4-octene ozonide (**69**) from **64** and **66**. Likewise, ozonolysis of a mixture of 3-hexene (**60**) and 4-octene (**61**) should afford the same four moieties and, thus, the same three ozonides. However, Criegee (*110*) was able to detect only 3-heptene ozonide (**67**) from ozonolysis of 3-heptene (**62**) and a mixture of only 3-hexene ozonide (**68**) and 4-octene ozonide (**69**) from ozonolysis of a mixture of 3-hexene (**60**) and 4-octene (**61**). Criegee (*110*) explained these surprising results on the basis that the cleavage of the primary ozonide and recombination of the carbonyl oxide and carbonyl moieties must occur in a solvent cage.

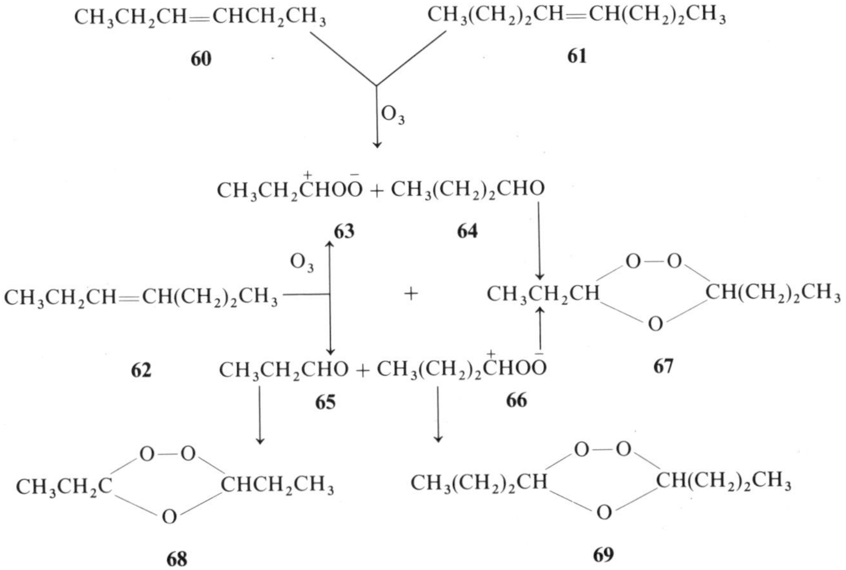

SCHEME 11. "Cross" ozonides from unsymmetrical olefins.

Subsequent to these efforts by Criegee, however, numerous papers have reported the formation of cross ozonides during ozonolyses (e.g., *128,148, 188,228,229,231,289,293,309*). Most of these were the result of advanced chromatographic techniques not available to Criegee at the time of his attempt (*14,27,187,188,227–229,310*). Also, reaction conditions, especially the concentration of the olefin solution, have been shown to play an important role in determining the yields of cross ozonides (*27,229,293,310,311*). Indo (*312*) appears to be the first to report, in 1959, the formation of a cross ozonide. He ozonized isoeugenol in chloroform and obtained, besides vanillin and acetaldehyde, two peroxides, one of which analyzed correctly for 2-butene ozonide and reduced to acetaldehyde. Vrbaski and Cvetanović (*313*), a year later, reported that a small amount of ethylene ozonide was produced from both isobutene and 1-butene. Early in 1963 Lorenz and Parks (*314*) reported a considerable quantity of 2-butene ozonide from ozonolysis of 2,6-dimethyl-2,6-octadiene. The first study of the formation of cross ozonides during ozonolysis was made by Rieszebos *et al.* (*229*). They obtained the six theoretically possible ozonides (cis and trans isomers of **71a**, **71b**, and **71c**) from ozonolysis of methyl oleate (**70**). These were obtained in approximately the statistical proportions of 2:1:1 for **71a**, **71b**,

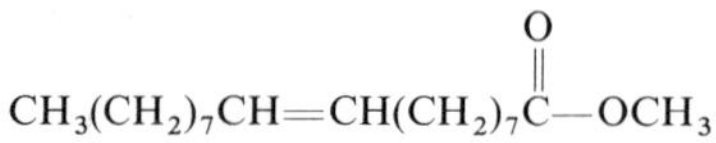

70

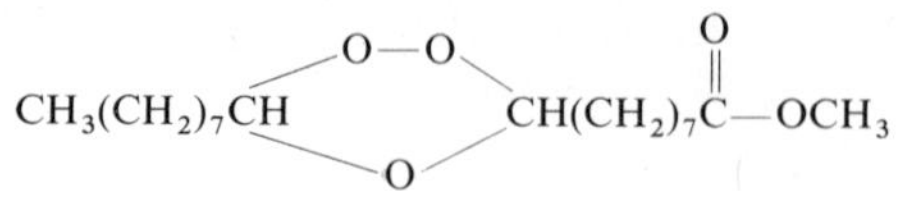

71a

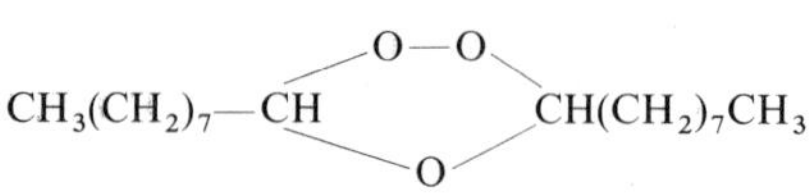

71b

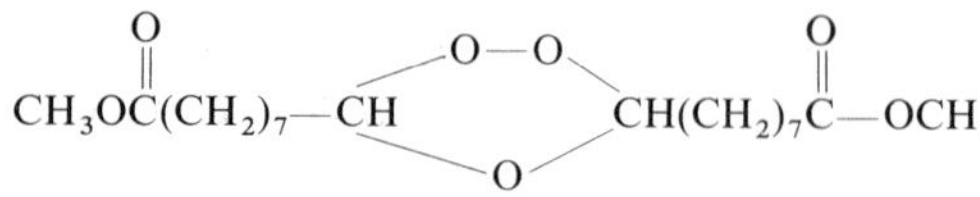

71c

and **71c**, respectively, when a 10% solution of methyl oleate in petroleum ether was ozonized at 10°C. Shortly afterward, Privett and Nickell (*228*) also, independently, reported the above six ozonides from ozonolysis of methyl oleate. These results, coupled with the demonstration that no exchange occurred between an ozonide and ozonolysis intermediates, appeared to confirm the Criegee mechanism completely (*229*).

After the work of Riezebos *et al.* (*229*) and Privett and Nickell (*228*), Murray and Williams (*293*) successfully performed the experiment at- by Criegee (*110*) (Scheme 11). They obtained the ozonides of 3-hexene, 3-heptene, and 4-octene, in cis and trans forms, from ozonolysis of 3-heptene alone and of mixtures of 3-hexene and 4-octene. Their success apparently was due to the use of more concentrated solutions of the olefins than those used by Criegee.

Not only have cross ozonides been produced as described above, but also cross ketone diperoxides (Chapter V, Section II) analogously have been

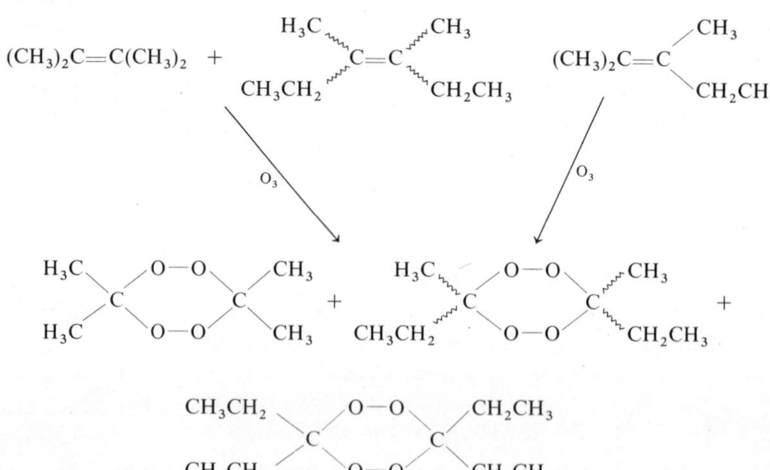

SCHEME 12. Cross ketone diperoxides.

synthesized by ozonolysis of either an unsymmetrical tetrasubstituted ethylene or a mixture of two symmetrical tetrasubstituted ethylenes (*251*). Examples are shown in Scheme 12. It is difficult to imagine any route to these diperoxides not involving Criegee carbonyl oxides.

No spectral evidence has been obtained for the Criegee carbonyl oxide from ozonolysis of an olefin. However, DeMore and Lin (*315*) observed, via low-temperature IR spectra, an unstable intermediate from ozonation of 2-butyne (and similar alkynes), which they believe to be some form of the expected carbonyl oxide, e.g., **72** from 2-butyne (this will be discussed further in Vol. II. This is also the carbonyl oxide expected from an olefin of type **73**.

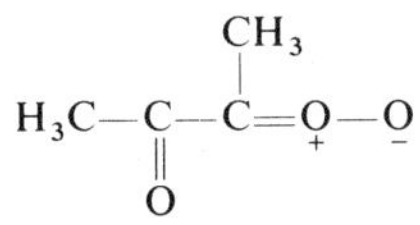

72

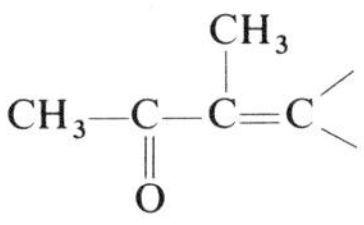

73

II. Stereochemistry. Flaws in Simple Criegee Mechanism

Concurrently with the accumulation of the convincing evidence described in Section I of this chapter favoring the Criegee mechanism, other evidence began to appear that cast doubt on the Criegee mechanism as the sole pathway to ozonides during ozonolysis of olefins. First was the discovery of Briner and co-workers (*316–319*) that certain *cis*- and *trans*- olefins give quite different ratios of ozonide to polymeric peroxides and/or free aldehydes upon ozonolysis. If both stereoisomeric olefins react with ozone to give identical pairs of carbonyl and carbonyl oxide moieties, as implied by the simple Criegee mechanism, they should give identical ratios of products.

Likewise, the simple Criegee mechanism assumed that *cis*- and *trans*-olefins should yield the same ozonide or mixture of ozonides (*109,110,112*). At first, available data seemed to confirm this. Tits and Bruylants (*178*) reported such from ozonolyses of *cis*- and *trans*-2-octenamide, but since their solvent was acetic acid, a participating solvent, they probably had no ozonide at all. More significant were the reports of identical ozonides or mixtures of ozonides from ozonolyses of the following cis-trans pairs in nonparticipating solvents: the dimethyl and diethyl esters of fumaric and maleic acids (*225*) and *cis*- and *trans*-stilbene (*109*). That these reports were incorrect and due to imprecise isolation methods was demonstrated by the elegant experiment of Schröder (*187*) who was the first to show that ozonides, including cis and trans isomers, could be separated from each other by vapor phase chromatography.

Although Criegee *et al.* (*189*) appear to have been the first to separate *cis*- and *trans*-ozonides from each other (by distillation), Schröder (*187*) was the first to determine *cis*- and *trans*-ozonide ratios accurately and to show a definite stereoselectivity in the formation of ozonides from *cis*- and *trans*-olefins. Schröder found that ozonolysis of *trans*-1,2-di-*tert*-butylethylene gave a 58% yield of ozonide, which consisted largely, if not solely (cf. *187*, *148*), of the *trans*-ozonide, while the *cis*-olefin gave an 82% yield of a mixture of 70% *cis*- and 30% *trans*-ozonides. After these results were published, many other such findings were reported (e.g., *128,141,148,193,231,238,291, 293,309,320*). Some examples are shown in Table VII.

Of even more significance was the exciting discovery of Murray *et al.* (*128,289*) and of Greenwood (*309*) that the cis:trans ratios of cross ozonides obtained from cis and trans unsymmetrical olefins often differ. Since that time, many other examples have been reported (e.g., *148,231,238,291,293*), some of which are shown in Table VIII. In addition, it has been shown that temperature (*148,309,320,321*), solvent (*148,238,309,320,322,323*), concentration of olefin (*322*), and other reaction conditions (*293*) affect cis:trans ozonide ratios, overall yields of ozonides, and the ratio of cross to normal ozonides. Some of these findings are illustrated in Tables VII and VIII. (*128,148,238,309,324*). A completely satisfactory mechanism must explain these phenomena.

Most of the yield and cis:trans ratio data given in Tables VII and VIII were obtained by gas chromatographic techniques. Rubinstein (*325*) has pointed out that the type of column and the gas chromatographic conditions can greatly affect ozonide stabilities and, thus, cis:trans ratios. The data in Tables VII and VIII appear to have been obtained under conditions where ozonides are stable.

Of prime importance in reconciling mechanistically the stereoselectivity observed in ozonide formation is the establishment of the validity of the configurational assignments made for the *cis*- and *trans*-ozonides. In general, *cis*-ozonides have a larger dipole moment than *trans*-ozonides (*189*) and have the longer retention time in gas chromatography (*128,148,187,188, 231*) or column chromatography (*326*) and the lower R_f value in thin-layer chromatography (*193,228,229*). As discussed earlier in Chapter V, Section I, B, IR and NMR spectra can often be used to distinguish between *cis*- and *trans*-ozonides. *Cis*-ozonides have a characteristic IR peak around 830 cm^{-1}, compared to 1320 cm^{-1} for *trans*-ozonides, and have ring proton absorption at lower field than have *trans*-ozonides. The NMR characteristic peaks are in the range of $\delta = 4.87 - 6.20$ ppm. Another criterion is that *cis*-ozonides usually are more rapidly reduced by chemical reducing agents than are *trans*-ozonides (*187,231*). None of these methods alone, however, are completely reliable. Murray *et al.* (*327*) and Criegee and Korber (*326*)

TABLE VII

CIS:TRANS RATIOS OF OZONIDES FROM CIS AND TRANS SYMMETRICAL OLEFINS[a]

Example	R in RCH=CHR	Configuration	Solvent	Concentration (M)	Temperature (°C)	Ozonide yield (%)	Ozonide cis:trans ratio	References
1	*t*-Butyl	Cis	Pentane	0.3	−75	82	70:30	*148,187*
2	*t*-Butyl	Trans	Pentane	0.3	−75	58	0:100	*187*
3	*t*-Butyl	Trans	Pentane	0.3	−75	51	28:72	*148*
4	Isopropyl[b]	Cis	Pentane	1	−70	85	66:34	*128*
5	Isopropyl[b]	Cis	Pentane	1	−75	[c]	62:38	*148*
6	Isopropyl	Cis	CH_2Cl_2	1	−70	58	63:37	*322*
7	Isopropyl	Cis	EtOAc	1	−70	81	53:47	*322*
8	Isopropyl	Cis	Et_2O	1	−70	84	56:44	*322*
9	Isopropyl[b]	Trans	Pentane	1	−70	47	53:47	*128*
10	Isopropyl[b]	Trans	Pentane	1	−75	52	45:55	*148*
11	Isopropyl	Trans	CH_2Cl_2	1	−70	56	49:51	*322*
12	Isopropyl	Trans	EtOAC	1	−70	57	35:65	*322*
13	Isopropyl	Trans	Et_2O	1	−70	73	35:65	*322*
14	Ethyl	Cis	Pentane	1	−70	81	53:47	*128*
15	Ethyl	Trans	Pentane	1	−70	49	53:47	*128*
16	Ethyl	Cis	Pentane	0.2	−30	88	48:52	*309*
17	Ethyl	Trans	Pentane	0.2	−30	22	41:59	*309*
18	Ethyl	Cis	Et_2O	0.2	−70	80	38:62	*309*
19	Ethyl	Trans	Et_2O	0.2	−70	42	33:67	*309*
20	Methyl[b]	Cis	Pentane	1	−70	48	39:61	*128*
21	Methyl[b]	Trans	Pentane	1	−70	36	38:62	*128*
22	Methyl[b]	Cis	Butane	0.2	−30	72	35:65	*309*
23	Methyl[b]	Trans	Butane	0.2	−30	36	34:66	*309*
24	Methyl	Cis	Me_2O	0.2	−70	62	29:71	*309*
25	Methyl	Trans	Me_2O	0.2	−70	34	30:70	*309*

[a] Reprinted, in part, with permission, from Bauld *et al.* (*148*). Copyright by the American Chemical Society.

[b] Cis:trans ratios for these ozonations were also reported by Lorenz and Parks (*231*), but the values were so different from those reported by the other sets of authors (whose values were within reasonable areas of agreement) that they appear to be highly questionable.

[c] Not determined.

TABLE VIII

CIS:TRANS RATIOS (C:T) OF OZONIDES FROM CIS AND TRANS UNSYMMETRICAL OLEFINS[a]

Example	R and R′ in RCH=CHR′	Configuration	Solvent[b]	Temperature (°C)	Pair A[c] C:T	Pair A[c] (%[d])	Pair B[c] C:T	Pair B[c] (%[d])	Pair C[c] C:T	Pair C[c] (%[d])	References
1	R = *i*-Pr, R′ = CH_3	Cis[e]	Pentane	−75	71:29	(11)	48:52	(86)	43:57	(3)	*148*
2	R = *i*-Pr, R′ = CH_3	Trans[f]	Pentane	−75	40:60	(12)	46:54	(85)	31:69	(3)	*148*
3	R = *i*-Pr, R′ = CH_3	Cis	Pentane	−30	66:34	(20)	50:50	(72)	43:57	(8)	*148*
4	R = *i*-Pr, R′ = CH_3	Trans	Pentane	−30	46:54	(17)	52:48	(77)	41:59	(6)	*148*
5	R = *i*-Pr, R′ = CH_3	Cis	AcOEt	−75	57:43	(30)	47:53	(58)	34:66	(12)	*148*
6	R = *i*-Pr, R′ = CH_3	Cis	AcOEt	0	60:40	(27)	51:49	(59)	40:60	(14)	*148*
7	R = *i*-Pr, R′ = CH_3	Trans	AcOEt	−75	49:51	(17)	48:52	(75)	32:68	(8)	*148*
8	R = *i*-Pr, R′ = CH_3	Cis	CH_2Cl_2	−75	71:29	(31)	63:37	(56)	56:44	(13)	*148*
9	R = *i*-Pr, R′ = CH_3	Cis	CH_2Cl_2	0	62:38	(24)	54:46	(58)	49:51	(18)	*148*
10	R = *i*-Pr, R′ = CH_3	Trans	CH_2Cl_2	−75	60:40	(25)	55:45	(60)	41:59	(15)	*148*
11	R = Et, R′ = CH_3	Cis	Pentane	−70	56:44	(14)	42:58	(68)	44:56	(18)	*128*
12	R = Et, R′ = CH_3	Trans	Pentane	−70	41:59	(8)	38:62	(82)	32:68	(10)	*128*
13	R = Et, R′ = CH_3	Cis	Me_2O[b]	−70	40:60	(22)	34:66	(53)	29:71	(25)	*309*
14	R = Et, R′ = CH_3	Trans	Me_2O[b]	−70	40:60	(17)	36:64	(52)	27:73	(31)	*309*
15	R = *t*-Bu, R′ = CH_3	Cis[g]	Pentane	−70	74:26	(3)	54:46	(88)	46:54	(9)	*128*
16	R = *t*-Bu, R′ = CH_3	Trans	Pentane	−70	ND[h]		30:70	(ND)[h]	(ND)[h]		*128*
17	R = *t*-Bu, R′ = Et	Cis[i]	Pentane	−70	82:18	(11)	67:33	(77)	62:38	(12)	*128*
18	R = *t*-Bu, R′ = Et	Trans[j]	Pentane	−70	68:32	(ND)[h]	38:62	(ND)[h]	(ND)[h]		*238*
19	R = CH_3, R′ = H	~	Isobutane	−95	81:19	(4)	~	(85)	~	(11)	*324*

[a] Reprinted, in part, with permission, from Bauld *et al.* (*148*). Copyright by the American Chemical Society.

[b] The solution concentrations were 1 *M* except in examples 13 and 14, where the concentrations were 0.2 *M*.

[c] Pair A is the higher molecular weight cross-ozonide pair, pair B is the normal ozonide pair, and pair C is the lower molecular weight cross-ozonide pair.

[d] This represents the percentage of each ozonide pair in the mixture of ozonides.

[e] Murray *et al.* (*128*) reported 66:34, 49:51, and 49:51 for pairs A, B, and C, respectively.

[f] Murray *et al.* (*128*) reported 50:50 (A), 48:52 (B), and 38:62 (C).

[g] Fliszár and Carles (*238*) reported 76:24 (A), 49:51 (B), and 46:54 (C).

[h] ND = not determined, usually because of some interference.

[i] Fliszár and Carles (*238*) reported 73:27 (A), 60:40 (B), and 51:49 (C).

[j] Murray *et al.* (*128*) reported 29:71 (B).

have made unequivocal configurational assignments on the basis that, with the ozonides studied, the trans isomer exists in racemic form and should be resolvable, whereas the cis isomer is meso. The experiment of Murray *et al.* (*327*) involved reduction of each isomeric ozonide with less than the stoichiometric amount of brucine and checking the unreacted ozonide for optical activity. Only one ozonide showed such activity and was accordingly assigned the trans configuration. Reduction by optically active brucine destroyed one enantiomer at a faster rate than the other. The method of Criegee and Korber (*326*) involved chromatography over an optically active substrate (cellulose acetate), on which one enantiomer of the *dl* pair was selectively absorbed. Their results coincided with some of the previous criteria, but not all. Comparison of retention times in gas and column chromatography was shown to be a reliable criterion with ozonides of 1,2-disubstituted ethylenes. For further discussion of cis:trans ozonide assignments and for gas chromatographic substrates and techniques, the reader is referred elsewhere (*14,237–239,326,327*).

III. The SMY Mechanism

From the data of Tables VII and VIII it can be seen that there is a definite stereoselectivity in the production of ozonides from *cis*- and *trans*-olefins. The stereoselectivity expresses itself in terms of a higher cis:trans ozonide ratio from a *cis*-olefin than from a *trans*-olefin, both in normal and cross ozonides. In fact, where R in a symmetrical ozonide pair is isopropyl or *tert*-butyl (Table VII, examples 1–13 and Table VIII, examples 1–10, 15–18), the cis:trans ozonide ratio reported is generally greater than one from a *cis*-olefin and less than one from a *trans*-olefin, especially in nonpolar solvents. This cannot be explained by the original Criegee mechanism, which proposed identical carbonyl oxide (**13**) and carbonyl (**14**) moieties (Scheme 2, Chapter III) as intermediates from *cis*- and *trans*-olefins (*110,112*).

For this reason, Story, Murray, and Youssefyeh (*127,128,289*) proposed in 1966 that there are at least three competing pathways for ozonide formation: the Criegee mechanism, which they at that time considered to be minor in most cases; a new mechanism, or working hypothesis, which will be referred to in this book as the SMY mechanism and which they considered to be important for many *trans*-olefins; and rearrangement of a σ complex (or peroxy epoxide), which they hypothesized might be major for *cis*-olefins with bulky substituents at the double bond. In this proposal, which is outlined in Schemes 13 and 14, it was suggested that a π complex (**74**) forms first. With *trans*-olefins and sterically unhindered *cis*-olefins, at least, the proposal considered the π complex to collapse to the primary ozonide, **75** (Scheme 13) and **85** (Scheme 14), as its major fate. The primary ozonide,

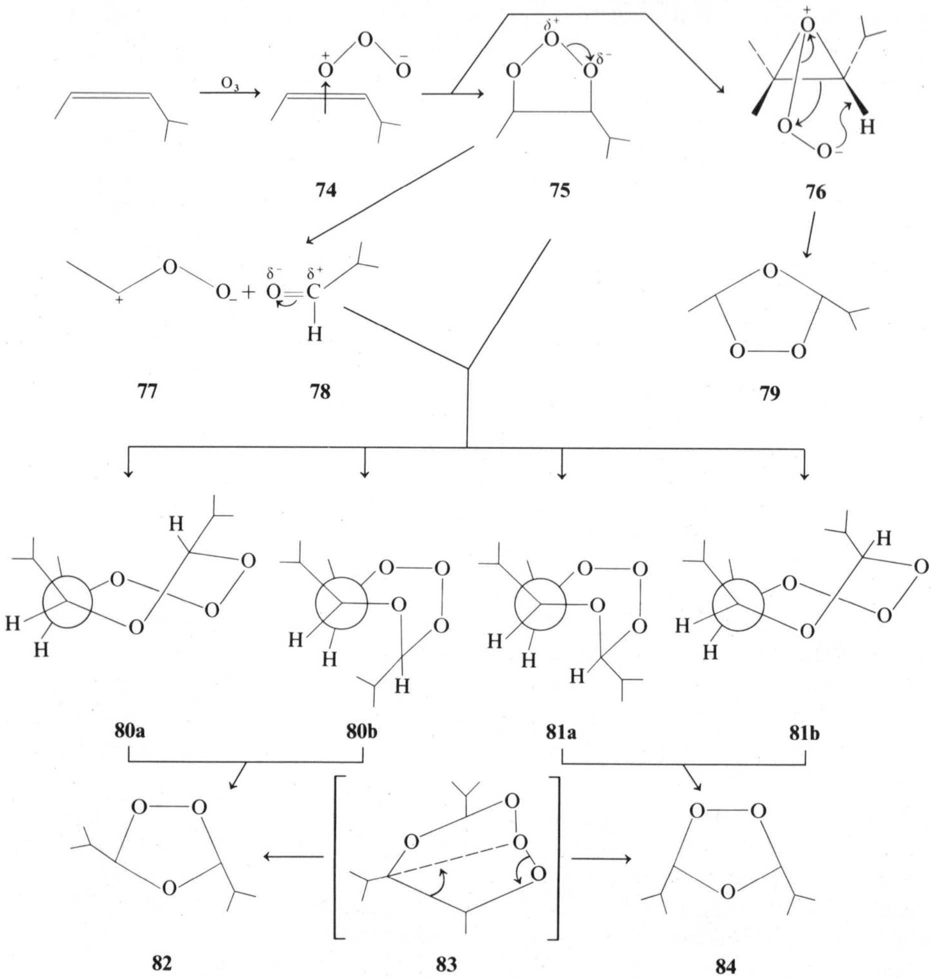

SCHEME 13. SMY mechanism for *cis*-olefins; illustrated with *cis*-4-methyl-2-pentene. Reprinted, in part, with permission, from Murray *et al.* (*128*). Copyright by the American Chemical Society.

in turn, might either break down to the Criegee moieties (**77** and **78**) or it might be attacked by an aldehyde to give an adduct (**80**, **81**, **86**, or **87**) the preferred configuration and conformation of which determined the predominant stereochemical course of normal ozonide or cross ozonide (**82** or **84**) formation. The conversion of the adduct to ozonide and aldehyde is illustrated in **83**. According to the conformational analysis of Murray *et al.* (*27,128*), conformation **81a** of Scheme 13 would be favored in the case of *cis*-olefins, since it has only H,H,H nonbonded repulsions, but there would

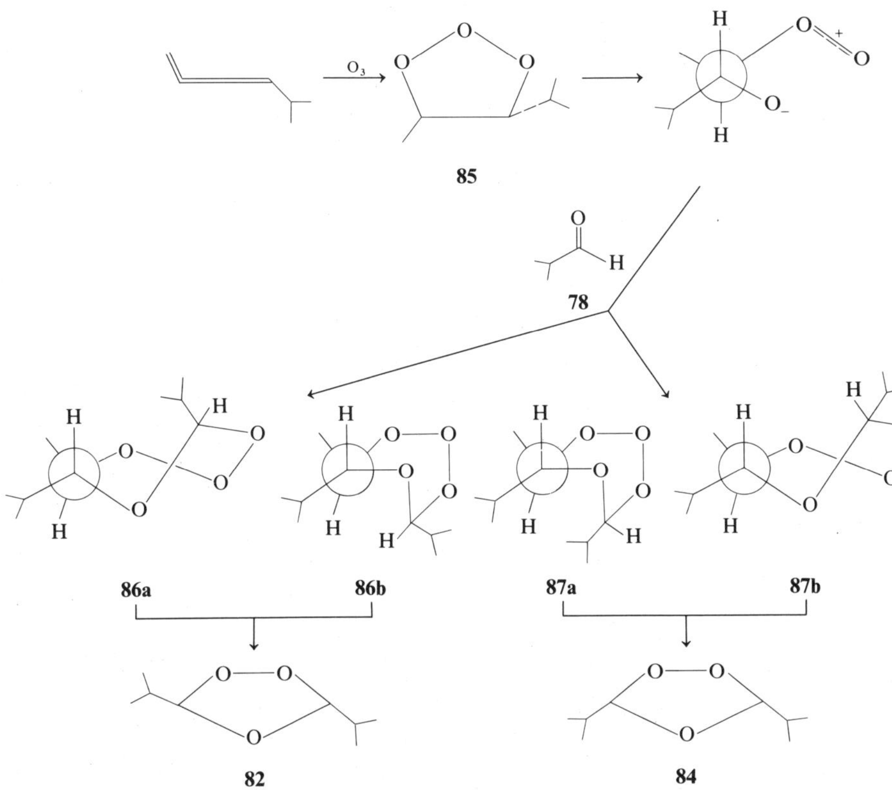

SCHEME 14, SMY mechanism for *trans*-olefins; illustrated with *trans*-4-methyl-2-pentene Reprinted, in part, with permission, from Murray *et al.* (*128*). Copyright by the American Chemical Society.

be no preference among the configurations and conformations **86** and **87** for *trans*-olefins. Thus, the effect of this mechanism, in competition with the Criegee mechanism and the rearrangement of the peroxy epoxide, **76** → **79**, would be to produce a higher cis:trans ratio in ozonides from a *cis*-olefin than in ozonides from a *trans*-olefin (*27,128*). This mechanism has now been shown not to occur to any measurable extent (*297,298*) (see Section VI, A of this chapter).

IV. Refinements of Criegee Mechanism

A. BAULD–BAILEY SCHEME. *SYN* AND ANTI CARBONYL OXIDES

To this reviewer, one apparent criticism of the above-described SMY mechanism is in the conformational analysis for *trans*-olefins (Scheme 14). Conformation **87b** has only H,H,CH_3 nonbonded repulsions and should be favored over the others. Therefore, there should be a preference for the *cis*-ozonide with both *cis*- and *trans*-olefins if the SMY mechanism prevails (*148*). This, however, is not the case, at least under the conditions employed in Tables VII and VIII.

For this reason and also because of the strong and varied evidence already discussed for the Criegee carbonyl oxide, Bauld, Bailey *et al.* (*148*) in 1968 suggested refinements for the Criegee mechanism in order to account for the stereochemistry of ozonide and cross-ozonide formation. Most important of the refinements were that (a) the 1,2,3-trioxolane primary ozonide and the 1,2,4-trioxolane final ozonide exist in nonplanar, probably half-chair, conformations; (b) the Criegee carbonyl oxide exists in syn and anti configurations (Scheme 15, structures **89** and **91a**), which are produced in a preferential manner depending on the geometry of the primary ozonide; (c) the recombination of the carbonyl oxide and carbonyl moieties to give a final ozonide (1,2,4-trioxolane) occurs stereoselectively, depending on the configuration of the carbonyl oxide and on the preferred conformation of the final ozonide.

These refinements emphasize suggestions of Huisgen (*80,150*) and Criegee (*112*) that ozonide formation is the result of two concerted 1,3-dipolar cycloadditions with a concerted 1,3-dipolar cleavage sandwiched in between. Criegee (*112*) had earlier suggested the possibility of stereoisomeric carbonyl oxide zwitterions, but did not tie them in with the stereoselectivity of ozonide formation.

Structures such as **91a** and **91b** are contributing structures to the resonance hybrid of the carbonyl oxide. Since **91a** should be more stable, owing to a lesser separation of charge, the zwitterion should have considerable double-bond character, making syn and anti forms of carbonyl oxides possible (*148*).

Calculations by Wadt and Goddard (*78*) indicate that **91b** is too unstable to make any appreciable contribution. As discussed earlier, Wadt and Goddard (*78*) believe the Criegee carbonyl oxide to be purely a singlet diradical. Even if this viewpoint prevails, their planar structure, **91c**, should be able to exist in syn and anti forms, particularly in view of the broad spectrum of character associated with singlet diradicals as discussed earlier (Chapter II, Section II; Section I of this chapter).

The Bauld–Bailey ideas are outlined in Scheme 15 and in Table IX. The half-chair conformations proposed in this scheme are such that the carbon atoms are maximally staggered in both the 1,2,3- and the 1,2,4-trioxacyclopentane rings. The transformations depicted were formulated by Bauld and Bailey (*148*) into three general rules. *Rule 1** states that *in the concerted*

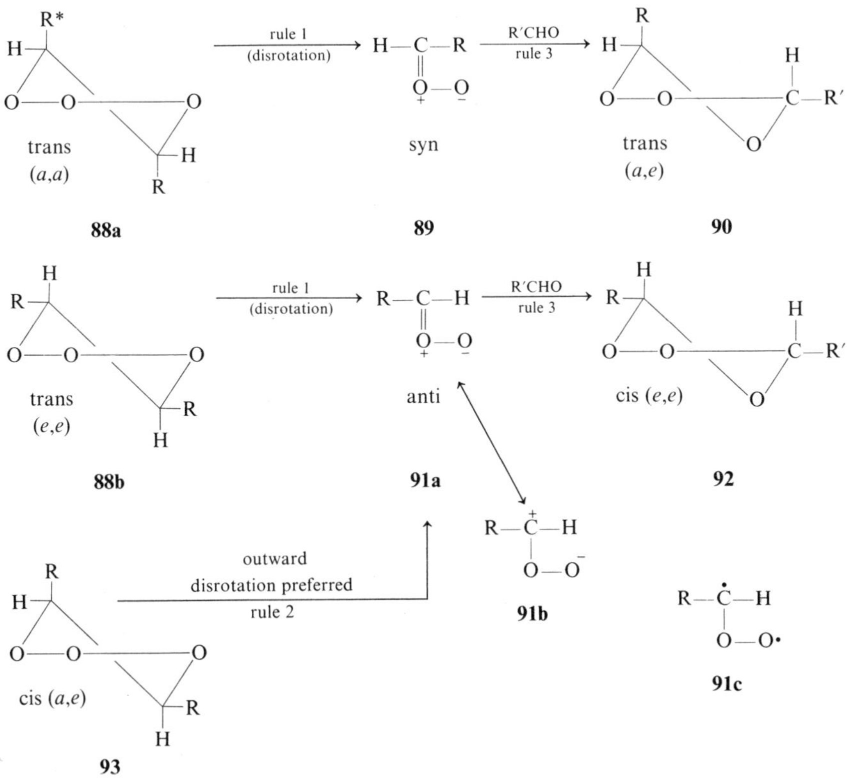

SCHEME 15. Bauld–Bailey scheme for stereoselectivity of ozonide formation. Reprinted, in part, with permission, from Bauld *et al.* (*148*). Copyright by the American Chemical Society. *a* = axial; *e* = equatorial; *-diaxial preferred with large substituents.

disrotatory 1,3-dipolar cleavage of primary ozonides, equatorial substituents are preferentially converted into anti and axial substituents into syn carbonyl oxide zwitterions (Scheme 15). This is a result of the fact that in the primary ozonide these groups are already approaching such geometry, and less rotation is required during cleavage in order to achieve this geometry. *Rule 2** states that *an equatorial substituent is incorporated into a carbonyl oxide zwitterion in preference to an axial substituent.* Such a choice arises only in the case of the cis primary ozonide. Here an outward cleavage movement (away from the substituents) should certainly be sterically preferred. Such a process would favor the incorporation of the equatorial substituent into the carbonyl oxide, again through the principle of least rotation. *Rule 3** states

* Reprinted, with permission, from Bauld *et al.* (*148*). Copyright by the American Chemical Society.

that *aldehydes preferentially interact with anti carbonyl oxides so as to orient bulky substituents diequatorially* (*cis*) *and with syn carbonyl oxides so as to orient bulky substituents into an equatorial–axial conformation* (*trans*). This results from the reverse of Rule 1. An anti carbonyl oxide preferentially installs its substituent into an equatorial position in the final ozonide, causing the aldehyde also to prefer to place its substituent equatorially, in order to avoid unfavorable interactions. A syn carbonyl oxide preferentially places its substituent axially, causing the aldehyde to install its substituent equatorially; this avoids 1,3-diaxial interactions between the substituents.

Table IX summarizes the effects of these rules. *Cis*-olefins having large substituents attached to the carbons of the double-bond yield, preferentially but not exclusively, *anti*-carbonyl oxides and cis final ozonides, whereas *trans*-olefins yield *syn*-carbonyl oxides and trans final ozonides, predominantly. This preference in the case of a trans olefin is due to the fact that in a trans primary ozonide with large substituents the diaxial conformation should be more stable. According to the Bauld–Bailey proposal, these trends should decrease as the bulk of the substituents decreases and might disappear entirely when neither substituent is larger than methyl or ethyl. On this basis, the Bauld–Bailey rules fit most of the data of Table VII and many of the data of Table VIII in regard to ozonides having both substituents larger than ethyl. Otherwise, most of the anomalies occur with polar solvents. Solvent and other effects are discussed in Chapter VII, Section I. It should be pointed out here, however, that even minor differences in experimental conditions and techniques can cause different results. Perhaps this is the reason for the different values in examples 9 and 10 in Table VII and in the footnotes of Table VIII.

Extended Hückel linear combination of atomic orbital–molecular orbital (LCAO–MO) calculations by Fliszár (*328*) and Rouse (*329*) for primary ozonides (1,2,3-trioxacyclopentanes) of ethylene, and methyl, ethyl, isopropyl, and *tert*-butyl mono- or disubstituted ethylenes confirmed the hypothesis of Bauld and Bailey (*148*) that primary ozonides have a carbon–carbon half-chair conformation (e.g., **88** and **93**).

Rouse (*329,330*), however, also calculated the stabilities of various conformations of primary ozonides by the complete neglect of differential overlap (CNDO/2) method, which he considered to be more accurate for such heterocycles. These calculations, contrary to the extended Hückel calculations, indicated that with primary ozonides having substituents such as methyl, ethyl, isopropyl, and *tert*-butyl, several conformers lie close together in energy and no one conformer would be expected to be major in a sample of the primary ozonide. Substituent lone-pair interactions appeared to be more important than substituent–substituent interactions in dictating ring conformation. The results did indicate, however, that as substituents

TABLE IX

ORIGINAL BAULD–BAILEY RULES: PREFERRED STEREOCHEMICAL COURSE OF OZONIDE FORMATION WITH OLEFINS HAVING BULKY SUBSTITUENTS[a]

Olefin configuration	Primary ozonide		Predominant carbonyl oxide configuration	Predominant final ozonide	
	Configuration	Conformation[b]		Configuration	Conformation[c]
Cis	Cis	*a,e*	Anti	Cis	*e,e*
Trans	Trans	*a,a*[d]	Syn	Trans	*a,e*
Trans	Trans	*e,e*	Anti	Cis	*e,e*

[a] Reprinted, with permission, from Bauld *et al.* (*148*). Copyright by the American Chemical Society.
[b] C–C half-chair conformation with substituents axial (*a*) or equatorial (*e*) as indicated.
[c] C–O half-chair conformation with substituents axial or equatorial as indicated.
[d] Axial–axial conformation preferred with primary ozonides having large substituents.

become larger the stability of the equatorial position decreases, especially with disubstituted primary ozonides. In contrast to Rouse's conclusions, Blint and Newton (*331*) pointed out that the carbon–carbon half-chair conformation proposed for a primary ozonide by Bauld and Bailey (*148*) is the same type of conformation found to exist for H_2O_3, of which a 1,2,3-trioxacyclopentane is, in essence, a dialkyl derivative.

1,2,4-Trioxacyclopentanes (final ozonides), with small substituents at least, have been shown experimentally to have an oxygen–oxygen half-chair, **94**, rather than the carbon–oxygen half-chair, **90** or **92**, conformation hypothesized by the Bauld–Bailey mechanism (*148*). This was established by means of electron diffraction studies for ethylene ozonide by Almenningen *et al.* (*332*) and by means of microwave spectra for ethylene, propylene, and *trans*-2-butene ozonides by Kuczkowski *et al.* (*324,333,334*). Electron-diffraction studies with the *trans*-ozonide of methyl *p*-methoxycinnamate (i.e., 3-*p*-anisyl-5-carbomethoxy-1,2,4-trioxacyclopentane) were not decisive, but indicated that the oxygen–oxygen half-chair conformation was probably predominant (*335*). Rouse (*336*) has made CNDO/2 calculations that are in agreement with these results, including ozonides that have substituents as large as *tert*-butyl. These findings, therefore, require modification of the Bauld–Bailey rules (*148*).

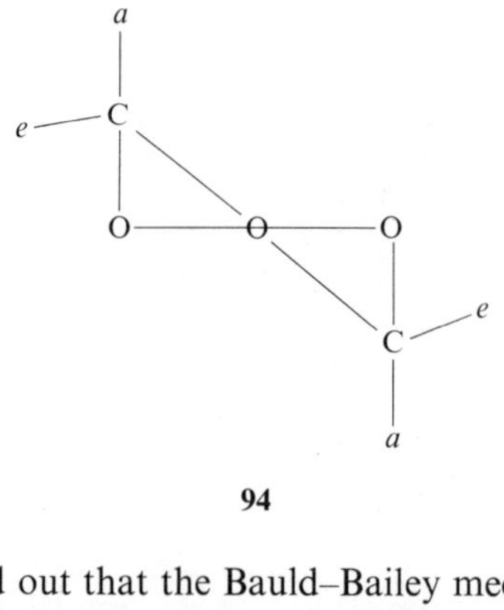

94

It should be pointed out that the Bauld–Bailey mechanism (*148*) was envisioned as the first step in reconciling the Criegee zwitterion mechanism with the new stereochemical findings and was intended only for olefins with large substituents at the double bond. It was realized that refinements would have to be made as new data and facts emerged, especially in regard to conformations of the primary and final ozonides. It was stated that any nonplanar model of these ring systems would lead to similar predictions (*148*). The contribution that was expected to stand the test of time was the basic concept that ozonide stereochemistry is largely dependent on the stereoselective formation and reaction (with an aldehyde or ketone) of *syn*- and *anti*-carbonyl oxides. This appears to be true. A discussion of various refinements of the Bauld–Bailey mechanistic ideas follows. All retain the *syn*- and *anti*-carbonyl oxide concept.

Kuczkowski *et al.* (*324,334*) proposed revisions of the Bauld–Bailey Rules 2 and 3, but in so doing they destroyed much of the original rationale of the proposal (*148*), as they admit (*324*). In the mind of this reviewer, only Rule 3 needs altering, at least for olefins with bulky substituents at the double bond, in order to conform to the oxygen–oxygen half-chair conformation of the final ozonide, **94**. In **94**, a trans final ozonide places its substituents either diaxially or diequatorially, whereas the substituents of a cis ozonide have an equatorial–axial arrangement. The *revised Rule 3* should then be stated as follows: *Aldehydes preferentially interact with anti carbonyl oxides to orient bulky substituents on the 1,2,4-trioxolane ring into an axial–equatorial conformation* (*cis*), *while aldehydes interact with syn carbonyl oxides to orient bulky substituents diaxially* (*trans*). The rationale is much the same as before. *Anti*-carbonyl oxides prefer to place their substituent equatorially and *syn*-carbonyl oxides to place their substituent axially for reasons given previously. When aldehydes and *anti*-carbonyl oxides interact, the aldehyde group generally goes in axially, producing a cis ozonide (*e*,*a*) since this gives less substituent-oxygen lone-pair interaction; this appears to be more important than substituent–substituent interactions (*324,330*). With *syn*-carbonyl oxides, the aldehyde places its group axially in order to minimize both substituent–substituent and substituent–lone-pair interactions, thus yielding a trans (*a*,*a*) final ozonide. These modified Bauld–Bailey rules, outlined in Table X, are consistent with the data of Table VII, at least for large substituents, just as were the original Bauld–Bailey rules.

B. KUCZKOWSKI SCHEME

Kuczkowski *et al.* (*324*) have explained ozonolysis results, such as those shown in Tables VII and VIII, on the basis of orbital symmetry considerations. Both ozone and the carbonyl oxide are 4π electron systems and carbon–carbon and carbon–oxygen double bonds are 2π electron systems. Therefore, the three reactions involved in the formation of final ozonides (formation and decomposition of the primary ozonide and addition of the carbonyl oxide to an aldehyde or ketone) are symmetry-allowed, concerted $[\pi^4 s + \pi^2 s]$ processes. The additions should be *supra*, *supra*, going through an exaggerated envelope transition state, as illustrated in Scheme 16.

The reaction between carbonyl oxide and carbonyl moieties to give the final ozonide is considered first, for convenience, followed by decomposition of the primary ozonide to *syn*- and *anti*-carbonyl oxides.

TABLE X

REVISED BAULD–BAILEY PREDICTIONS[a]

Olefin configuration	Primary ozonide		Predominant carbonyl oxide configuration	Predominant final ozonide	
	Configuration	Conformation[b]		Configuration	Conformation[c]
Cis	Cis	*a*,*e*	Anti	Cis	*a*,*e*
Trans	Trans	*a*,*a*	Syn	Trans	*a*,*a*,

[a] For olefins with large substituents at the double bond. For others, the conformations of Table XII can be used.
[b] C–C half-chair conformation with substituents axial (*a*) or equatorial (*e*) as indicated.
[c] O–O half-chair conformation with substituents pseudo-axial or pseudo-equatorial as indicated.

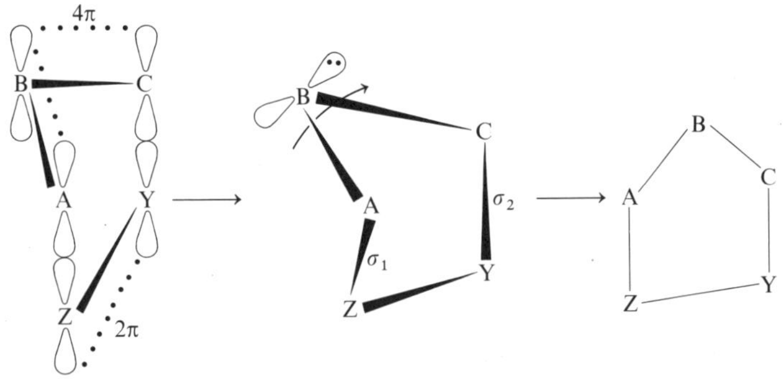

SCHEME 16. *Supra, supra* addition of 4π and 2π systems. Reprinted, with permission, from Lattimer *et al.* (*324*). Copyright by the American Chemical Society.

Two orientations each are possible for reactions of syn (A and B) and anti (C and D) carbonyl oxides with an aldehyde, as illustrated in Scheme 17 (*324*). Of the two orientations leading to final ozonide from a *syn*-carbonyl oxide, orientation A is suggested as being preferable since the bulky R groups would interact less with lone pairs of the terminal carbonyl oxide oxygen. Likewise, orientation C is said to be preferable for the *anti*-carbonyl oxide since interactions of the bulky group of the aldehyde would be less with lone pairs of the terminal carbonyl oxide oxygen. Thus, *syn*-carbonyl oxides should produce more trans final ozonide and *anti*-carbonyl oxides more cis final ozonide by this scheme, just as with the Bauld–Bailey scheme.

In the cycloreversion of a primary ozonide to carbonyl oxide and aldehyde, the transition state also would be an envelope, as shown in Scheme 16. The oxygen adjacent to the carbon that incorporates into the carbonyl oxide must move up or down as shown in Scheme 18. The downward movement (pathway I) would lead to a *syn*-carbonyl oxide and the upward movement (pathway II) to an *anti*-carbonyl oxide. Separate analyses are necessary for 1-alkenes, *trans*-alkenes, and *cis*-alkenes (*324*).

For 1-alkenes it can be seen that pathway I (Scheme 18) would put the R group in an axial position, where it would be gauche to ring bonds (Scheme 19), whereas pathway II (Scheme 18) would place R equatorially and trans to the ring bonds (Scheme 19). The latter is assumed to be the more stable transition state; therefore an *anti*-carbonyl oxide (Scheme 18) and a cis final ozonide (Scheme 17) should predominate in the higher molecular weight cross ozonide, as found in example 19, Table VIII (*324*).

For *trans*-alkenes with bulky groups, pathway I (Scheme 18) is chosen (*324*) because the R group of the carbonyl oxide (to be) would be gauche to the ring bonds and trans to the other large R group (Scheme 19) in the transition state, rather than vice versa with pathway II. This also is assumed

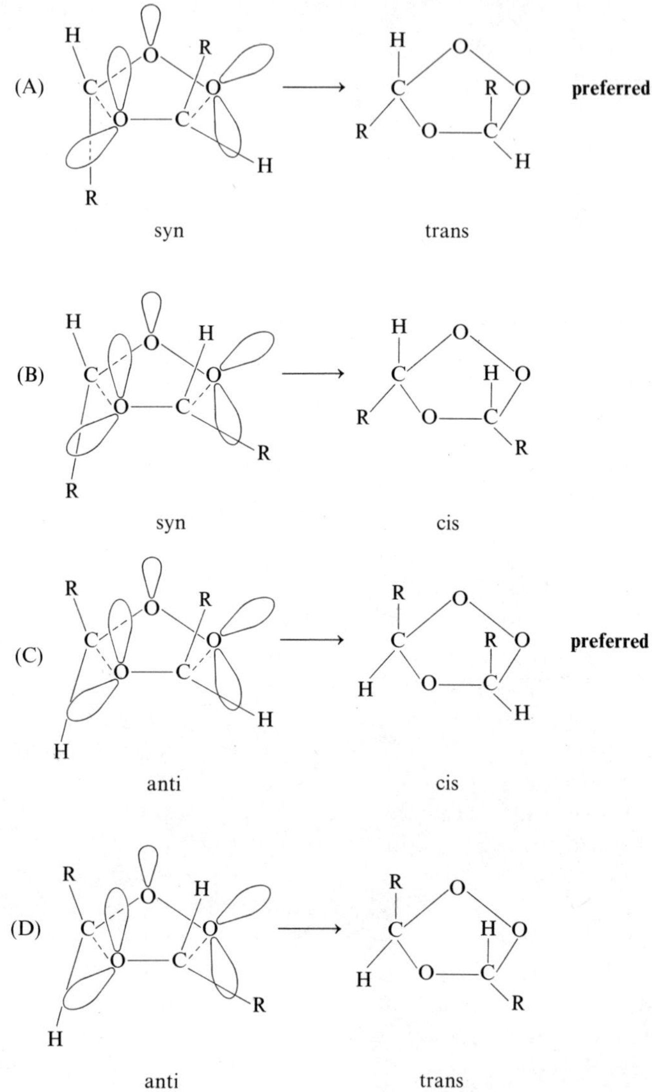

SCHEME 17. Reaction of syn and anti carbonyl oxides to produce final ozonides. Reprinted, with permission, from Lattimer *et al.* (*324*). Copyright by the American Chemical Society.

In (A)–(D), the plane of the carbonyl compound is approaching the plane of the zwitterion. The following representation makes clear the geometric relations of the pertinent bonds and orbitals of the two approaching moieties.

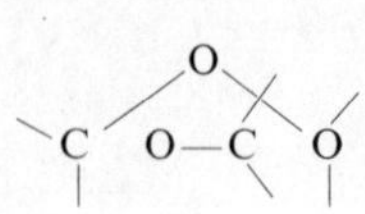

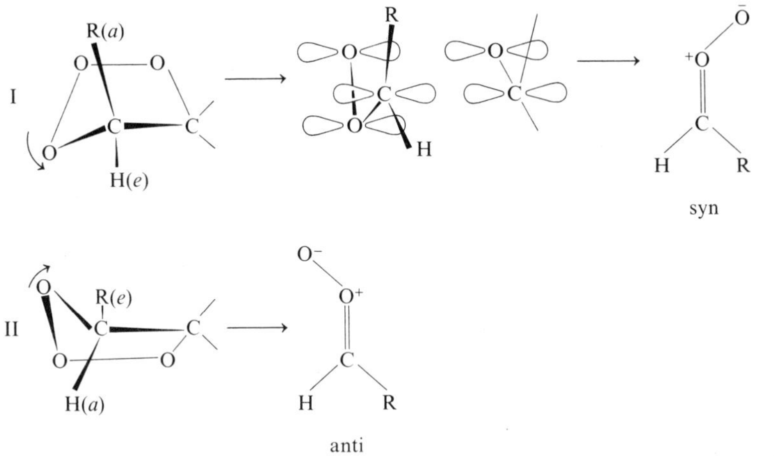

SCHEME 18. Kuczkowski primary ozonide decomposition scheme. Reprinted, in part, with permission, from Lattimer *et al.* (*324*). Copyright by the American Chemical Society.

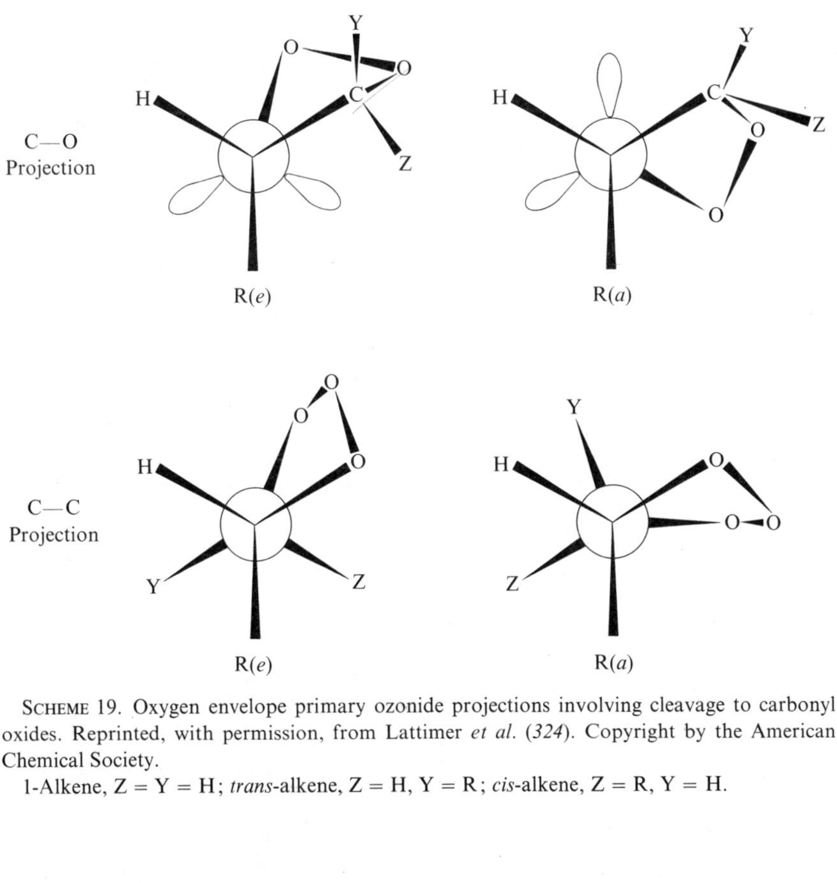

SCHEME 19. Oxygen envelope primary ozonide projections involving cleavage to carbonyl oxides. Reprinted, with permission, from Lattimer *et al.* (*324*). Copyright by the American Chemical Society.

1-Alkene, Z = Y = H; *trans*-alkene, Z = H, Y = R; *cis*-alkene, Z = R, Y = H.

to prevail if the substituents are small, causing *trans*-olefins to give a predominance of *syn*-carbonyl oxides (Scheme 18) and trans final ozonide (Scheme 17), as found in Tables VII and VIII (*324*).

For *cis*-alkenes the R groups will be gauche to each other by both pathways. However, pathway II (Scheme 18) is said to be preferable (*324*) for symmetrical *cis*-alkenes because this places the R group of the incipient carbonyl oxide trans to the opposite ring bond (Scheme 19). Thus, *anti*-carbonyl oxide (Scheme 18) and *cis*-ozonide (Scheme 17) predominate, as found in Tables VII and VIII, at least for bulky substituents. This does not, however, explain the results with *cis*-2-butene (Table VII, example 20). For unsymmetrical *cis*-alkenes, route II (Scheme 18) is also favored (*324*), the bulkier of the two groups taking an equatorial position in the transition state leading to the larger carbonyl oxide, which then becomes anti. Thus, more cis larger cross ozonide would be obtained. In the transition state leading to the smaller carbonyl oxide, it is suggested that the smaller substituent takes an axial position (*324*), since this keeps the larger substituent trans to the ring bond (Scheme 19). Thus, more syn smaller carbonyl oxide and more trans smaller cross ozonide are predicted. The normal ozonide would be expected to be in between. This fits to some extent the data of Table VIII (examples 1, 11, 15, and 17).

The stereochemical consequences of the Kuczkowski scheme are outlined in Table XI.

TABLE XI

STEREOCHEMICAL COURSE OF NORMAL AND CROSS OZONIDE FORMATION BY THE KUCZKOWSKI SCHEME[a]

Olefin configuration	Primary ozonide: Configuration	Primary ozonide: Conformation[b]	Carbonyl oxide configuration	Final ozonide configuration[c]
1-Alkene	—	*e*	Anti	Cis[d]
Trans	Trans	*a*[e]	Syn	Trans
Cis	Cis	*e*[f]	Anti	Cis
Cis	Cis	*a*[g]	Syn	Trans

[a] Reprinted, with permission, from Lattimer *et al.* (*324*). Copyright by the American Chemical Society.

[b] Oxygen envelope transition state with *a* and *e* referring to the axial or equatorial position of the R group going to the carbonyl oxide.

[c] Predominant.

[d] Obtained as a cross ozonide.

[e] Preferred by symmetrical or unsymmetrical alkenes.

[f] Preferred by symmetrical alkenes with bulky groups or when the larger substituent of an unsymmetrical alkene is incorporated into the carbonyl oxide.

[g] Preferred by symmetrical alkenes with small groups or when the smaller substituent of an unsymmetrical alkene is incorporated into the carbonyl oxide.

C. SYMMETRICAL AND UNSYMMETRICAL OLEFINS WITH SMALL SUBSTITUENTS AT DOUBLE BOND

The consequences of the modified Bauld–Bailey and the Kuczkowski schemes differ little from each other. Both have difficulty with symmetrically substituted ethylenes bearing small substituents.

The Bauld–Bailey scheme was not originally intended for ethylenes with small substituents. However, if one incorporates into the Bauld–Bailey scheme (Table X) some calculations and experimental results of Fliszár *et al.* (*328,337,338*), most of which are outlined in Table XII, the scheme affords a rationale for data pertaining to symmetrical and unsymmetrical ethylenes with one or both substituents small (Tables VII and VIII), which is just as reasonable as that of Kuczkowski. The conformational preferences shown in Table XII are based on the Bauld–Bailey primary ozonide model, which Fliszár's (*328*) calculations favored, and on the experimental studies of Fliszár *et al.* (*337,338*) which determined the preferentially formed carbonyl oxide from an unsymmetrical olefin (see Chapter VII, Section III for further discussion).

At first sight the *cis*- and *trans*-2-butenes appear to be anomalous because both give the *trans*-ozonide predominantly (Table VII). However, for the trans isomer the *a,a* and *e,e* conformations are equal in energy (*328*), and for the cis isomer the axial substituent forms the carbonyl oxide preferentially (*338*). Thus, both *cis*- and *trans*-2-butene might be expected to produce a *syn*-carbonyl oxide preferentially, from which a trans final ozonide would result (Table X).

The surprising result with propene (example 19, Table VIII) can also be rationalized readily by the Bauld–Bailey scheme. The methyl group is equatorial and becomes the carbonyl oxide preferentially (Table XII). This should result predominantly in the *anti*-carbonyl oxide and the cis isomer (Table X) of the 2-butene cross ozonide, as found (Table VIII).

Example 15 of Table VIII (*cis*-1-methyl-2-*t*-butylethylene) is also an interesting example. Accordingly to Table XII the *tert*-butyl group is equatorial and preferentially forms the carbonyl oxide. Therefore, the larger cross ozonide should and does have a cis:trans ratio greater than 1 (Table X). The normal ozonide would be, and is, a borderline case, since the major carbonyl oxide would be reacting with a small aldehyde. The smaller cross ozonide would come from the minor carbonyl oxide, which should be syn, causing more *trans*-ozonide, as found (Table VIII). This analysis would also explain the results of example 1, Table VIII (*cis*-1-methyl-2-isopropylethylene), except for the strange, anomalous, conformational analysis and results of Fliszár (*328,337*), which place the isopropyl group in an axial position, in contrast to ethyl and *tert*-butyl (Table XII). Here, perhaps, the conformational analysis is in error; possibly several conformations are close in energy, as claimed by Rouse (*330*). Likewise, the results of Fliszár and

TABLE XII

PREFERRED CONFORMATIONS (R AND R′) OF SUBSTITUENTS IN PRIMARY OZONIDES ACCORDING TO FLISZÁR (*328*, *337*, *338*)

RCH=CHR′ configuration	R and R′ in RCH=CHR′		Preferred R in carbonyl oxide	Percent preferred
	R[a]	R′[a]		
—	Me (*e*)	H (*a*)	Me	62
—	Et (*e*)	H (*a*)	Et	63
—	*n*-Pr (*e*)	H (*a*)	*n*-Pr	60
—	*i*-Pr (*e*)	H (*a*)	*i*-Pr	64
—	*t*-Bu (*e*)	H (*a*)	*t*-Bu	70
Cis	Me (*a*)	Et (*e*)	Me	52
Cis	Me (*e*)	*n*-Pr (*a*)	Me	54
Cis	Me (*e*)	*i*-Pr (*a*)	Me	57
Cis	Me (*a*)	*t*-Bu (*e*)	*t*-Bu	62
Cis	Et (*a*)	*t*-Bu (*e*)	*t*-Bu	63
Trans	Me (*e*)	Et (*e*)	Me	56
Trans	Me (*e*)	*n*-Pr (*e*)	Me	60
Trans	Me (*a*)	*i*-Pr (*a*)	Me	71
Trans	Me (*a*)	*t*-Bu (*a*)	Me	88
Trans	Et (*a*)	*i*-Pr (*a*)	Et	66
Trans	Et (*a*)	*t*-Bu (*a*)	Et	87

[a] *e* = equatorial; *a* = axial.

Renard (*337*) in regard to the major carbonyl oxide produced, which were obtained in methanol solution, may not carry over to ozonolyses in inert solvents. With these assumptions, the results in Table VIII pertaining to cis olefins in hydrocarbon solvents can be reconciled. For a trans olefin, Ex. 2 of Table 8 will be taken as an illustration. According to Fliszár (328,338) the conformation is diaxial. Therefore, following Tables 10 and 12, the predominant final ozonide configuration should be trans for the normal and both cross ozonides, as observed. The analysis fails, however, for ex. 18, Table 8, in the case of cross ozonide A. The Kuczkowski scheme also fails to account for this anomaly.

D. FURTHER COMPARISONS

Both the Bauld-Bailey and the Kuczkowski schemes are based on the *syn*- and *anti*-carbonyl oxide concept and assume a concerted decomposition of the primary ozonide. With both schemes a nonplanar primary ozonide is involved, an axial substituent incorporates itself into a syn carbonyl oxide, and an equatorial substituent becomes an anti carbonyl oxide. *Syn*-carbonyl oxides form *trans*-final ozonides and *anti*-carbonyl oxides form *cis*-final ozonides, predominantly. In fact, all predictions made for bulky olefins by the two schemes are the same (cf. Schemes 10 and 11 and the preceding discussion). This results from the fact that the Kuczkowski scheme (*324*) is a modification of the Bauld–Bailey scheme (*148*) in that it adapts the principles laid down by the Bauld–Bailey scheme to different conformations of the primary and final ozonides (or transition states thereof). This was clearly stated by Bauld, Bailey *et al.* (*148*) to be possible.

The Bauld–Bailey (*148*) and the Kuczkowski (*324*) schemes are similar in regard to the transition state for primary ozonide decomposition, which resembles the primary ozonide. This is as it should be, since decomposition of the primary ozonide is a highly exothermal reaction (*134,145,182,234*). In the recombination of the Criegee moieties to produce the final ozonide, however, the transition states of the two schemes appear to differ, that of Kuczkowski *et al.* resembling the reactants and that of Bauld–Bailey the product. Kuczkowski *et al.* (*324*) have criticized the Bauld–Bailey scheme in this respect. Whereas this may be justified, certain other criticisms (*324*) are not.

Both schemes are based on orbital symmetry considerations. The transition states in the Bauld–Bailey scheme utilize the supra-supra orbital alignment just as surely as do those of the Kuczkowski scheme. This point was not emphasized in 1968 (*148*), however, when such considerations were in their infancy. Kuczkowski *et al.* (*324*) stated that the physical bases for axial substituents to produce *syn*-carbonyl oxides (Bauld–Bailey Rule 1) and for equatorial substituents, in a *cis*-primary ozonide, to be preferentially incorporated into the carbonyl oxide (Bauld–Bailey Rule 2) are not clear. Actually, one need only take a model and look down the carbon-oxygen bond of the potential carbonyl oxide and apply the rule of least motion to see why a *syn*-carbonyl oxide should be preferentially produced from an axial substituent and an *anti*-carbonyl oxide from an equatorial substituent (Rule 1). For Rule 2, one need only consider, in addition, that greater steric repulsions are encountered as an axial substituent, relative to an axial hydrogen, moves inward toward the ring during the cleavage and rehybridization processes. It is of interest that the calculations and experimental results of Fliszár *et al.* (*328,337,338*), as illustrated in Table XII, confirm Bauld–Bailey Rule 2 (Table X) in an independent manner.

In contrast to the Bauld–Bailey and Kuczkowski schemes, Benson (*183*) and Wadt and Goddard (*78*) have assumed that the decomposition of the primary ozonide is stepwise, rather than concerted, involving first an oxygen bond and a diradical intermediate. Whereas this certainly must be the mechanism in vapor-phase decompositions (see Chapter XII), it is not unreasonable to assume that a concerted process, resulting in the zwitterion form of the carbonyl oxide, occurs in solution, where solvation can stabilize otherwise unlikely structures and processes. It is difficult to explain the stereochemistry of ozonide formation by the stepwise process. It is conceivable, however, that some contribution from such a process accounts to some extent for the variations observed under different experimental conditions.

Recent studies concerning differences in stereochemical results due to differences in the rate of warm-up of an ozonolysis reaction mixture (*321*) or to the presence of a reagent capable of complexing with the carbonyl oxide (*320*) clearly indicate that although the *syn*- and *anti*-carbonyl oxide concept is the best rationalization for the observed results in ozonide formation, the details still require further refinement. For further discussion, see Chapter VII and the Epilogue at the end of this volume.

V. The "Unified Concept"

One other mechanistic proposal needs comment in order to avoid confusion in the literature. This is the so-called "unified concept" outlined in Scheme 20 (*97,98*).

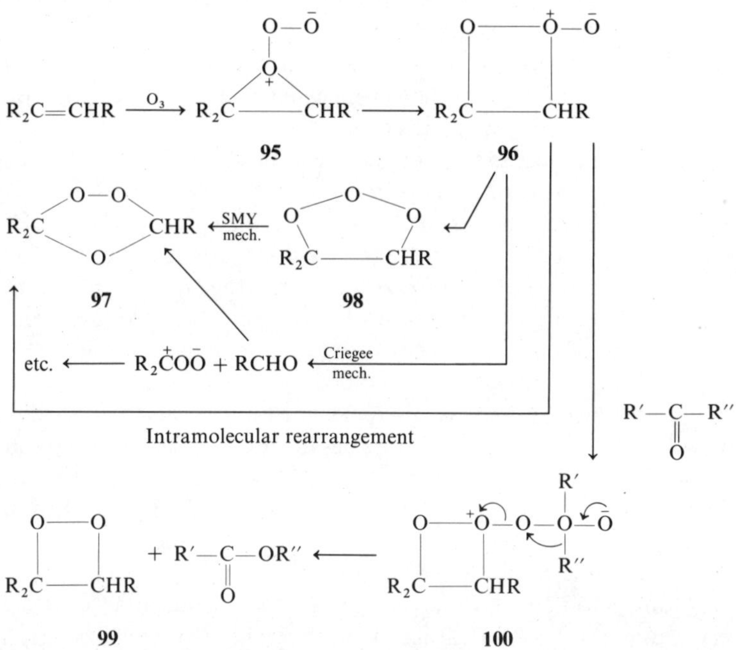

SCHEME 20. The Story "unified concept" of ozonolysis.

The key intermediate in this mechanism is the Staudinger molozonide (**96**), for which it was suggested that a peroxy epoxide (**95**) might be a precursor (*97,98*). Several fates were then proposed for the molozonide intermediate (**96**), including a breakdown into the Criegee moieties, rearrangement to a 1,2,3-trioxolane (**98**), a primary ozonide, followed by the

SMY mechanism, rearrangement to a final ozonide (**97**), and/or Baeyer–Villiger oxidation of certain aldehydes and ketones via **100** (*97,98*). The latter reaction was the principal evidence for the mechanism initially. The authors (*96,98*) reported that added propionaldehyde or butyraldehyde was oxidized to the corresponding carboxylic acid, cyclohexanone was oxidized to ϵ-caprolactone (or 6-hexanolide, **101**), and methyl *tert*-butyl ketone (pinacolone) was oxidized to *tert-butyl* acetate. The other product in these Baeyer–Villiger oxidations should be a dioxetane (**99**), which, later, the authors (*98*) reported that they had isolated.

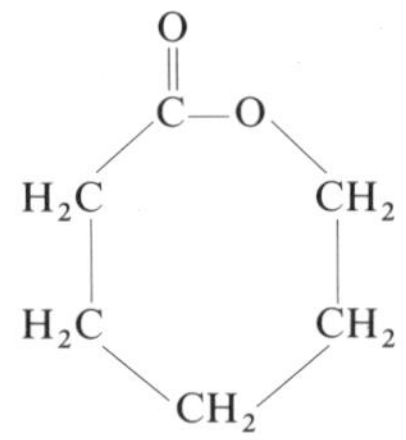

101

The mechanism attracted much attention, and several investigators tried to use it as a synthetic method for dioxetanes, one with reported success (*339*). Work from several different laboratories, however, has shown that these reports have no basis in fact, but resulted from compounded errors. The oxidation of propionaldehyde to propionic acid was shown to occur in the presence of a 1,2,3-trioxolane (**98**), making the proposal of the Staudinger molozonide unnecessary (*99*). At the same time it was shown (*99–101*) that no oxidation of either pure cyclohexanone or pure pinacolone occurs under the reported conditions (*96,98*). It is well known, however, that both of these substances are oxidized to the products reported while sitting on the shelf for a period of time. It was further found that the report (*98*) of dioxetane formation was in error (*99,100*). Kopecky (*100*) made one of the dioxetanes claimed and found it to have quite different properties from those reported by Story *et al.* (*98*). The products actually obtained (*98*) were conventional peroxidic ozonolysis products (*99,101*). Likewise, the reported synthesis of 3,4-bisbiphenylene-1,2-dioxetane by the Story method (339) was shown to be in error (*339a,340*).

VI. Weighing the Evidence: Criegee Carbonyl Oxide Confirmed

After the "unified concept," described in the preceding section, was shown to be invalid, the only mechanisms left by 1973 to be given the final test were the Criegee carbonyl oxide mechanism, with necessary refinements as described in Section IV of this chapter, and the SMY mechanism, described in Section III.

A. ^{18}O STUDIES

It was realized early that one way to decide between these mechanisms was to ozonize an olefin in the presence of an excess of an aldehyde labeled with oxygen-18. By the Criegee carbonyl oxide mechanism the ^{18}O should be incorporated into the ether bridge of the final ozonide, whereas the SMY mechanism should cause it to appear in the peroxide bridge, as outlined in

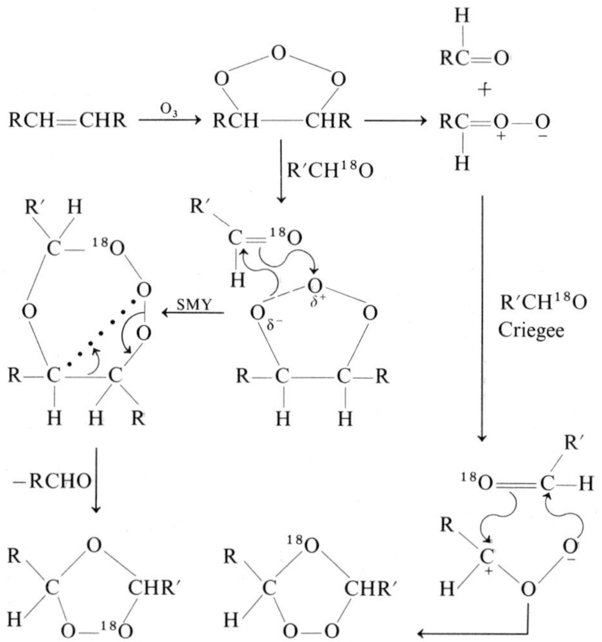

SCHEME 21. ^{18}O Incorporation by SMY and Criegee mechanisms.

Scheme 21. This was tested early by Story *et al.* (*292*), who ozonized *trans*-1,2-diisopropylethylene in the presence of ^{18}O-labeled acetaldehyde and reported that approximately 75% of the reaction proceeded by the SMY mechanism, placing the ^{18}O in the peroxide bridge. Their method of analysis involved reduction of the ozonide with either lithium aluminum hydride or methyllithium (*292*). They had previously attempted to show that such reductions resulted in loss of ^{18}O from a peroxide but not from an ether bridge (*341*). Studies by Carles and Fliszár (*342*), however, cast doubt on the validity of this assumption and indicated that, at most, 48% of the labeling was in the peroxide bridge.

Next, Fliszár *et al.* (*290,291,343*) ozonized *cis*- and *trans*-stilbene, styrene, *trans*-β-methylstyrene, and triphenylethylene in the presence of ^{18}O-labeled benzaldehyde and found that the ^{18}O label appeared exclusively in the ether bridge, showing that only the Criegee mechanism operated in this case. Their analytical method involved direct mass spectrometry of the labeled ozonides. This was confirmed by Story *et al.* (*344*) who explained the discrepancy between their results and those of Fliszár *et al.* on the basis that resonance stabilization of carbonyl oxides of phenyl-substituted olefins should cause the Criegee mechanism to be favored.

Murray and Hagen (*294*), also using mass spectrometry as their analytical method, ozonized *cis*- and *trans*-1,2-diisopropylethylene in the presence of ^{18}O-labeled isobutyraldehyde. They found ^{18}O incorporation in both the ether and the peroxide bridges for both the *cis*- and the *trans*-olefins, but they found the degree of incorporation to be temperature dependent. Their reported peroxide bridge ^{18}O value, indicating the SMY mechanism, was 40% at −122°C, but only 20% at approximately 0°C. Thus, their results indicated that, at best, the SMY mechanism is minor.

A year later, Kuczkowski *et al.* (*295,296*) ozonized ethylene, propylene, *cis*- and *trans*-2-butene, and *cis*- and *trans*-1,2-diisopropylethylene in the presence of an excess of ^{18}O-labeled formaldehyde or acetaldehyde at temperatures as low as −126°C, and analyzed the resulting ozonides for ^{18}O content by means of microwave spectroscopy and/or mass spectrometry. They found that ^{18}O incorporation occurred either exclusively or at least 90% at the ether bridge. Most recently, both Gallaher and Kuczkowski (*297*) and Higley and Murray (*298*) have shown in an unambiguous and conclusive manner that the SMY mechanism (Section III of this chapter) makes no significant contribution to final ozonide formation, leaving, of the present-day theories, only a carbonyl oxide-type mechanism capable of explaining the results.

The studies of Higley and Murray (*298*) included syntheses of peroxide-labeled and ether-labeled ozonides and ^{18}O-position determinations both by mass spectrometry and by triphenyl phosphite reductions. The latter was shown to be consistently reliable, whereas mass spectrometry was shown to give a misleading fragmentation pattern with certain ozonides (*298*), thereby causing erroneous conclusions (*294*). Obviously, the lithium aluminum hydride and methyllithium analyses (*292,341*) led to even greater misinterpretations. By the triphenyl phosphite method of analysis, Higley and Murray (*298*) and Gallaher and Kuczkowski (*297*) showed that essentially no ^{18}O was incorporated into the peroxide linkage of an ozonide during ozonolysis of an olefin in the presence of ^{18}O-labeled aldehydes.

Kinetic evidence against the SMY mechanism had been reported earlier by Greenwood and Durham (*145*) and by Razumovskii (*182*). They showed that the disappearance of primary ozonide during ozonation of *trans*-3-hexene is clearly first order even at −100°C and that aldehyde is formed immediately. If aldehyde thereafter participated to any appreciable degree in the primary ozonide decomposition, the kinetics should have reflected it.

During the time that the ^{18}O evidence still appeared to support the SMY mechanism, attempts were made to find other possible explanations for the apparent incorporation of ^{18}O into the peroxide linkage of an ozonide during ozonolyses in the presence of ^{18}O-labeled aldehydes. These are outlined in Schemes 22, 23, and 24.

Concerning Scheme 22, Fliszár *et al.* (*345*) felt that this possibility was eliminated when they made polarographic studies of the products formed between carbonyl oxides and alcohols and found no evidence for the hydroxyalkyl peroxides (**104**) expected from carbonyl oxide **103a**. Wadt and

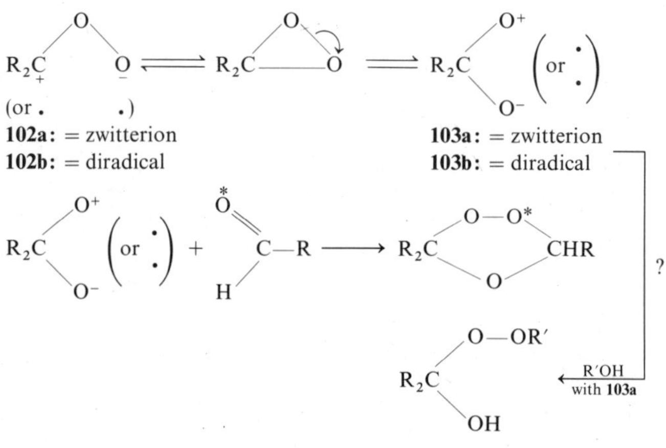

104

* Denotes O^{18}

SCHEME 22.

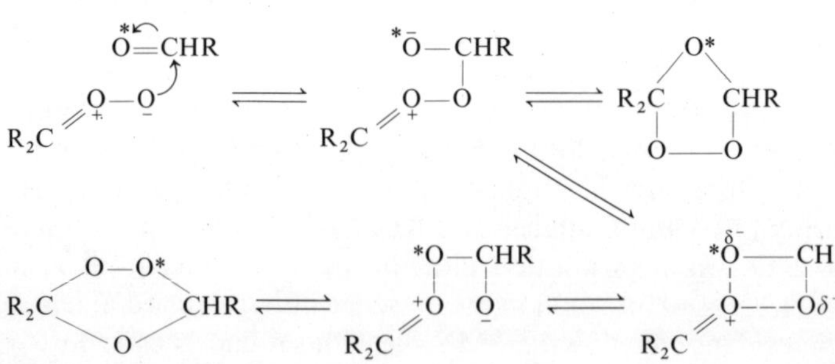

* Denotes O^{18}

SCHEME 23. Reproduced by permission of the National Research Council of Canada from Fliszár and Carles (238).

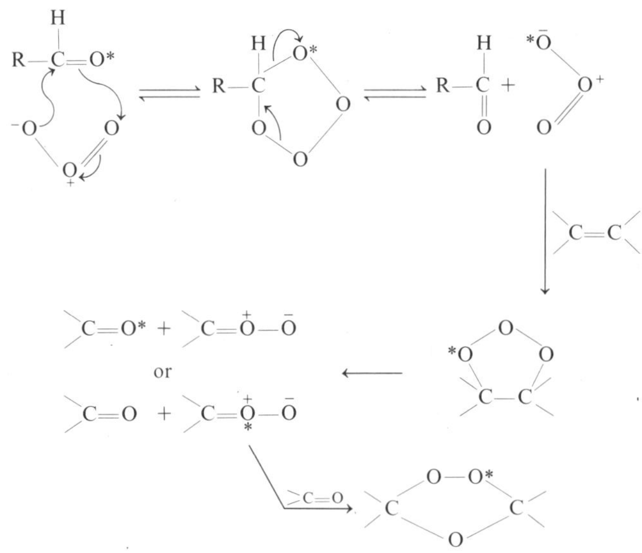

* Denotes O^{18}.

SCHEME 24. Reprinted, with permission, from Klopman and Joiner (346). Copyright by the American Chemical Society.

Goddard (*78*), however, on the basis of their singlet diradical structure for **102**, believe that isomerization of **102b** to **103b** is thermodynamically likely and that singlet diradical **103b** should react with methanol to form a methoxyalkyl hydroperoxide rather than **104**.

Scheme 23 (*238*) appears to be invalidated by the work of Higley and Murray (*298*), which showed that no appreciable amount of ^{18}O appears in the peroxide bridge and, thus, no explanation (*238*) is necessary.

Scheme 24 was proposed by Klopman and Joiner (*346*). The possibility of 1,3-dipolar cycloaddition of ozone with a carbonyl group seems attractive. Indeed, Klopman and Joiner (*346*) reported a rapid loss of ^{18}O-labeled isobutyraldehyde upon treatment with ozone. They did not follow through, however, with the identification of ^{18}O in ozone (*346*). Scheme 24 is not necessarily invalidated by the work of Higley and Murray (*298*), which found no appreciable ^{18}O in the peroxide bridge of ozonides, as discussed. On the contrary, it is likely that the Klopman–Joiner (*346*) results are significant, but that 1,3-dipolar cycloaddition between ozone and a carbon–carbon double bond is faster than 1,3-dipolar cycloaddition between ozone and a carbonyl group.

B. CRIEGEE'S LAST CONTRIBUTION

During the period that the SMY mechanism was still under consideration and of great interest, a definitive testing of ozonolysis mechanisms was renewed by Criegee* and co-workers (*340,347–349*). These experiments, which involved competitions between intramolecular normal and cross-ozonide formation, have elegantly and decisively demonstrated the necessity of the Criegee carbonyl oxide intermediate in ozonolyses producing ozonides. The most important of these are illustrated in Schemes 25–30.

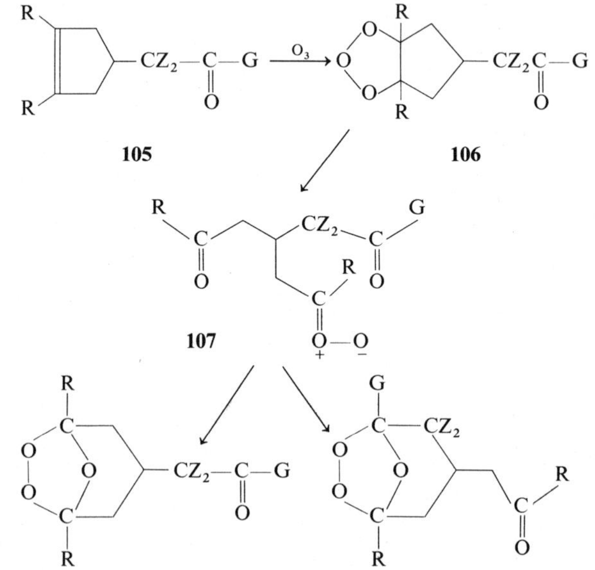

105 **106** **107** **108** **109**

a: $R = CH_3$, $Z = D$, $G = CD_3$; **b**: $R = CH_3$, $Z = H$, $G = C_6H_5$;
c: $R = CH_3$, $Z = H$, $G = H$; **d**: $R = C_6H_5$, $Z = H$, $G = CH_3$.

SCHEME 25. Competitions between intramolecular normal and cross-ozonide formation.

It does not seem reasonable that the keto group of primary ozonides **106**, **111**, **116**, and **122** (in Schemes 25–28) could get close enough to the trioxolane ring to produce cross ozonides **109**, **114**, **119**, and **126** via the SMY mechanism, especially in Schemes 26–28. Further, in the one case where this might be possible (primary ozonide **129**, Scheme 29), the cross ozonide (**131**) is not produced. This is in line with earlier findings that ozonides of cyclohexenes are produced only in very low yields if at all (Chapter V, Section I, A) (*62*). No known mechanism other than the Criegee mechanism involving carbonyl oxides **107**, **112**, **117**, **124**, and **130** can explain these results.

The experiment shown in Scheme 28 is especially elegant (*349*). The normal ozonide **127**, or the exo–endo isomer thereof, is obtained only in

* The author feels that it is noteworthy that Professor Criegee did much of this work himself, during the period when illness was sapping his strength rapidly.

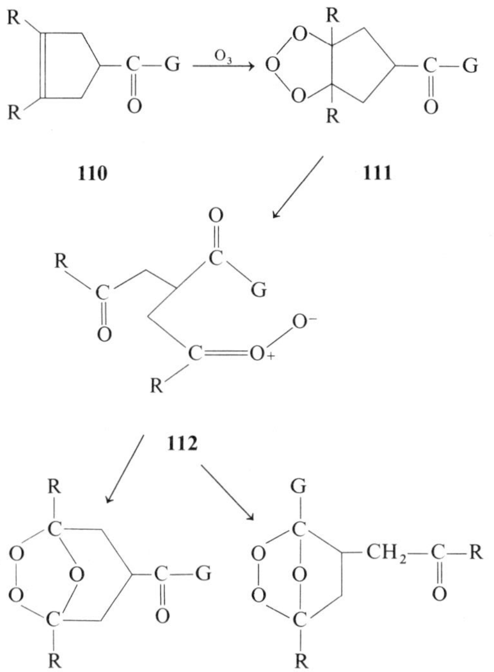

110 **111** **112** **113** **114**

a: R = G = CH_3; **b**: R = CH_3, G = C_6H_5; **c**: R = C_6H_5, G = CH_3

SCHEME 26. Further intramolecular ozonide competitions.

very low yield. Instead, carbonyl oxide **124b** rotates to give conformation **124a**, after which the carbonyl oxide group reacts with the foreign acetyl group to give cross ozonide **126**, with the remaining acetyl groups in the trans configuration rather than the less stable cis configuration (**127**, or its endo–exo isomer). Another product, in low yield, was **128**, produced from another conformation, **125**, of the Criegee intermediate. It is also interesting, in view of the discussions in Chapter V, Section I, A, that ozonide **123** was not produced (*349*). This would be the ozonide of a cyclopropene derivative; such are as yet unknown (cf. *350*).

Some additional aspects of the experiments of Schemes 25–29 are also interesting. In Scheme 25 (*340,348*) ozonides **108a** and **109a** were produced in equal amounts. This is not likely by any route other than one involving Criegee intermediate **107**; this further indicates that the half-life of **107** is more than infinitesimal. From **105b** and **105d**, however, ozonides **108b** and **109d**, respectively, were the major products, and from **105c** only ozonide **109c** was obtained, demonstrating that the 1,3-dipolarophilic character of the carbonyl groups

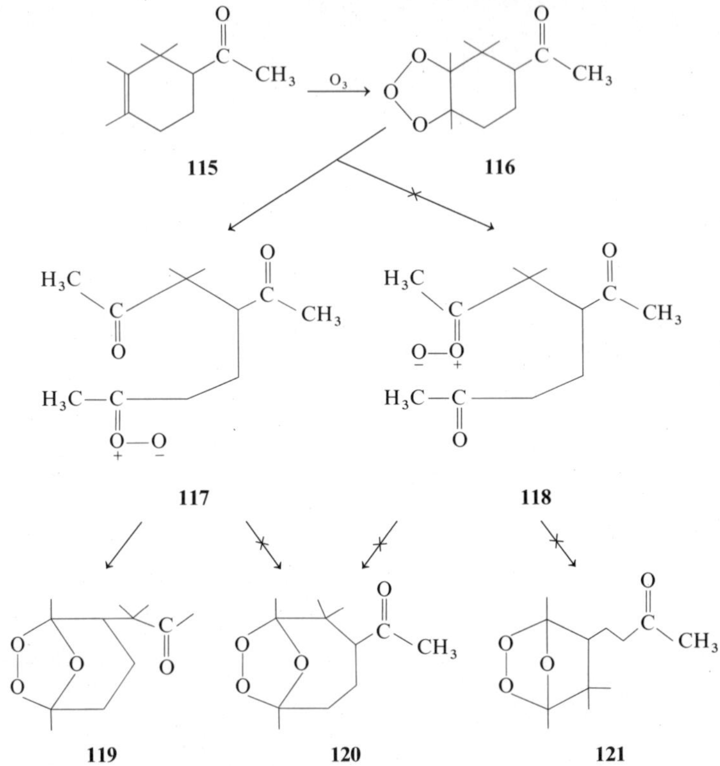

SCHEME 27. Intramolecular cross ozonide formation.

involved was in the order: formyl > acetyl > benzoyl. In Scheme 26 (348) **110a** and **110c** gave ozonides **114a** (exo and endo) and **114c** (exo and endo) exclusively, and **110b** yielded ozonide **114b** (exo and endo) as the major product and **113b** as a very minor product. This demonstrates that ring closure to a "cyclobutene" ozonide ring system is faster than ring closure to an ozonide of a cyclopentene.

In Scheme 27 (*347*) two carbonyl oxide intermediates are possible, **117** and **118**. None of the normal cyclohexene-type ozonide, **120**, was produced, as expected (Chapter V, Section I, A) (*62*), but of the other two possibilities only ozonide **119**, a cyclopentene type, was obtained, rather than ozonide **121**, a cyclobutene type. In view of the preceding discussion, it is evident that intermediate **117** was formed to the exclusion of **118**. The controlling factor was shown to be the presence of methyl groups alpha to the double bond in **115** (*347*).

Finally, as shown in Scheme 30 (*348*), a competition between an alcohol group and an acetyl group for the carbonyl oxide group was tested. The product resulted from interaction of the carbonyl oxide group with the alcohol function rather than with the carbonyl function of intermediate **132**.

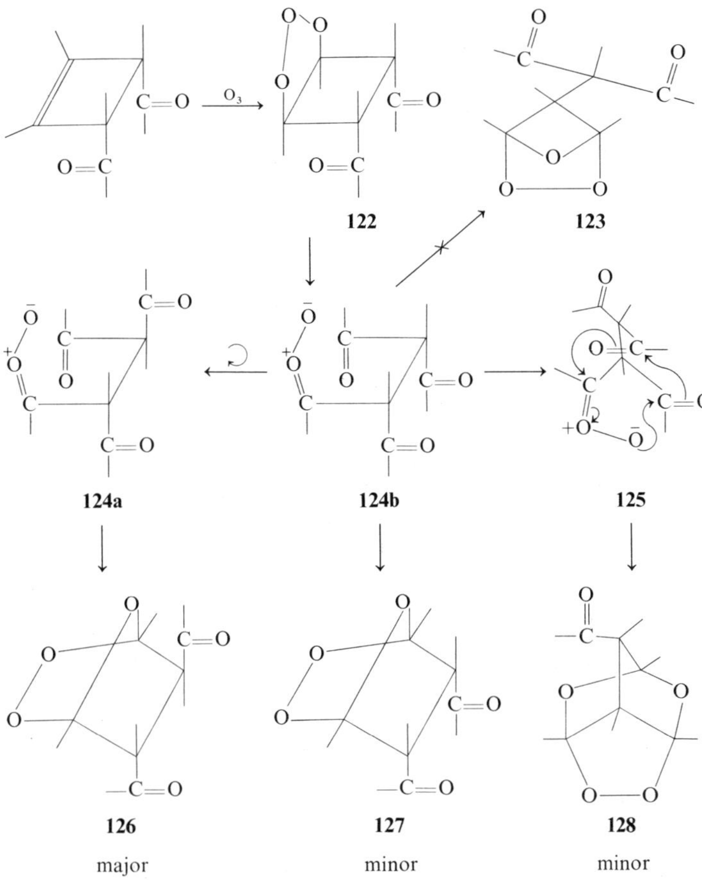

SCHEME 28. Cis-trans isomerization in cross ozonide formation.

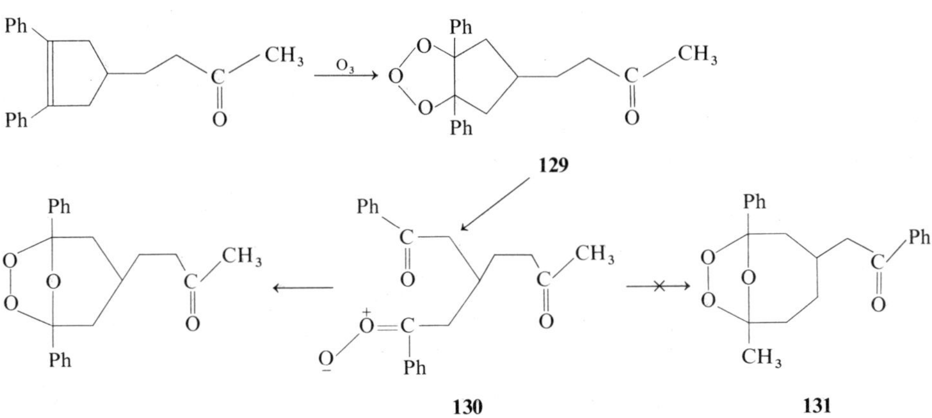

SCHEME 29. Competition between intramolecular cyclopentene and cyclohexene ozonide formation.

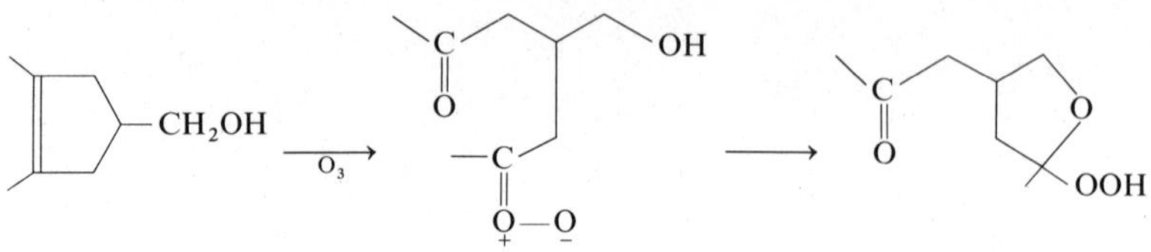

132

SCHEME 30. Competition between intramolecular ozonide and alkoxyalkyl hydroperoxide formation.

The combined force of these reactions, coupled with the O^{18} evidence discussed earlier (Section VI, A of this chapter), is to negate any mechanism of ozonolysis except some type of refined Criegee carbonyl oxide mechanism. Further refinements continue to challenge the investigator.

CHAPTER VII

Ozonolysis of Olefins: Competitions in Peroxidic Product Formation

In Chapter VI it become evident that the only important mechanism for the ozonolysis reaction involves the intermediacy of the Criegee carbonyl oxide and that the stereochemistry of ozonide formation is related to the syn and anti geometry of the carbonyl oxide.

The activation energies of the various steps leading to the different peroxidic ozonolysis products all appear to be low (*78,182*); thus, relatively minor factors can influence the course that these reactions take and the proportions of the peroxidic ozonolysis products thus obtained. This is especially true concerning the fates of the carbonyl oxide intermediate. The purpose of this chapter is to point out and discuss these various competitions and to see whether or not our present concepts are sophisticated enough to rationalize them. In this connection the author wishes to make it clear that, although he will attempt to explain as much as possible through the concepts developed in the preceding chapters, especially Chapters IV and VI, he does not mean to imply that present-day knowledge is sufficient or that these are necessarily the best explanations. To the contrary, it will become evident that much research is still needed before all of the many subtleties of the ozonolysis mechanism are uncovered and understood. Nevertheless, it seems worthwhile to do the best we can with what we have at present.

I. Competitions in Nonparticipating Solvents

The competitions referred to in the preceding paragraph are especially important in nonparticipating solvents. A nonparticipating solvent is one that does not interact with the carbonyl oxide intermediate, in contrast to solvents, such as methanol, that do (Section II of this chapter). In nonparticipating solvents, the principal competitive routes are: (a) 1,3-dipolar cycloaddition of a carbonyl oxide and an aldehyde (or ketone in some cases) to give an ozonide (examples in Chapter V, Section I); (b) dimerization of the carbonyl oxide to give an aldehyde or ketone diperoxide (examples in

Chapter V, Section II); (c) polymerization of the carbonyl oxide, either alone or with an aldehyde or ketone, to give oligomers (examples in Chapter V, Section III). Even within the confines of the reaction between the carbonyl oxide and an aldehyde to give an ozonide, various factors, such as temperature, type of solvent, olefin concentration, source of the carbonyl oxide, complexing agents, etc., affect cis:trans ozonide ratios, etc.

A. OZONIDE VS. DIPEROXIDE AND OLIGOMER FORMATION

Accompanying ozonide formation, in varying degrees, is almost always dimerization and polymerization of the carbonyl oxide intermediate (e.g., *105, 109,134,187,189*, and references to follow). These generally become the major reactions when the moiety formed along with the carbonyl oxide is a ketone (e.g., *60,194*, and references to follow). The failure of the carbonyl oxide moiety to add to the carbonyl group of a ketone, along with exceptions to the rule, was discussed in Chapter V, Section I, A. This inability appears to be due to the decreased 1,3-dipolarophilic and electrophilic character of a ketone carbonyl in comparison to that of an aldehyde carbonyl.

Fliszár and co-workers (*248,254,351–353*) have made detailed studies of the peroxidic and nonperoxidic products obtained from ozonolyses of various di- and tetrasubstituted ethylenes in carbon tetrachloride at temperatures in the range of 0° to 45°C. For tetraphenylethylene, they found that no ozonide was produced, but for each mole of olefin ozonized, 1 mole of benzophenone and 1 mole of diphenylcarbonyl oxide were formed; 57% of the carbonyl oxide dimerized (to **132a**, R = R′ = C_6H_5), and 43% formed an oligomer which they assumed was open chain, e.g., **134** (*254*). Similar results were reported for tetramethylethylene and biscyclohexylidene (*352*). For 1,2-disubstituted ethylenes, however, Fliszár *et al.* (*248,351–353*) found that considerably more than 1 mole of aldehyde was produced per mole of olefin ozonized. The other products were ozonide and aldehyde diperoxide (**132a**, R′ = H). For example, *trans*-stilbene yielded ozonide, benzaldehye, and diperoxide in the ratio 0.677:0.468:0.089, respectively. Fliszár *et al.* (*248,351,352*) suggested that the excess aldehyde is produced by the interaction of two or more carbonyl oxides, in competition with oligomer formation, as shown by **133** in Scheme 31. They concluded that whereas oligomers of disubstituted carbonyl oxides were stable up to 80°C, those of monosubstituted carbonyl oxides decompose spontaneously to aldehyde and oxygen, at least at temperatures above 0°C (*352*).

This probably explains reports in the older literature concerning the determination of considerable amounts of aldehydes during ozonolyses of various olefinic compounds of type RCH=CHR′ (*221–223,316–318*) and of "spontaneous decompositions" of ozonides to aldehydes, ketones, etc., during ozonolyses (*57,354–356*).

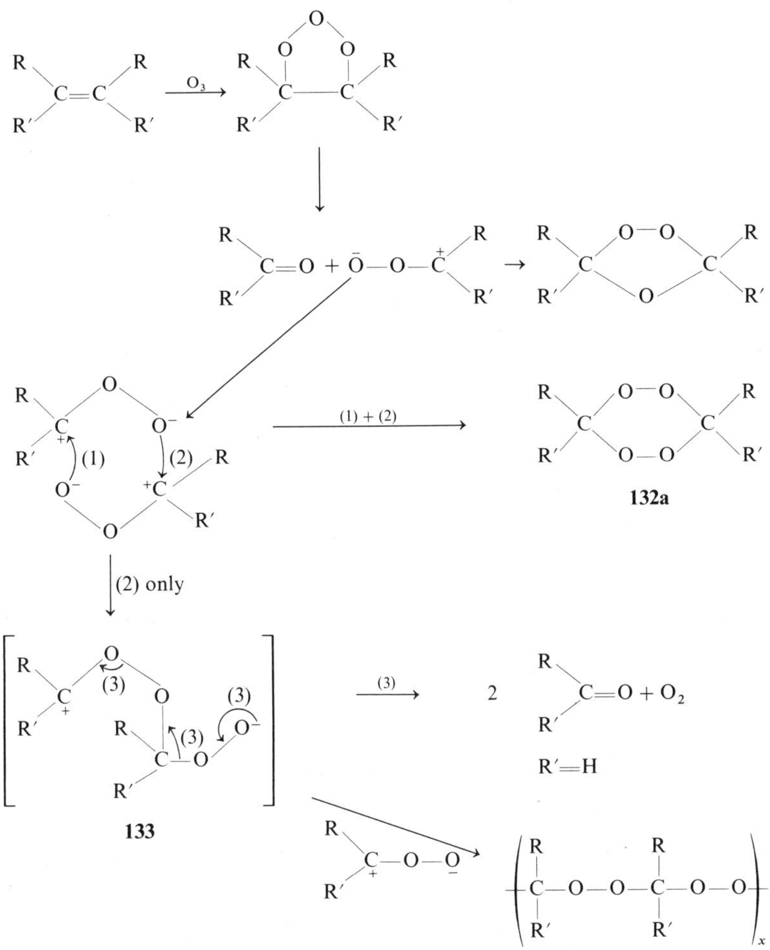

SCHEME 31. Mechanisms of diperoxide and oligomer formation.

Whereas Fliszár's studies were carried out at approximately room temperature, Greenwood (*263*), Razumovskii (*234,261*), Menyailo (*114,230*), and Milas (*250*) *et al.* studied oligomer formation during ozonolyses of various olefins at $-78°C$ or lower. They found, in contrast to Fliszár *et al.*, that the oligomer was composed not only of carbonyl oxide, but also of carbonyl moieties (see Chapter V, Section III). In fact, Razumovskii *et al.* (*261*) believe that oligomers of simple olefins are largely polyozonides of essentially the same elemental composition as monomeric ozonides (see Chapter V, Section III). Greenwood *et al.* (*142,143,145,263*) have followed the decomposition of primary ozonides to ozonides, oligomer, and aldehydes by NMR at low temperatures and suggest (*263*) that oligomer formation might occur more by the attack of carbonyl oxides on primary ozonides than by copolymerization of carbonyl oxides and aldehydes or ketones. Their proposal is illustrated in Eq. (11). They suggested several fates for **135** by which polymerization could continue. These included (a) attack of **135** on the primary ozonide, and (b) loss of aldehyde from **135** and attack of the remaining carbonyl oxide on the primary ozonide or on aldehyde or other carbonyl oxides. This, coupled with linear combinations of the Criegee moieties, would give rise to oligomers of the various structures found by Greenwood and Rubinstein (*263*) (see Chapter V, Section III). Carbonyl oxide attack on an oxygen, rather than a carbon of the primary ozonide [Eq. (12)], as well as aldehyde attack on a primary ozonide, might also be at least minor pathways to oligomer. Such mechanisms might explain the differences in results reported by Fliszár and by other investigators. At low

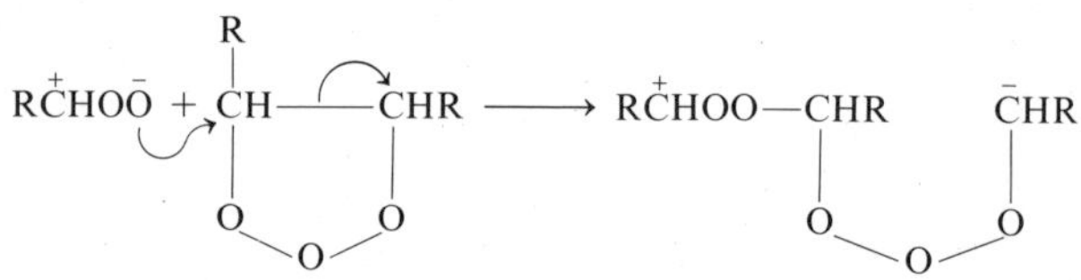
(11)

135

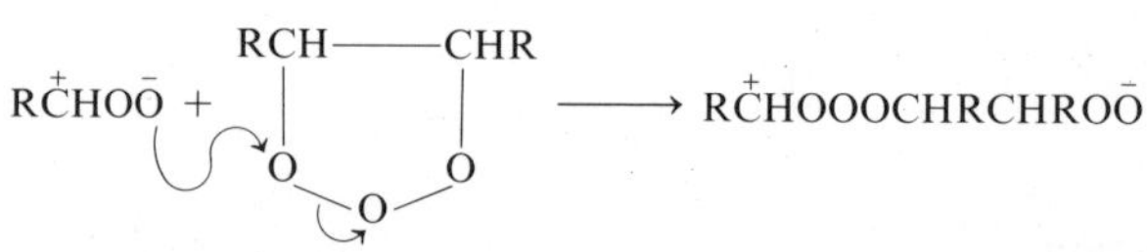
(12)

temperatures such interactions between a primary ozonide and a carbonyl oxide or an aldehyde might occur more readily than thermal decomposition of the primary ozonide, whereas at higher temperatures the latter reaction should prevail. Another factor is the structure of the olefin. The primary ozonide of tetraphenylethylene (*254*) should certainly decompose readily, and the resulting moieties (diphenylcarbonyl oxide and benzophenone) would not be expected to interact to give either mono- or polyozonides.

One of the principal arguments of Greenwood and Rubinstein (*263*) for their mechanism for oligomer formation was the fact that *trans*-olefins yield much more oligomer than do *cis*-olefins; conversely, *cis*-olefins give much higher yields of ozonide than do *trans*-olefins (see Table VII, Chapter VI, Section II). 1-Alkenes also afford high ozonide yields (*237*,*357*). These results would be expected by Greenwood's mechanism, since trans primary ozonides are much more stable than *cis*- or 1-alkene (*137*,*357*) primary ozonides and, consequently, are available for attack by a carbonyl oxide or aldehyde. As further evidence, Greenwood *et al.* (*143*,*263*) found that with both *cis*- and *trans*-olefins, oligomer formation increased as the ozonation temperature decreased; the primary ozonides are longer lived at the lower reaction temperature.

A possible argument against the Greenwood–Rubinstein oligomer mechanism is that Greenwood also has reported that decomposition of a primary ozonide is a first-order reaction (*145*) (see Chapter VI, Section VI, A). This would appear to preclude attack by an outside species. In fact, in dimethyl ether at −100°C or below, oligomer formation was the major reaction with *trans*-3-hexene even while first-order kinetics were being observed (see, however, discussion on solvent effects in Section I, B, 2).

Competitions between ozonide and oligomer formation are also interesting with cycloolefins. As pointed out in Chapter V, Section I, A, 1,2-disubstituted cyclobutene and cyclopentene types generally give good yields of ozonides (*62,104,106,109,268*) (see references in Table III), whereas cyclobutene and cyclopentene types with only one substituent, or none, at the double bond, as well as cyclohexenes and higher cycloolefins (*62,109,206,358,359*), give largely oligomer. The chances for intramolecular ring closure are not as good with the Criegee intermediates of cyclohexene or larger cycloolefins as with those of cyclobutenes and cyclopentenes. Further, the absence of substituents around the carbonyl oxide and/or carbonyl ends of the Criegee intermediate from mono- and unsubstituted cyclobutenes and cyclopentenes must allow intermolecular interactions to be favored over intramolecular 1,3-dipolar cycloaddition.

Ozonolysis of tetramethylethylene appears to be somewhat special. The various peroxidic ozonolysis products obtained were shown in Chapter V, Section III. In addition, Milas and Belic (*250*) and Story and Burgess (*220*) have cited evidence for the isomerization of dimethylcarbonyl oxide (**136**) to the corresponding vinyl hydroperoxide (**137**) during ozonolysis of tetramethylethylene in nonparticipating solvents. In their first report, Story and Burgess (*219*) thought that the carbonyl oxide (**136**) then added to the vinyl hydroperoxide (**137**), to give compound **139**. This was disproved, however, and it was later shown (*220*) that the vinyl hydroperoxide (**137**), in turn, isomerized to hydroxyacetone (**138**), which then reacted with the carbonyl oxide (**136**), to give compound **141** via **140**. These reactions are shown in Scheme 32. The ozonide of tetramethylethylene has never been obtained, even when ozonolyses were carried out in excess acetone (*113,218*). Criegee (*113,218*) suggested that this is due to the ease with which dimethylcarbonyl oxide (**136**) isomerizes.

In passing, it is noteworthy that addition of a carbonyl oxide to an olefin is a possible reaction route by which certain epoxides have been obtained (*360,361*), but the evidence is not clear-cut and can be explained in other ways (*360*).

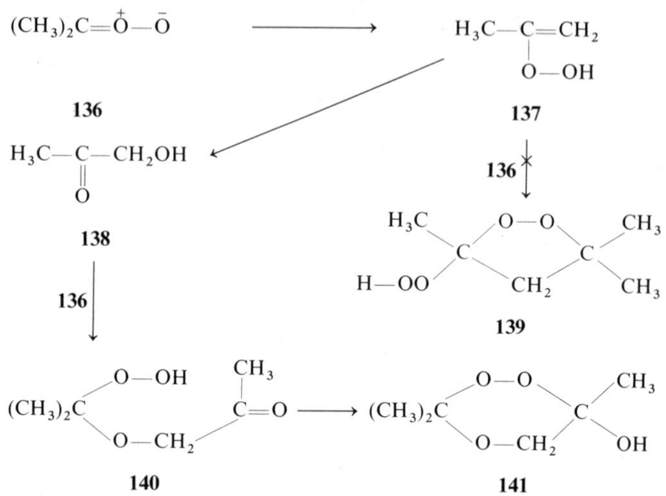

SCHEME 32. Isomerization of dimethylcarbonyl oxide.

B. EFFECTS OF VARIOUS EXPERIMENTAL CONDITIONS ON OZONIDE FORMATION

As pointed out earlier, various changes in experimental conditions can have relatively large effects on cis:trans ozonide ratios, ozonide yields, and the proportions of cross to normal ozonides. It is here, especially, that suggested explanations are, admittedly, speculative, and that further research into subtle details is needed.

First, some generalities need to be taken care of. Throughout Chapter VI, the importance of cross ozonide formation in arguments for and against various mechanisms has been discussed frequently. It is important also in the present chapter. Essential to these discussions is the fact that ozonide (1,2,4-trioxolane) formation is irreversible in the sense that one ozonide can neither be converted to another nor break down to Criegee moieties, which could then recombine to cross ozonides. The irreversibility of ozonide formation in this sense has been established by Murray and Williams (*237*) and by Kerur and Diaper (*362*).

Also, in the following discussions frequent reference will be made, either directly or by inference, to reactions of the Criegee moieties in a "solvent cage." Loan *et al.* (*188*) have been able to estimate in a clever manner the extent of the solvent cage reaction in the ozonolysis of 2-pentene.

1. MIXTURES OF OLEFINS

Ozonolyses of equimolar mixtures of the four possible combinations of the geometrical isomers of 3-hexene and 4-octene have been studied by

Murray and Williams (*293*); the results were similar to those obtained from ozonolyses of the individual olefins (Chapter VI, Section II). Ozonolysis of a mixture of the cis isomers afforded 3-heptene ozonide and 3-hexene ozonide with cis:trans ratios greater than unity, whereas ozonolysis of a mixture of the trans isomers gave ratios considerably less than unity. The cis:trans ratios of the ozonides obtained from equimolar mixtures of the *cis*- and *trans*-olefins (3-hexene and 4-octene) fell in between these values. Reliable cis:trans ratios for 4-octene ozonide were not obtained (*293*). These results are consistent with *syn*- and *anti*-carbonyl oxide rules (Bauld–Bailey or Kuczkowski, Chapter VI). It is not surprising that the values obtained were somewhat different from those obtained with a pure 3-hexene isomer and a pure 3-heptene isomer alone, since the conformations and steric requirements of each primary ozonide play a role in determining the final cis:trans ozonide ratios.

2. Solvent Effects

The nature of the solvent can have striking effects on the total ozonide yield, the relative yields of normal and cross ozonides from unsymmetrical olefins, and cis:trans ozonide ratios from *cis*- and *trans*-olefins.

The ozonide yield from *cis*-olefins is usually somewhat lower with polar than with nonpolar solvents, but this effect is greater when solvents such as methylene chloride, rather than nucleophilic solvents, are used (see Table XIII, examples 1–4, 7–9, 14, and 15). Similar results have been reported for methyl oleate (*228*), 1-alkenes (*230*), and certain tetrasubstituted olefins (*189*). This is understandable on the basis that the carbonyl oxide and aldehyde (or ketone) moieties from decomposition of the unstable primary ozonide diffuse apart more readily in polar solvents, thus increasing the opportunity for side reactions to occur at the expense of geminate recombination to the ozonide (*148*). The fact that the ozonide yield is not lowered as much (or sometimes not at all) with nucleophilic solvents may be due to solvation of the carbonyl oxide [Eq. (13)]. This should stabilize it and slow down polymerization (or rearrangement to acid) and give time for interaction with the aldehyde moiety. The ultimate result of this stabilization of the carbonyl oxide by nucleophilic solvents is, of course, the actual reaction with a protic nucleophilic solvent to give a hydroperoxide (Chapter V, Section IV; Section II of this chapter).

In contrast to the results just discussed, the ozonide yields from *trans*-olefins (at least those with small substituents on the double-bond carbons) are often greater in polar than nonpolar solvents. The effect is especially pronounced with dimethyl and diethyl ether, which are relatively small nucleophiles (see Table XIII, examples 10–13 and 16–18). A possible reason for this is that decomposition of primary ozonides is enhanced by polar

TABLE XIII

SOLVENT EFFECTS ON OZONIDE YIELDS FROM VARIOUS OLEFINS

Example	R in RCH=CHR	Configuration	Solvent	Conc. (*M*)	Temp. (°C)	Total yield (%)	References
1	*t*-Bu	Cis	Pentane	0.3	−75	82	*148,187*
2	*t*-Bu	Cis	Et_2O	0.1	−75	77	*323*
3	*t*-Bu	Cis	AcOEt	0.15	−75	57	*323*
4	*t*-Bu	Cis	CH_2Cl_2	0.15	−75	27	*323*
5	*t*-Bu	Trans	Pentane	0.3	−75	51–58	*148,187*
6	*t*-Bu	Trans	Et_2O, AcOEt, CH_2Cl_2 $CFCl_3$ or acetone	0.1–0.15	−75 to −110	0–10	*148,323*
7	*i*-Pr	Cis	Pentane	1.0	−70	84	*322*
8	*i*-Pr	Cis	Et_2O or AcOEt	1.0	−70	84, 81	*322*
9	*i*-Pr	Cis	CH_2Cl_2	1.0	−70	58	*322*
10	*i*-Pr	Trans	Pentane	1.0	−70	47	*322*
11	*i*-Pr	Trans	CH_2Cl_2	1.0	−70	56	*322*
12	*i*-Pr	Trans	AcOEt	1.0	−70	57	*322*
13	*i*-Pr	Trans	Et_2O	1.0	−70	73	*322*
14	Et	Cis	Pentane	0.2	−70	81	*309*
15	Et	Cis	Et_2O	0.2	−30, −110	80	*309*
16	Et	Trans	Pentane	0.2	−30, −70	23, 21	*309*
17	Et	Trans	CH_2Cl_2	0.2	−30, −70	24, 23	*309*
18	Et	Trans	Et_2O	0.2	−30, −70	44, 41	*309*

solvents, especially if they are nucleophilic. Thus, the opportunities for side reactions that lead to oligomer formation through attack on the primary ozonide by outside species (carbonyl oxides, aldehydes, etc., as discussed in Section I, A of this chapter) are decreased, and the chances of ozonide formation are, thereby, increased (*137,148,263*).

An exception to this generalization concerning *trans*-olefins is *trans*-1,2-di-*tert*-butylethylene (Table XIII, examples 5 and 6) and, perhaps, other *trans*-1,2-disubstituted ethylenes with large substituents. The ozonide yield decreased with increases in solvent polarity and nucleophilicity. Here, the bulky substituents hinder outside attacks on the primary ozonide, thus, this route to oligomer and the help of nucleophilic solvents in preventing it are unimportant. The increased separation of the carbonyl and carbonyl oxide moieties from each other in the more polar solvents is the more important factor. Interestingly, Greenwood and Durham (*145*) found this also to be true with smaller *trans*-olefins at low temperatures, where the primary ozonide is more stable. This, perhaps, explains the discrepancy noted in Section I, A concerning the Greenwood–Rubinstein oligomer formation mechanism.

Another effect of increasing the polarity of the ozonation solvent is to increase the ratio of cross to normal ozonides from unsymmetrical olefins (Table XIV, example 1 vs. examples 4 and 6; example 8 vs. examples 10 and 11). This effect was also observed and discussed by Privett and Nickell (*228,311*). This is understandable on the basis of previous discussion. The increased solvent polarity increases the separation of the carbonyl oxide and carbonyl moieties, allowing them to react more independently rather than in a solvent cage (*148,311*). It is noteworthy that an increase in reaction temperature has no effect with polar solvents (Table XIV, examples 4–7) whereas it has with nonpolar solvents, because even at low temperatures in polar solvents the solvent cage effect is minor (*148*).

Finally, regarding solvent effects, the cis:trans ozonide ratio tends to be higher in pentane or methylene chloride than in ether or ethyl acetate with either the *cis*- or the *trans*-olefin (see Table VII, Chapter VI, examples 4–13, 16–19, and 22–25). The same trend appears to occur with both normal and cross ozonides of unsymmetrical olefins, although not as clear-cut; the principal differences are between methylene chloride and ethyl acetate solvents

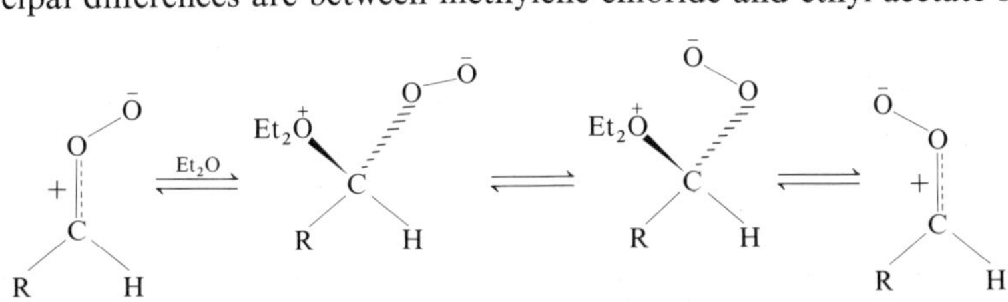

(13)

TABLE XIV

Relative Normal and Cross Ozonide Yields

Example	R and R′ in RCH=CHR′	Configuration	Solvent	Conc. (*M*)	Temp. (°C)	Relative yields (%) Pair A[a]	Pair B[a]	Pair C[a]	References
1	*i*-Pr, CH_3	Cis	Pentane	1	−75	11	86	3	*148*
2	*i*-Pr, CH_3	Cis	Pentane	1	−30	20	72	8	*148*
3	*i*-Pr, CH_3	Cis	Pentane	2.5	−75	20	70	10	*148*
4	*i*-Pr, CH_3	Cis	AcOEt	1	−75	30	58	12	*148*
5	*i*-Pr, CH_3	Cis	AcOEt	1	0	27	59	14	*148*
6	*i*-Pr, CH_3	Cis	CH_2Cl_2	1	−75	31	56	13	*148*
7	*i*-Pr, CH_3	Cis	CH_2Cl_2	1	0	24	58	18	*148*
8	*i*-Pr, CH_3	Trans	Pentane	1	−75	12	85	3	*148*
9	*i*-Pr, CH_3	Trans	Pentane	1	−30	17	77	6	*148*
10	*i*-Pr, CH_3	Trans	AcOEt	1	−75	17	75	8	*148*
11	*i*-Pr, CH_3	Trans	CH_2Cl_2	1	−75	25	60	15	*148*
12	Et, CH_3	Cis	Butane	0.2	−30	9	87	4	*309*
13	Et, CH_3	Cis	Me_2O	0.2	−70	22	53	25	*309*
14	Et, CH_3	Trans	Butane	0.2	−30	15	80	5	*309*
15	Et, CH_3	Trans	Me_2O	0.2	−70	17	52	31	*309*

[a] Pair A is higher molecular weight cross ozonide pair; pair C is lower molecular weight cross ozonide pair; and pair B is normal ozonide pair.

(see Table VIII, Chapter VI, examples 1, 2, 5, 7, 8, 10, and 11–14). There appears to be more correlation with the nucleophilicity than with the polarity of the solvent (*309,322*). A possible explanation is that nucleophilic solvents solvate the carbonyl oxide zwitterion and either cause equilibration of the syn and anti isomers [Eq. (13)] (*148*) or alter the steric requirements in the transition state leading to the ozonide.

Fliszár and Carles (*238*) reported that cis:trans ozonide ratios do not vary appreciably starting from a *cis*- or a *trans*-olefin when the ozonolysis is carried out in benzene solution at 25°C, whereas they do vary in pentane solution even at 25°C. They attributed this to equilibration of the *syn*- and *anti*-carbonyl oxides at the relatively high temperature in benzene solution. They suggested that in pentane solution the equilibration is prevented by a selective complexation of an *anti*-carbonyl oxide with a *cis*-olefin and a *syn*-carbonyl oxide with a *trans*-olefin. Thus, *cis*-olefin would still give more *cis*-ozonide, and *trans*-olefin more *trans*-ozonide. Fliszár and Carles (*238*) suggested that in benzene solution π complexes are formed between benzene (which is in excess) and the *syn*- and *anti*-carbonyl oxides, thus preventing the complexation with the olefins and allowing equilibration to occur. This explanation appears unlikely, however, because (a) recent studies with aromatic complexing agents have shown that complexation at 25°C does not occur appreciably; and (b) when complexation of carbonyl oxides with aromatics does occur, it appears to slow down equilibration (*320*) (see also Section I, B, 6 of this chapter).

3. Excess Aldehyde

To some extent, results obtained from ozonolyses of olefins in the presence of excess aldehyde parallel those obtained by increasing the polarity of the solvent. The *trans*-ozonide content of the ozonide mixture obtained increased markedly, in the case of *cis*-alkenes and the cross ozonide of 1-alkenes, as the concentration of added aldehyde in the reaction mixture increased from 0 to about 1 *M*, after which the cis:trans ozonide ratio remained constant (*237,293*). Specifically, during ozonolyses of *cis*-3-hexene (0.5 *M* in pentane), the *trans*-ozonide content of the 3-hexene ozonide (normal ozonide) increased from 43% to 60% (Fig. 1), and that of the 3-heptene ozonide (cross ozonide) increased from 47% to 55%, as the butyraldehyde content was increased from 0 to 1 *M* (*293*). Under the same conditions ozonolyses of *trans*-3-hexene resulted in only a slight increase in trans content of the normal (Fig. 1) and cross ozonides as the butyraldehyde content was increased. Further, in the case of the normal ozonide, the final cis:trans ozonide ratios reached at high aldehyde concentrations (1.75–2.0 *M*) were approximately the same starting from either *cis*- or *trans*-3-hexene (Fig. 1).

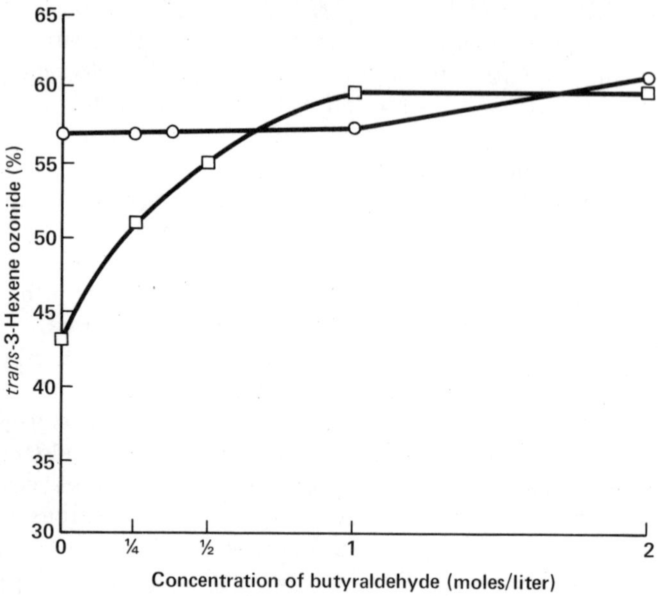

FIG. 1. 3-Hexene ozonide cis:trans ratios as a function of added butyraldehyde in the ozonolysis of *cis*-3-hexene (□) and *trans*-3-hexene (○). Reprinted, with permission, from Murray and Williams *in* "Oxidation of Organic Compounds-III" (*293*). Copyright by the American Chemical Society.

Since butyraldehyde cannot participate chemically in the 3-hexene ozonide formation, these effects must be largely solvent effects. This suggests an explanation similar to that proposed in the preceding subsection for the higher trans content in ozonides obtained by ozonolyses in nucleophilic solvents than in nonnucleophilic solvents, i.e., solvation of the carbonyl oxide zwitterion by the excess aldehyde. However, it is difficult to see how solvation of the zwitterion by the aldehyde could occur without the two reacting to give the cross ozonide. Thus, the effect of excess aldehyde on cis:trans ozonide ratios appears to be unexplained.

Murray and Williams (*237*) have determined the cis:trans cross ozonide ratios obtained upon ozonolyses of 3,3-dimethyl-1-butene (i.e., *tert*-butylethylene) in the presence of increasing amounts of the following aldehydes: propanal, 2-methylpropanal, and 2,2-dimethylpropanal. First, it was found that in each case the trans content of the ozonides increased as the added aldehyde concentrations were increased from 0 to about 1 *M* (*237*), just as in the cases discussed in the preceding paragraph. It was further found that the bulkier the added aldehyde, the greater the cis:trans ratio over the entire range of the aldehyde concentration curve (Fig. 2) (*237*). This is understandable on the basis of either the Bauld–Bailey (Table X) or the Kuczkowski (Table XI) refinements of the Criegee mechanism (Chapter VI, Section IV), keeping in mind that with 1-alkene primary ozonides the substituent is equatorial (*324,328*). This leads to a predominance of anti-carbonyl oxide and cis final ozonide; this tendency should increase as the bulk of the added aldehyde increases (Chapter VI, Section IV).

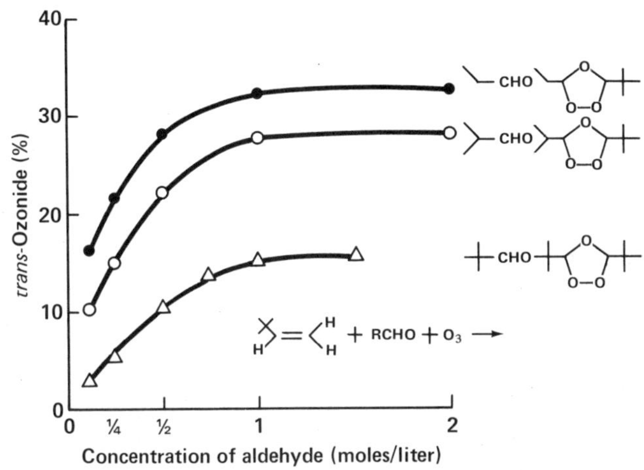

FIG. 2. The percentage of *trans*-ozonide obtained as a function of added aldehyde concentration in the ozonolysis of 3,3-dimethyl-1-butene. Reprinted, with permission, from Murray and Williams (*237*). Copyright by the American Chemical Society.

Another interesting observation made by Murray and Williams (*237*) came from a comparison of the cis : trans cross-ozonide ratios obtained from ozonolysis of 1-pentene (propylethylene) in the presence of varying amounts of 2,2-dimethylpropanal on the one hand and of 3,3-dimethyl-1-butene (*tert*-butylethylene) in the presence of propionaldehyde on the other. The curves, showing the percentage of *trans*-ozonide vs. concentration of added aldehyde in the two cases, were very nearly superimposable, indicating that steric requirements in the olefin and the aldehyde are equally important. This is not surprising in view of the discussion in the preceding paragraph. It seems likely, however, that the near superimposability of the two curves was coincidental, especially since the cross ozonides in the two cases were not identical.

The variation in overall ozonide yield from ozonolyses of the 3-hexenes in the presence of increasing concentrations of butyraldehyde is almost as difficult to understand as the variation in cis:trans ozonide ratios under the same conditions. With *cis*-3-hexene the total ozonide yield remained constant up to about a 1 *M* concentration of butyraldehyde, after which it decreased slightly as the concentration increased to 3 *M*, as shown in Fig. 3 (*293*). This, perhaps, is caused by some attack of the excess aldehyde on the primary ozonide, even though its stability is very slight at the temperature employed, −70°C; however, the effect discussed is also slight, and it occurs only in the presence of a considerable excess of aldehyde. Along with the slight decrease in total ozonide yield was, of course, a dramatic increase in cross ozonide (3-heptene ozonide) yield, at the expense of normal ozonide, as the butyraldehyde content was increased (Fig. 3) (*293*).

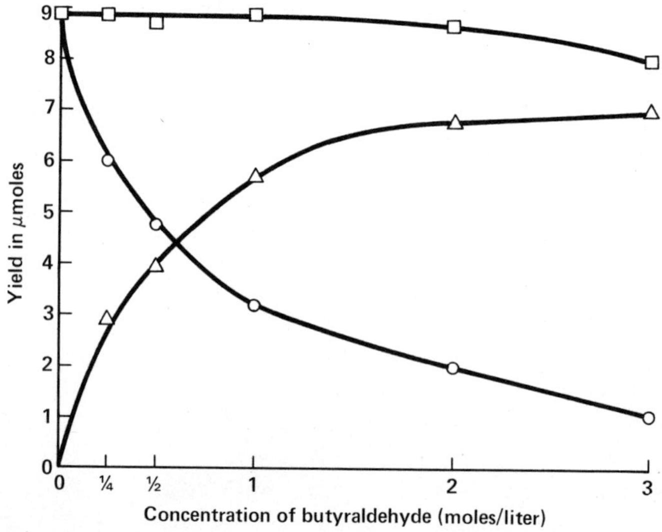

FIG. 3. Variation of 3-hexene ozonide (○), 3-heptene ozonide (△), and total ozonide (□) yields as a function of added butyraldehyde concentration in the ozonolysis of *cis*-3-hexene. Reprinted, with permission, from Murray and Williams, *in* "Oxidation of Organic Compounds-III" (*293*). Copyright by the American Chemical Society.

With *trans*-3-hexene, the total ozonide yield first increased from 50% to 77% as the added butyraldehyde concentration was increased from 0 to 0.5 *M*, then fell to 60% as the butyraldehyde concentration was increased to 2.0 *M*, after which it remained constant (Fig. 4) (*293*).

This, again, reminds one of the effect of increased solvent polarity on ozonide yields from *trans*-olefins, discussed in the preceding subsection. The fact that the ozonide yield first increased and then decreased as the butyraldehyde concentration increased, causes one to suspect that two competing effects are in operation. One is the solvent effect. Increased polarity caused

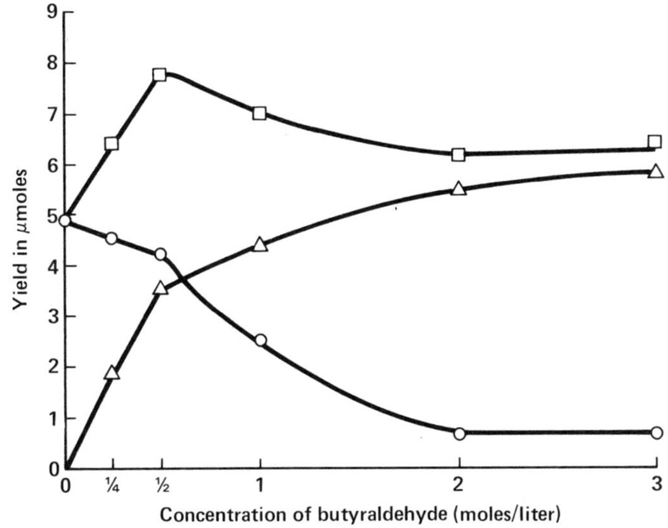

FIG. 4. Variation of yields of 3-hexene ozonide (○), 3-heptene ozonide (△), and total ozonide (□) as a function of added butyraldehyde concentration in the ozonolysis of *trans*-3-hexene. Reprinted, with permission, from Murray and Williams, *in* "Oxidation of Organic Compounds-III" (*293*). Copyright by the American Chemical Society.

by the presence of butyraldehyde increases the rate of primary ozonide decomposition, thereby decreasing the chances of oligomer formation by attack of aldehyde or carbonyl oxide on the primary ozonide. When, however, the foreign aldehyde concentration gets above 0.5 *M*, the oligomer formation begins to increase owing to attack of aldehyde on the primary ozonide. As with *cis*-3-hexene, the yield of cross ozonide increased and that of normal ozonide decreased as the butyraldehyde concentration was increased. The cross ozonide yield increase was dramatic up to the 0.5 *M* butyraldehyde concentration mark, while the normal ozonide yield decrease was slow; above 0.5 *M* butyraldehyde concentration just the opposite was true (Fig. 4) (*293*). This is consistent with the above explanation. It is interesting that no matter how much the butyraldehyde concentration was increased some normal ozonide was always produced (*293*). Similar results were reported by Privett and Nickell (*228*) from the ozonolysis of methyl linoleate in the presence of nonanal.

4. Concentration Effects

Murray *et al.* (*322*) have studied the effect of changes in olefin concentration upon the cis:trans ozonide ratios obtained upon ozonolyses of the *cis*- and *trans*-1,2-diisopropylethylenes in pentane at −70°C. There was very little effect as olefin concentrations were varied between 1 and 4 *M*, especially with the *cis*-olefin (Fig. 5). However, as the concentration was decreased from 1 *M*, the percentage of *trans*-ozonide increased rapidly with both isomeric olefins. In the case of the *trans*-olefin, the percentage of *trans*-ozonide appeared to reach a maximum at about 1/16 *M* (Fig. 5), after which it began to decrease again. With the *cis*-olefin the percentage of *trans*-ozonide continued to increase, and it appeared that the cis: trans ozonide ratios might become equal (at about 48% trans) at extremely low concentrations. The decrease in the percentage of *trans*-ozonide from the maximum as the concentration of *trans*-olefin is increased is not readily explainable at the present time. The approach of the cis: trans ozonide ratios from the *cis*- and *trans*-olefins toward each other at very low olefin concentrations, however, is understandable on the basis of a time factor. At high dilutions, more time is required for the carbonyl oxide to find an aldehyde moiety with which to react and form an ozonide. During this time, equilibration of the *syn*- and *anti*-carbonyl oxides begins to occur. At some point of high dilution the thermodynamic equilibrium mixture should be reached, starting from either the *cis*- or the *trans*-olefin.

The effect of changes in olefin concentration on the total ozonide yields from ozonolyses of *cis*- and *trans*-1,2-diisopropylethylene in pentane at

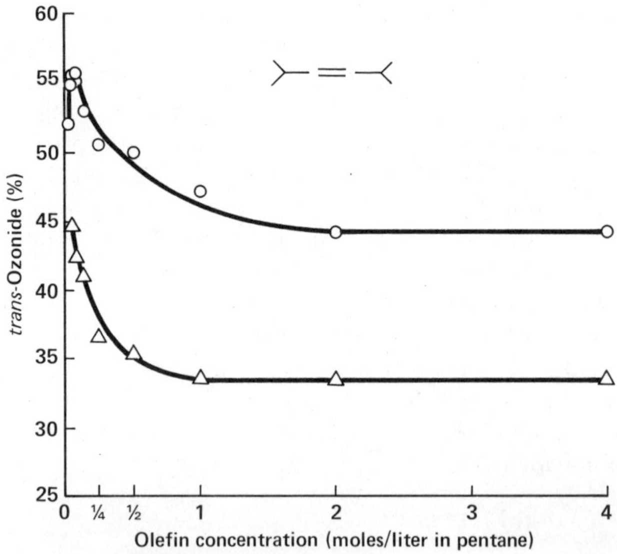

FIG. 5. Percentage of *trans*-ozonide from (△) *cis*- and (○) *trans*-1,2-diisopropylethylene as a function of concentration. Reprinted with permission of author and publisher from Murray *et al.* (*322*). Copyright by Permagon Press, Ltd.

−70°C (*322*) perhaps can be similarly rationalized. At all concentrations the *cis*-olefin gave higher ozonide yields than the *trans*-olefin. The maximum ozonide yield in each case occurred at an olefin concentration of about 0.25 *M* (Fig. 6). Below this concentration the ozonide yields dropped rapidly, and at higher concentrations they dropped more slowly. The yields from the *trans*-olefin dropped more rapidly than those from the *cis*-olefin, as the concentration was increased above 0.25 *M*, and leveled off at about 2 *M* (30% ozonide yield), whereas the drop in yield from the *cis*-olefin had not yet leveled off at 4 *M* (60% yield) (*322*). These data perhaps are understandable on the basis that as the olefin concentration is decreased below 0.25 *M*, the concentrations of carbonyl oxide and carbonyl moieties present at any instant are so low that other fates of the carbonyl oxide (polymerization or rearrangement) compete favorably with geminate recombination with the aldehyde moiety to produce ozonide. As the olefin concentration is increased above 0.25 *M*, two effects perhaps become possible. One is that the polarity of the medium increases (due both to olefin and to ozonolysis products), and this may well cause the ozonide yield from the *cis*-olefin to decrease for reasons already discussed under solvent effects. The other effect is that the

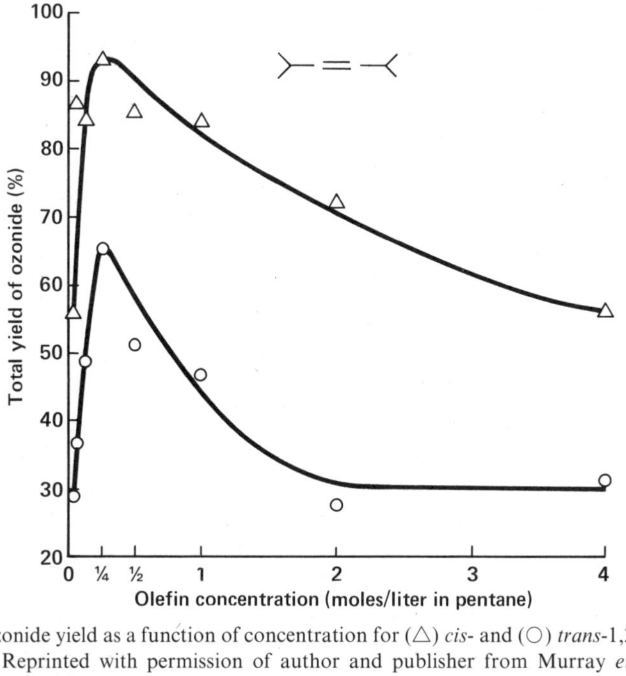

FIG. 6. Ozonide yield as a function of concentration for (△) *cis*- and (○) *trans*-1,2-diisopropylethylene. Reprinted with permission of author and publisher from Murray *et al.* (*322*). Copyright by Permagon Press, Ltd.

chances for outside species (aldehyde or carbonyl oxide) to attack the primary ozonide are increased. This, as discussed earlier, appears to increase oligomer formation at the expense of ozonide formation (*148,263*), especially with the *trans*-olefin because of the greater stability of its primary ozonide than that of the *cis*-olefin.

Olefin concentration changes also affect the proportions of cross to normal ozonides obtained upon ozonolysis of unsymmetrical olefins. An increase in olefin concentration in a nonpolar solvent increases the ratio of cross to normal ozonides (Table XIV, examples 1 and 3). This effect has also been observed by Riezebos *et al.* (*229*) in the ozonolysis of methyl oleate. The increased concentration brings the aldehyde and carbonyl oxide moieties closer together and thereby allows combinations outside the solvent cage to occur more often (*148*).

5. Temperature Effects

The most comprehensive study of temperature effects in ozonide formation has been made by Murray and Hagen (*321*) with the *cis*- and *trans*-1,2-diisopropylethylenes in pentane solution (0.2 *M*, 50% conversion). The most striking finding of the investigation was the effect of the rate of warming of the reaction mixtures on the cis:trans ozonide ratios. With a fast warm-up (plunging the reaction vessel into a water bath at room temperature), it was found that the isomeric olefins gave different, but temperature-independent, cis:trans ozonide ratios (56:44 for the *cis*-olefin and 39:61 for the *trans*-olefin). With a slow warm-up (reaction mixtures kept in the cooling bath, which was allowed to warm up slowly), however, the cis:trans ozonide ratios obtained from ozonolyses carried out below −100°C for the *cis*-olefin and below −80°C for the *trans*-olefin were different from those obtained at higher temperatures. For the *cis*-olefin the effect was less dramatic than for the *trans*-olefin, the cis:trans ozonide ratios varying from 56:44 for ozonolyses at −100°C to 61:39 for those at −123°C. The striking effect with the *trans*-olefin is shown in Fig. 7 (*321*). This effect has been verified by Bailey *et al.* (*320*). The fact that the variation of cis:trans ozonide ratios with temperature occurs only below −100°C for the *cis*-olefin and below −80°C for the *trans*-olefin suggests that it is associated with the stability of primary ozonides. It is interesting that under the slow warm-up, both the *cis*- and the *trans*-olefins give cis:trans ozonide ratios greater than unity at low temperatures. This would have been predicted by the SMY mechanism (Chapter VI, Section III) as interpreted by Bauld and Bailey (*148*) (see Chapter VI, Section IV, A), but, as discussed earlier, this mechanism has now been disproved (Chapter VI, Section VI). Current results (*363*) indicate that *trans*-1,2-diisopropylethylene may be unique in this respect; neither *trans*-

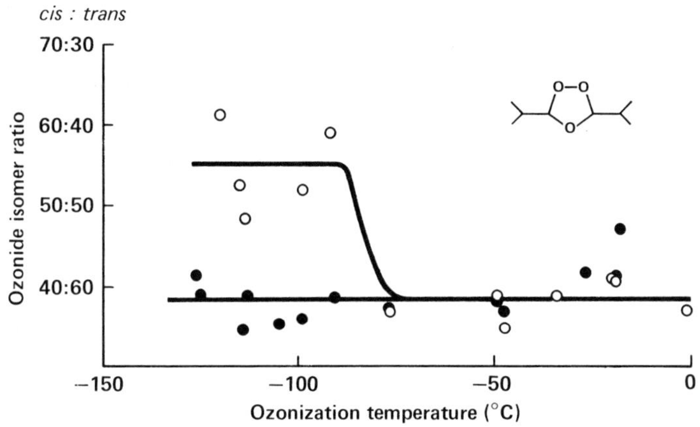

FIG. 7. Dependence of cis:trans ozonide ratio on warm-up rate of ozonolysis reaction mixture of *trans*-diisopropylethylene. ●, Fast warm-up; ○, slow warm-up. Reprinted, with permission, from Murray and Hagen, (*321*). Copyright by the American Chemical Society.

1,2-di-*tert*-butylethylene nor *trans*-3-hexene gave cis:trans ozonide ratios greater than unity under the same conditions. It is noteworthy that Fliszár *et al.* (*328,337*) found the isopropyl group to be anomalous in their studies with unsymmetrical olefins (Chapter VI, Section IV, C and Chapter VII, Section III).

These strange results concerning differences in cis:trans ozonide ratios obtained under fast and slow warm-up conditions show that neither the Bauld–Bailey nor the Kuczkowski schemes in their present forms are sophisticated enough to explain all findings. Perhaps the clue to further refinement lies in the calculations of Rouse (*329,330*), which indicate that several conformations for a primary ozonide lie close together in energy and no one should be major (Chapter VI, Section IV, A). It is possible that different conformations are involved under slow and fast warm-up conditions, due to the fact that energy is released suddenly in the latter case.

Temperature studies also were made with ozonolyses of *cis*- and *trans*-1,2-diisopropylethylene carried out in the presence of equimolar amounts of propionaldehyde (*321*). In a third set of runs, the propionaldehyde was added after the ozonations were complete (*321*). With the *trans*-olefin, ozonized in the presence of propionaldehyde and using the slow warm-up, curves similar to those of Fig. 7 were obtained for both the normal and the cross ozonides, with the exception that the normal ozonide gave higher cis:trans ratios in the presence of propionaldehyde than in its absence (cf. Figs. 7 and 8). In the third set of experiments, the low-temperature portion of the curve for the normal ozonide was the same as for the ozonolyses

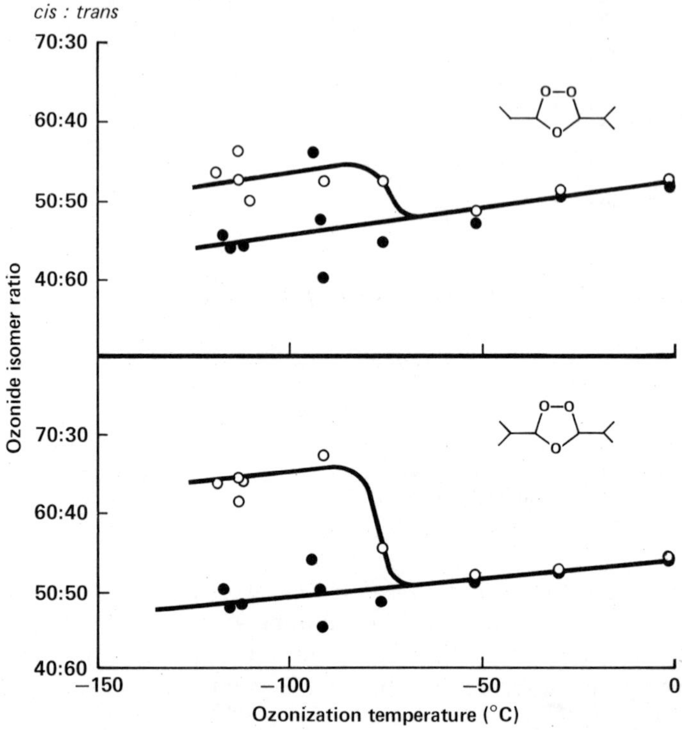

FIG. 8. Dependence of the ozonide cis:trans ratio on the ozonolysis temperature of *trans*-diisopropylethylene with propionaldehyde present during the ozonolysis. ●, Fast warm-up; ○, slow warm-up. Reprinted, with permission, from Murray and Hagen, (*321*). Copyright by the American Chemical Society.

carried out in the presence of propionaldehyde (Fig. 8), and the higher-temperature portion was similar to that of Fig. 7 (see Fig. 9). Both the *cis*- and the *trans*-olefins exhibited a temperature dependence in regard to cis:trans normal and cross ozonide ratios for ozonolyses performed in the presence of propionaldehyde followed by either the slow or the fast warm-up. For the *cis*-olefin the cis:trans ratios decreased with increasing ozonation temperature, and there was essentially no difference in the fast and slow warm-up (*321*). The *trans*-olefin showed a small but steady increase in cis:trans ratio with increasing reaction temperature in the fast warm-up experiments (Fig. 8). The temperature dependence in the presence of propionaldehyde, but not in its absence, is just as difficult to understand as the effect of added aldehyde on cis:trans ozonide ratios discussed in Section I, B, 3 of this chapter. In regard to the cross ozonide (2-methyl-3-hexene

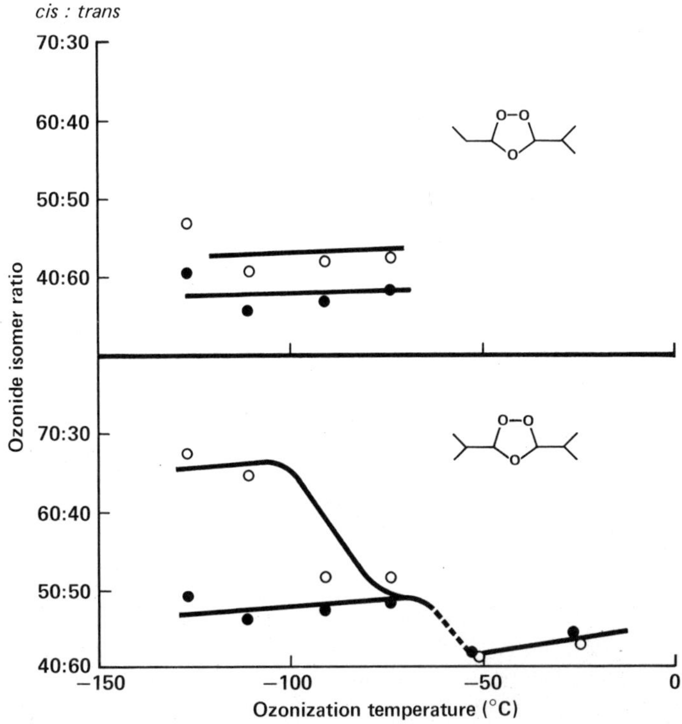

FIG. 9. Dependence of the ozonide cis : trans ratio on the ozonolysis temperature of *trans*-diisopropylethylene with propionaldehyde added after the ozonolysis. ●, Fast warm-up; ○, slow warm-up. Reprinted, with permission, from Murray and Hagen, (*321*). Copyright by the American Chemical Society.

ozonide) obtained during ozonolyses of *cis*-diisopropylethylene in the presence of propionaldehyde, Murray and Hagen (*321*) plotted the logarithms of the cis : trans ozonide ratios obtained against the reciprocals of the absolute temperatures employed, and, instead of a straight line, two intercepting straight lines were obtained. They interpreted this as indicating that a mechanism change occurred at about −60°C. Since, at present, all mechanisms other than the Criegee type have been disproved, with the possible exception of one involving rearrangement of a σ complex (Section I, C of this chapter), there must be some other explanation. Obviously, much more work is needed concerning the effects of fast and slow warm-up, particularly with a variety of solvents.

Murray and Hagen (*321*) found very little dependence of ozonide yields on temperature. In the absence of added aldehyde the total ozonide yield from *cis*-1,2-diisopropylethylene decreased from 70% to 60% for ozonolyses performed at −123°C and 0°C, respectively, whereas the yields from the *trans*-olefin remained fairly constant at 43% (Table XV, examples 5–8) throughout the temperature range. This is to be contrasted with the di-*tert*-butylethylenes, with which there were significant decreases in ozonide yield with increasing temperature, especially with the trans isomer (Table XV, examples 1–4), and with the 3-hexenes, with which there were slight increases in ozonide yield from both isomers as the ozonation temperature increased (Table XV, examples 9–15). In regard to experiments carried out by Murray and Hagen (321) in the presence of propionaldehyde, the overall ozonide

TABLE XV

TEMPERATURE EFFECTS ON OZONIDE YIELDS FROM VARIOUS OLEFINS[a]

Example	R in RCH=CHR	Configuration	Solvent	Temp. (°C)	Yield (%)	References
1	*t*-Bu	Cis	Pentane	−75	82	*148,187*
2	*t*-Bu	Cis	Pentane	RT	64	*323*
3	*t*-Bu	Trans	Pentane	−75	51–58	*148,187*
4	*t*-Bu	Trans	Pentane	RT	0	*323*
5	*i*-Pr	Cis	Pentane	−123	70	*321*
6	*i*-Pr	Cis	Pentane	0	60	*321*
7	*i*-Pr	Trans	Pentane	−123	43	*321*
8	*i*-Pr	Trans	Pentane	0	43	*321*
9	Et	Cis	Pentane	−70	81	*309*
10	Et	Cis	Pentane	−30	88	*309*
11	Et	Trans	Pentane	−70	21	*309*
12	Et	Trans	Pentane	−30	23	*309*
13	Et	Trans	Et_2O	−110	35	*309*
14	Et	Trans	Et_2O	−70	41	*309*
15	Et	Trans	Et_2O	−30	44	*309*

[a] In all cases the olefin concentration was 0.1 to 0.3 *M*.

yields from *cis*-1,2-diisopropylethylene fell to 35% (for ozonation at −100°C) and 40% (for ozonation at −1°C); the decrease was less with the trans isomer (to ca. 30% total yield).

These temperature effects seem to be related to the increased decomposition rate of the primary ozonide and to the greater dispersion of the Criegee moieties as the temperature increases (Table XV). With olefins having large substituents at the double bond (the di-*tert*-butylethylenes), the separation of the carbonyl oxide and aldehyde moieties is probably the more important factor. This leads to independent side reactions outside the solvent cage at the expense of the geminate recombination of the moieties to form ozonide. With olefins having small substituents (the 3-hexenes), however, the faster decomposition of the primary ozonide at higher temperatures is more important, decreasing the opportunity for attacks of outside species on the primary ozonide and, thus, decreasing the oligomer yield (*148,263*). The diisopropylethylenes should fall somewhere in between. Further evidence for this viewpoint was furnished by Durham and Greenwood (*143*), who showed that the amount of oligomer produced from small *cis*-olefins increased considerably when the ozonolyses were carried out at −130°C, and the primary ozonide was allowed to decompose slowly at this temperature.

Murray and Hagen (*321*) also studied two unsymmetrical olefins, *cis*- and *trans*-4-methyl-2-pentene (methylisopropylethylene). In regard to cis:trans ozonide ratios, there was very little dependence on temperature or configuration with the normal ozonides. In the case of the cross ozonides, however, the ratios from the cis and trans isomers got closer together as the ozonolysis temperature was increased (*321*) (see also Table VIII, examples 1–4). Similar results have been reported by Fliszár and Carles (*238*). This would appear to be indicative of equilibration of *syn*- and *anti*-carbonyl oxides at higher temperatures.

In regard to the effect of temperature changes on the ratio of cross to normal ozonides during ozonolyses of unsymmetrical olefins, this ratio increases as the ozonolysis temperature increases (*229,321*) (see also Table XIV). This almost certainly is due to the Criegee moieties separating from each other and reacting more independently rather than with each other at higher temperatures.

6. Complexing Agents

In Chapter XI, Section IV, the discovery of ozone π complexes, which are stable only at low temperatures (ca. −140°C or below), is discussed. Currently, studies involving the use of ozone complexes as ozonation agents are being conducted (*320,363*). Some results, comparing the cis:trans ozonide ratios obtained from *cis*- and *trans*-diisopropylethylene using first "free" and then complexed ozone in two different solvents, are shown in Table XVI (*320*). Several observations are important. First, the differences in cis:trans ozonide ratios observed between the *cis*-olefin on the one hand and the *trans*-olefin on the other, as well as differences observed in ozonolyses using "free" or using complexed ozone, are greater in isopentane than in Freon 12. Second, the differences in cis:trans ozonide ratios obtained using "free" or using complexed ozone were much greater with the *cis*- than with the *trans*-olefin (there were essentially no differences in Freon 12 with the *trans*-olefin).

TABLE XVI

CIS: TRANS OZONIDE RATIOS FROM OZONATION OF *cis*- AND *trans*-1,2-DIISOPROPYLETHYLENES WITH OZONE COMPLEXES[a,b]

			Freon 12		Isopentane	
Example	Temp. (°C)	Complexation agent	Cis isomer	Trans isomer	Cis isomer	Trans isomer
1	−78	None	50:50	49:51	54:46	49:51
2	−150	None	55:45	51:49	67:33	58:42
3	−150	Toluene	—[c]	—[c]	74:26	62:38
4	−150	*o*-Xylene	61:39	48:52	76:24	65:35
5	−150	Isodurene	61:39	51:49	75:25	57:43
6	−150	HEB[d]	—[e]	—[e]	75:25	65:35
7	−150	MPE[d]	64:36	50:50	78:22	54:46

[a] Reprinted, with permission, from Bailey *et al.* (*320*). Copyright by the American Chemical Society.

[b] A slow warm-up was used throughout these studies.

[c] Complex is stable in Freon 12 only at lower temperatures.

[d] HEB = hexaethylbenzene; MPE = 1-mesityl-1-phenylethane.

[e] The complexing agent is too insoluble in Freon 12 for use in the usual concentration.

Third, the results obtained by ozonizing with "free" ozone and then adding the complexing agent to the cold reaction mixture containing the primary ozonide were essentially no different from those obtained using complexed ozone (cf. normal and inverse addition results in Table XVII.

It was concluded from this study (*320*) that the differences observed using "free" or complexed ozone were not caused by different types of ozone attack, but, instead, resulted from complexation of the carbonyl oxide zwitterion, which is isoelectronic with ozone, as it was produced from decomposition of the primary ozonide in the presence of the complexing agent. The complexation of the carbonyl oxide slowed down or prevented equilibration of the syn and anti stereoisomers. In the absence of the complexing agent, the equilibration is evidenced by the facts that the cis: trans ozonide ratios from both isomeric olefins approach 50:50 as the ozonation temperature is increased (cf. examples 1 and 2, Table XVI) and that the ratios are lower (or closer to 50:50) in the more polar Freon-12 solvent than in isopentane. The carbonyl oxide complexation theory also explains the observation that it is only with the *cis*-olefin that the cis: trans ozonide ratios are different, depending on whether or not a complexing agent is present. Trans primary ozonides decompose at temperatures too high for complexation of the carbonyl oxide to be effective, whereas cis primary ozonides decompose

TABLE XVII

COMPARISON OF NORMAL AND INVERSE OZONATION WITH COMPLEXING AGENTS[a]

DIPE[b] isomer	Solvent	Complexation agent	Cis : trans ozonide ratios	
			Normal	Inverse[c]
Cis	Freon 12	*o*-Xylene	61:39	60:40
Cis	Freon 12	Isodurene	61:39	58:42
Cis	Freon 12	MPE[d]	64:36	62:38
Cis	Isopentane	*o*-Xylene	76:24	75:25
Cis	Isopentane	Isodurene	75:25	74:26
Cis	Isopentane	MPE[d]	78:22	77:23
Trans	Freon 12	*o*-Xylene	48:52	47:53
Trans	Freon 12	Isodurene	51:49	49:51
Trans	Isopentane	Toluene	62:38	64:36
Trans	Isopentane	*o*-Xylene	65:35	65:35
Trans	Isopentane	Isodurene	57:43	55:45
Trans	Isopentane	HEB[e]	65:35	65:35
Trans	Isopentane	MPE[d]	54:46	53:47

[a] Reprinted, with permission, from Bailey *et al.* (*320*). Copyright by the American Chemical Society.

[b] DIPE = 1,2-diisopropylethylene. Ozonation temperature = −150°C.

[c] Inverse ozonation refers to ozonation with "free" ozone followed by addition of the complexing agent before warm-up.

[d] MPE = 1-mesityl-1-phenylethane.

[e] HEB = hexaethylbenzene.

at much lower temperatures, and the resulting carbonyl oxide is immediately complexed.

Whereas these findings further emphasize the validity of the *syn*- and *anti*-carbonyl oxide concept, they also point again to the inadequacies of both the Bauld–Bailey and the Kuczkowski schemes in their present forms (Chapter VI, Section IV) in explaining some of the fine details of the stereoselectivity observed in ozonide formation. This is especially true in regard to the surprisingly high cis:trans ozonide ratios obtained from *trans*-1,2-diisopropylethylene in isopentane under slow warm-up conditions, particularly in the presence of a complexing agent (Tables XVI and XVII).

7. MISCELLANEOUS

In regard to the equilibration of *syn*- and *anti*-carbonyl oxides frequently mentioned in the preceding discussion, some work of Murray and coworkers (*239,305,307*) is of interest. This concerns the formation of ozonides from carbonyl oxides obtained from aryl- and alkyldiazo compounds in the presence of aldehydes. The carbonyl oxides afforded cis:trans ozonide ratios similar to those obtained from ozonolyses of *trans*-olefin in the presence of aldehydes. This implies that the carbonyl oxide produced synthetically is composed of the most thermodynamically stable equilibrium mixture of the syn and anti forms and that this is closer to that obtained from the *trans*-olefin, at least in the presence of excess aldehyde.

In this connection it also is of interest that autoxidation of benzaldehyde under alkaline conditions is reported to give, along with benzoic acid (the major product) and phenol (0.4% yield), a 0.05% yield of stilbene ozonide which consisted of 80% cis isomer (*364*). It would appear that the carbonyl oxide intermediate produced under these conditions consisted largely of the anti form (Chapter VI, Section IV).

C. PEROXY EPOXIDES

Rearrangement of peroxy epoxides was proposed by Murray *et al.* (*127,128*) as a competing route to ozonides. Is there other evidence for such? To explore this question is the purpose of this subsection.

1-Alkenes afford very low yields of cross ozonides, although the total ozonide yield is high (*237*). This, of course, could be explained on the basis that the primary ozonide breaks down almost exclusively to give only one pair, rather than two pairs, of Criegee moieties (see Section III of this chapter). Although the alkyl-substituted carbonyl oxide has been shown to be strongly favored (*337*), other factors also appear to be involved.

Murray and Williams (*237,357*) have shown that, even in the presence of a large excess of foreign aldehyde, 1-alkenes give no more than 30–40% of cross ozonide whereas with other olefins as much as 80–90% cross ozonide is obtained under analogous conditions. This could be interpreted to indicate that the major route to ozonides with 1-alkenes involves an intramolecular rearrangement of either the primary ozonide (cf. *91,92,187*) or a peroxy epoxide (*127,128*) (Chapter VI, Section III). However, Murray and Williams (*357*) rule out such mechanisms on the basis that they should be stereospecific, which is contrary to their findings. Separate ozonolyses of *cis*- and of *trans*-1-deutero-1-hexene gave 50:50 mixtures of the *cis*- and *trans*-deuteroozonides. Murray and Williams (*357*) suggest an extreme solvent cage reaction involving rapid cleavage of an unstable primary ozonide followed by rapid recombination of the Criegee moieties.

The formation of a peroxy epoxide has been suggested as an explanation for the differences in ease with which certain cyclobutenes form ozonides (*130*). As shown in Scheme 33, *cis*-1,2,3,4-tetraphenylcyclobutene (**142**) gave a single ozonide, which was believed to be the endo isomer (**146a**). In contrast, *cis*-3-methyl-1,2,3,4-tetraphenylcyclobutene (**148**) afforded both the endo (**146b**) and the exo (**150**) ozonides, in approximately equal amounts. *trans*-1,2,3,4-Tetraphenylcyclobutene (**151a**) of course, produced a single ozonide (**153a**), but *trans*-3-methyltetraphenylcyclobutene (**151b**) also gave only one ozonide, thought to be **153b**. Of these four cyclobutenes, three afforded ozonides even in the presence of the "participating" (Chapter V, Section IV; Section II of this chapter) solvent methanol; only *cis*-1,2,3,4-tetraphenylcyclobutene (**142**) yielded a product (**147**) derived by solvent participation. It was further found that hexasubstituted cyclobutenes such as **156** did not react with ozone at all.

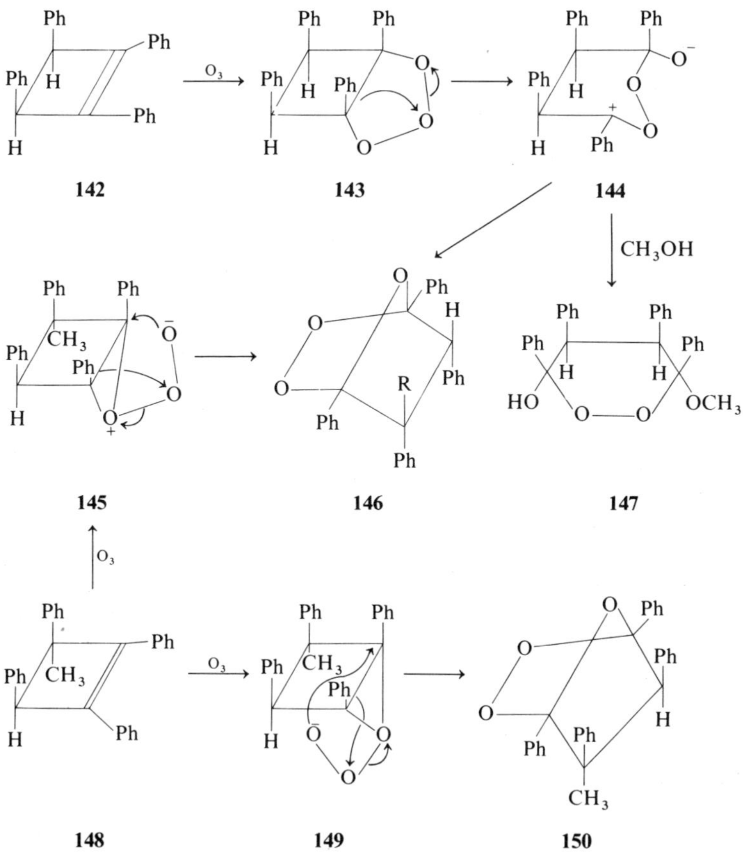

a R = H; **b** R = CH_3.

SCHEME 33. Ozonation of phenyl-substituted cyclobutenes.

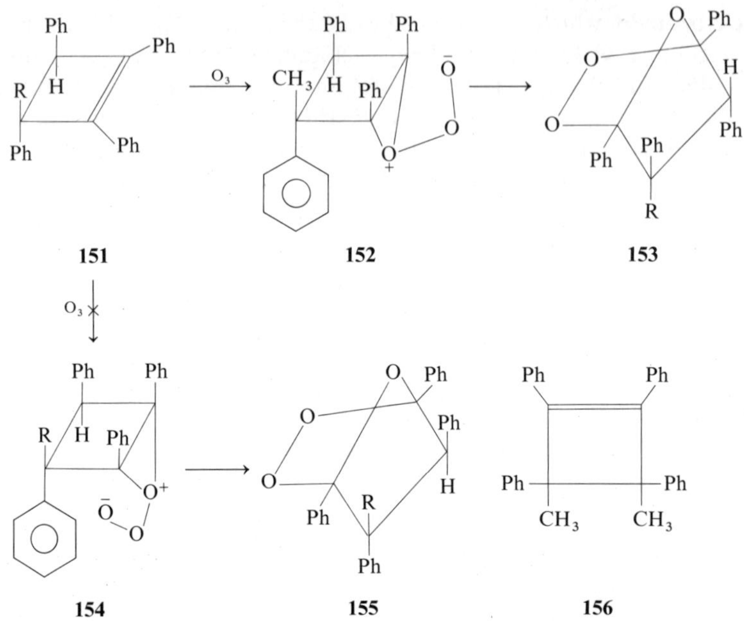

a R = H; b R = CH_3

SCHEME 33 (*Continued*)

On the basis of these results, Reinhardt *et al.* (*130*) concluded that at least one sp^3 carbon of the cyclobutene ring must be mono- rather than disubstituted, in order for ozone to react, and that initial ozone attack occurs only at that face of the cyclobutene ring to which a hydrogen atom is attached. They also suggested that only *cis*-tetraphenylcyclobutene (**142**), which has one face free of substituents, produces a primary ozonide, **143**, which then reacts through intermediate **144**, rather than a Criegee carbonyl oxide, to give ozonide **146a** in a nonparticipating solvent, or **147** in methanol. With the other tetra- or pentasubstituted cyclobutenes, which have at least one substituent on each face of the cyclobutene ring, it was suggested (*130*) that steric hindrance prevents the formation of a primary ozonide and a peroxy epoxide is produced instead. In the case of *cis*-methyltetraphenylcyclobutene, **148**, either peroxy epoxide, **145** or **149**, could be formed. On concerted rearrangement, peroxy epoxide **145** would yield the *endo*-ozonide, **146**, and **149** would yield the *exo*-ozonide, **150**. With the *trans*-tetraphenylcyclobutenes only a peroxy epoxide of type **152** can be produced, because the bulk of the phenyl group prevents the formation of **154** (*130*). This leads only to ozonide **153b** (not to **155b**) in the case of *trans*-methyltetraphenylcyclobutene (**151**b). According to the scheme of Reinhardt *et al.* (*130*), the peroxy epoxides lead only to ozonides and do not undergo reactions incorporating participating solvents. In the case of the hexasubstituted cyclobutenes, steric hindrance prevents both primary ozonide and peroxy epoxide formation. Further studies are necessary in order to test these mechanistic hypotheses and also to determine whether or not such mechanisms account for the peculiar ease with which many cyclobutene and cyclopentene types form ozonides, often even in the presence of participating solvents (Section II, C of this chapter).

D. SUMMATION

In summary, it can be said that much progress has been made in understanding the mechanism of ozonide formation during ozonolysis of olefins. Also, although all mechanisms seem to have been eliminated except the *syn*- and *anti*-carbonyl oxide mechanism, it appears that there still are subtleties that cannot yet be explained by either the Bauld–Bailey or the Kuczkowski variations in their present forms. Much work needs to be done in refining the *syn*- and *anti*-carbonyl oxide mechanism and also in ascertaining whether or not peroxy epoxides play a role in ozonide formation.

II. Competitions in Participating Solvents

A. α-OXYALKYL HYDROPEROXIDE FORMATION

Experience has shown (Chapter III, Section I, C and Chapter V, Section IV that ozonolyses in participating (i.e., protic, nucleophilic) solvents generally lead to α-oxyalkyl hydroperoxides, etc. (e.g., **160** or **164**) (*60,62,94 110–112,194,264,265,268,272,273*) rather than to the ozonides, polymeric ozonides, and dimeric and polymeric peroxides obtained from ozonolyses in nonparticipating solvents, as described in Section I of this chapter. Although it is possible that such hydroperoxides could be produced in either of two ways, reaction of a carbonyl oxide with the participating solvent (e.g., Scheme 34, **157** → **159** → **160**) or direct attack of the participating solvent on the primary ozonide (e.g., Scheme 34, **158** → **161** → **160**), the former, as described by the Criegee mechanism (Chapter III, Section I, C) appears to be the more likely. Addition of an excess of methanol to a cold solution of the primary ozonide of *trans*-1,2-di-*tert*-butylethylene did not change the NMR spectrum of the primary ozonide (Chapter IV, Section I) until the normal decomposition temperature of the primary ozonide had been reached (*365*).

The yields of α-alkoxyalkyl hydroperoxides (**160**) or of peroxides derived from them, are especially high from methanol or ethanol (*60,112,149,265, 268,272,273,281,345,366,367*) (see Chapter V, Section IV, Tables V and VI, for examples). In some cases high yields were reported from decanol, tetradecanol, isopropyl alcohol, and *tert*-butyl alchol (*94,272,273*). In general, however, yields appear to be lower when secondary or tertiary alcohols are used (*264,273*).

Carboxylic acids also add to carbonyl oxides to give α-acyloxyalkyl hydroperoxides (**162**) (*60,112,272,280*) (see Chapter V, Section IV). Although

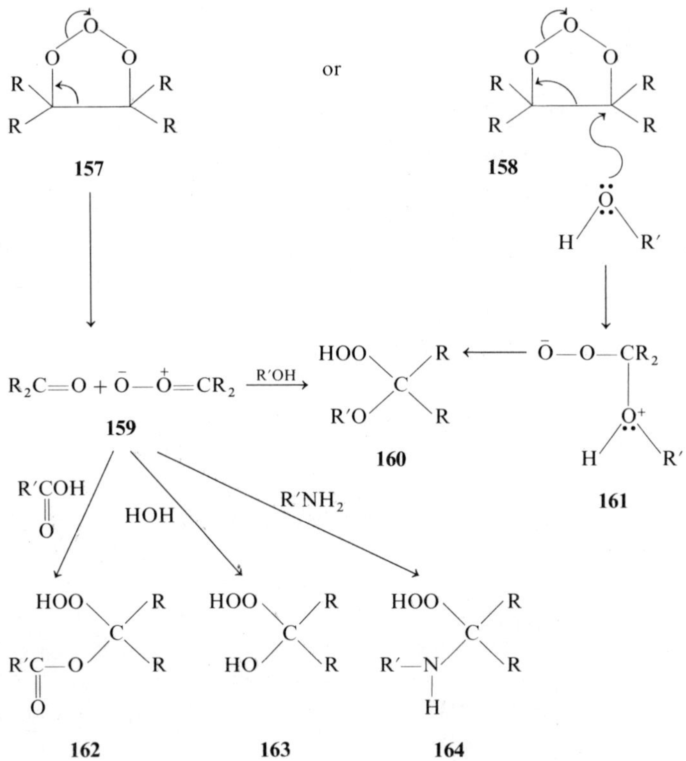

SCHEME 34. Roles of participating solvents.

these have been fully characterized only in the case of acetyloxy hydroperoxides (*60,272,280*), they have been reported in high yield also with butyric acid, pelargonic acid, and stearic acid as the participating solvent (*272*). Ozonolyses carried out in concentrated formic acid presumably give α-formyloxyalkyl hydroperoxides, but these have not been fully characterized (*368*).

Water has occasionally been used as a participating solvent in ozonolyses of olefins (*60,268,369,370*), but few hydroxyalkyl hydroperoxides (or peroxides derived from them) have ever been isolated (see Chapter V, Section

IV). It is likely that participation by water is general, but that the resulting hydroxyalkyl hydroperoxides (**163**), are generally too unstable to isolate. Instead, they either lose hydrogen peroxide and produce aldehydes or ketones (*60,369*), or else rearrange to carboxylic acids (*369*). Other examples of this will be seen with aromatic hydrocarbons (to be discussed in **Vol. II**).

Ammonia and amines also can behave as participating solvents, but, after the formation of **164**, other reactions immediately occur (see Section II, B of this chapter). The use of hydrogen cyanide and hydrogen peroxide as participating solvents also is discussed in the same subsection.

As pointed out in Chapter V, Section IV, intramolecular interaction tends to occur between the carbonyl and hydroperoxy functions of α-oxyalkyl hydroperoxides formed from cyclic olefins, provided that six- to eight-membered rings are produced. Some examples of such cyclic hemiperacetals are shown in Table VI (Chapter V). Further examples can be found with aromatic compounds, especially phenanthrene (*259,371*) and naphthalene (*372,373*) (to be discussed in **Vol. II**). From example 3 in Table VI, two different hemiperacetal structures, **165** and **166**, are possible. The one shown in Table VI is that assumed by Warnell and Shriner (281). Fliszár *et al.* (*367*), however, found that the product was a 50:50 mixture of both possible cyclic hemiperacetals.

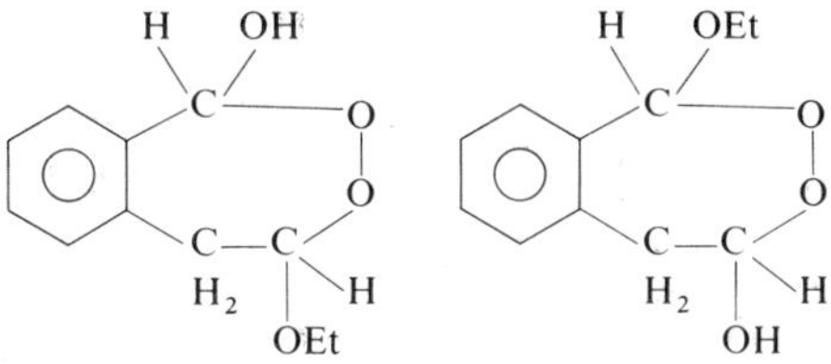

165 **166**

This addition of a hydroperoxy group to the carbonyl group of the aldehyde moiety also tends to occur intermolecularly, at least with cyclic olefins where intramolecular interaction is unfavorable. The principal example involves cyclohexene, which upon ozonolysis in methanol gave a polymeric peroxide (average $x = 1$) the composition of which appears to be between that of structure **167** and structure **168** (*374,375*). When the cold reaction mixture was immediately evaporated, the composition of the peroxidic ozonolysis product was closer to that of **167**, a simple hemiperacetal from addition of one molecule of **169** to another. After the ozonolysis mixture had been allowed to stand for several days, the composition of the peroxidic ozonolysis product was closer to that of **168** with, perhaps, some **170** also present. Similar results have been reported by Franz *et al.* (*262*) with a cyclohexene derivative. Peroxide **168** is a polyfunctional open-chain peracetal and acetal, produced by reaction of **167** with methanol.

Perry (*122*) studied the ozonolysis of norbornylene (**171**) under similar conditions and found that the peroxidic ozonolysis product was a mixture, apparently consisting largely of **174** but containing small amounts of **172**,

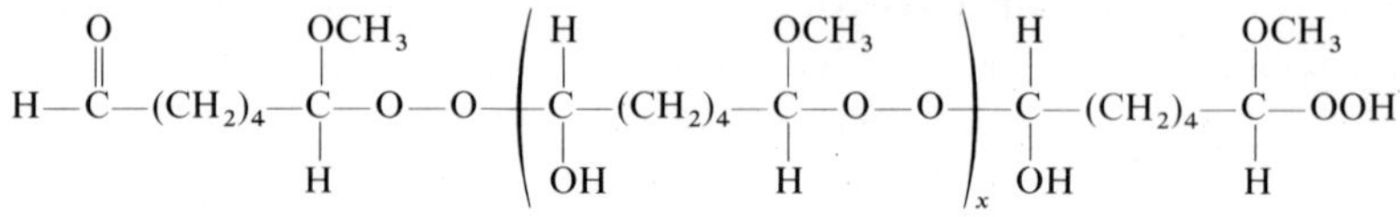

167

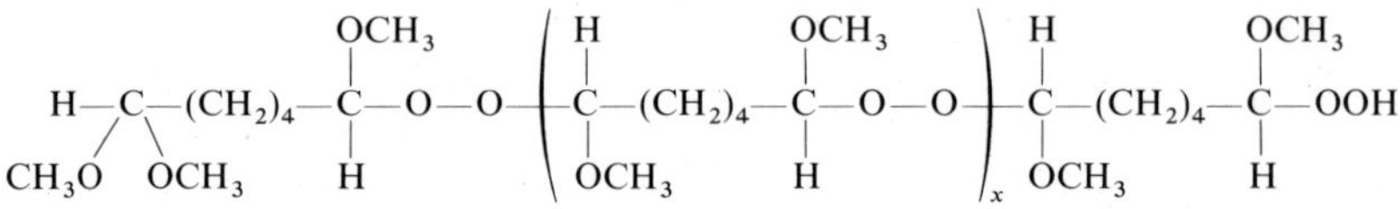

168

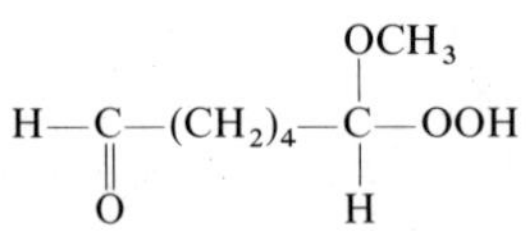

169

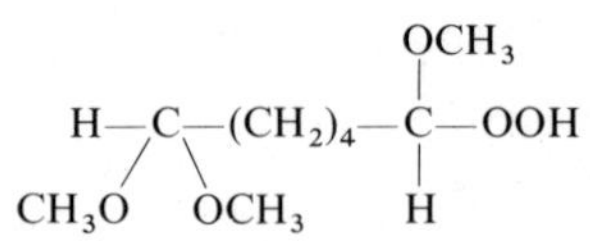

170

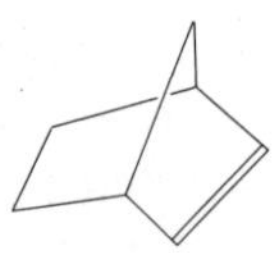

171

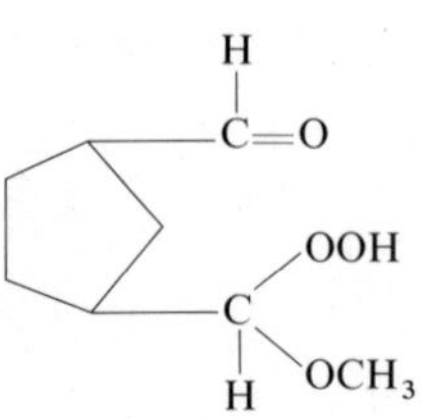

172

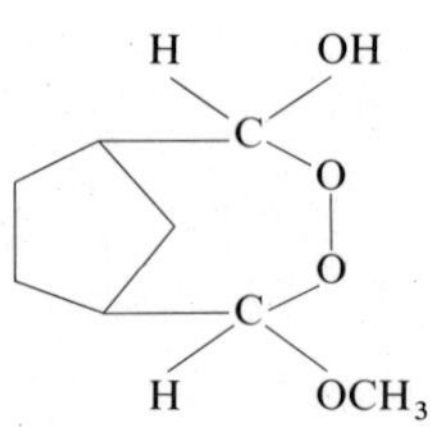

173

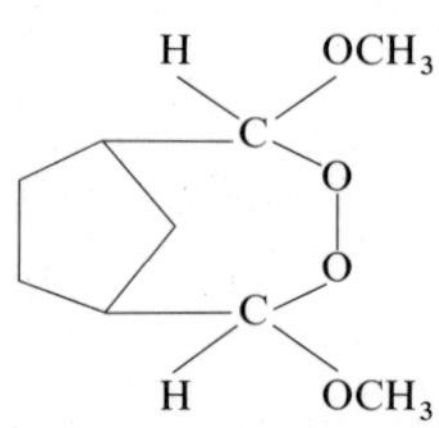

174

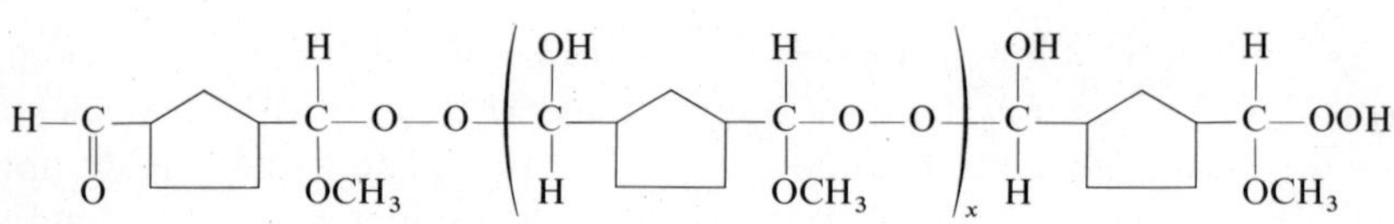

175

173, and polymeric peroxides similar to those obtained from cyclohexene (e.g., **175**). Peroxide **174** is a cyclic peracetal produced by cyclization of **172** to give hemiperacetal **173**, followed by reaction with methanol.

These additions of a hydroperoxy group to a carbonyl group followed by further reaction with alcohols (or hydroperoxides) have ample analogy in peroxide chemistry (*243,376,377*). Another example of peracetal formation during ozonolysis is the reaction of **165** or **166** with ethanol under acidic conditions to give the corresponding peracetal. Other examples can be found in the ozonolyses of phenanthrene (*259,371*) and of naphthalene (*373*) (to be included in Part II).

Similar additions of α-alkoxyalkyl hydroperoxides to aldehydes apparently also can occur during ozonolyses of open-chain olefins if catalytic amounts of acid are present (*271*). This also could explain the results of Milas *et al.* (*94*) during ozonolyses of olefins in *tert*-butyl alcohol–sulfuric acid solution.

B. AMOZONOLYSIS, CYANOZONOLYSIS

Until the early 1960s, the only participating solvents which had been used in ozonolyses and had been reported to add to carbonyl oxides were alcohols, carboxylic acids, and water. Between 1963 and 1965, however, Fremery and Fields ozonized several compounds in the presence of hydrogen peroxide (*370*), ammonia or primary amines (*378*), and hydrogen cyanide (*379*). Although the α-substituted alkyl hydroperoxides were not usually isolated, they definitely were intermediates in the reactions.

The hydrogen peroxide reactions were carried out in aqueous alkaline emulsions with cycloolefins and the products were dicarboxylic acids (*370*). Although Fremery and Fields (*370*) assumed that the nucleophile which attacked the carbonyl oxide was hydroxyl ion, the results could also be explained by attack by hydrogen peroxide or its anion. Robertson and Vergino (*380*) obtained diphenylmethyl bishydroperoxide (**178**) in 50% yield upon ozonolysis of tetraphenylethylene (**176**) in methanol at −78°C. As shown in Scheme 35, they suggested that it was produced by addition of hydrogen peroxide (from ozonation of methanol) to the carbonyl oxide (**177**). To check this hypothesis, they ozonized **176** in the presence of hydrogen peroxide and increased the yield of **178** to 71%. Milas (*381*) has reported in the patent literature the ozonolyses of several compounds in the presence of *tert*-butyl hydroperoxide and suggested that alkyl hydroperoxides participate in the same way as do alcohols.

$$CH_3OH + O_3 \longrightarrow HCOOH + HOOH, \text{etc.}$$

$$\underset{\mathbf{176}}{(C_6H_5)_2C{=}C(C_6H_5)_2} \xrightarrow{O_3} (C_6H_5)_2C{=}O + \underset{\mathbf{177}}{(C_6H_5)_2C{=}\overset{+}{O}{-}\overset{-}{O}}$$

$$\mathbf{177} \xrightarrow{HOOH} \underset{\mathbf{178}}{(C_6H_5)_2C(OOH)_2}$$

SCHEME 35. Hydrogen peroxide participation.

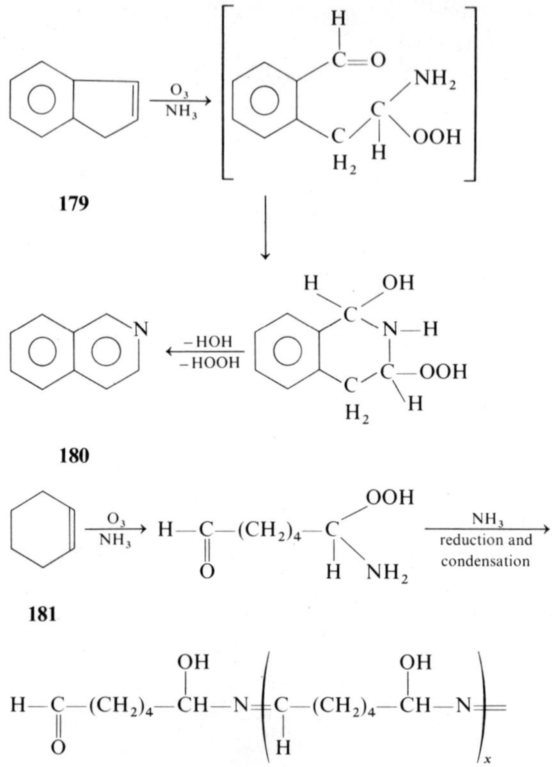

179

180

181

182

SCHEME 36. "Amozonolyses."

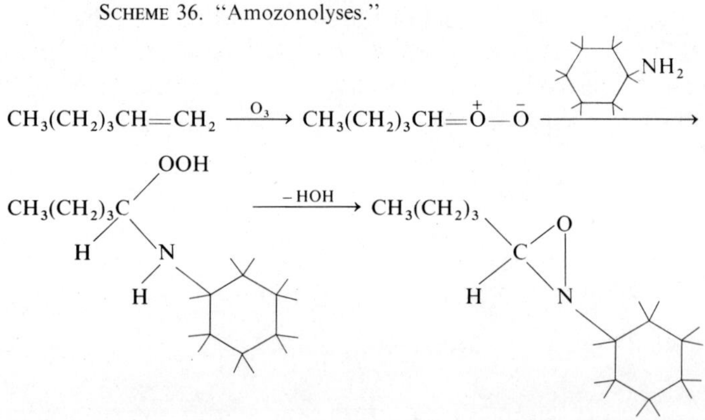

183 184

SCHEME 37. Amine participation to yield oxaziridines.

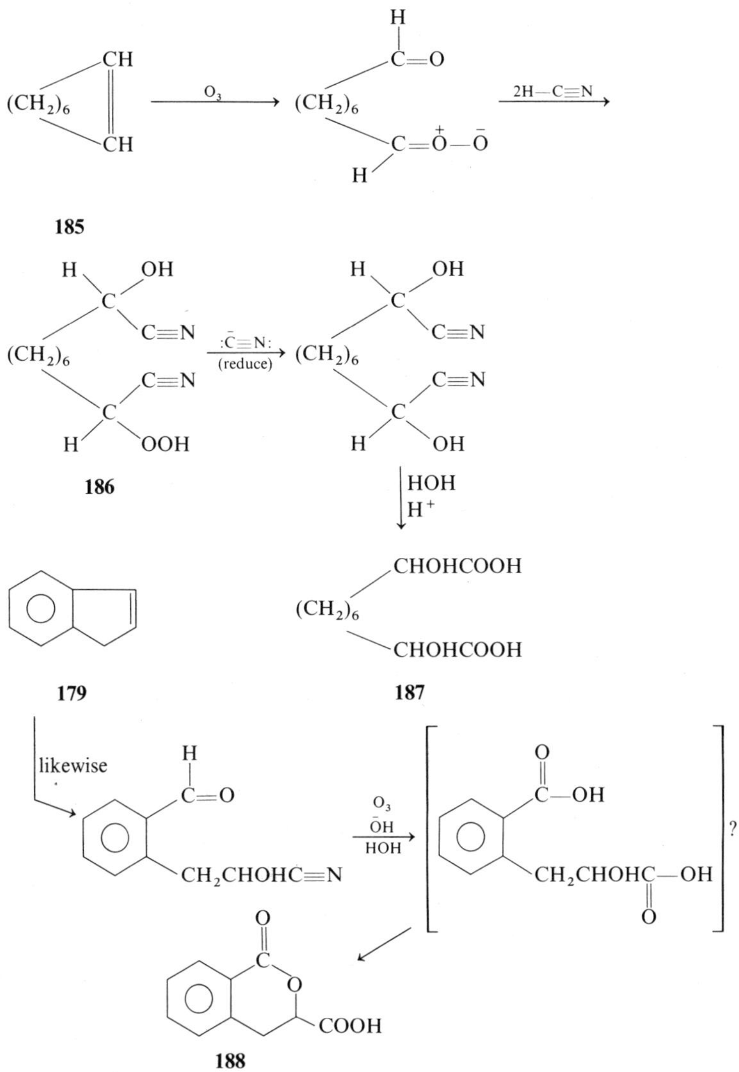

SCHEME 38. "Cyanozonolyses"

The ozonolyses in the presence of ammonia and amines were called "amozonolyses" by Fremery and Fields (*378*). With five-membered cyclo-olefins, six-membered heterocyclics were obtained. For example, indene (**179**) gave isoquinoline (**180**) in 62% yield as shown in Scheme 36. Six-membered or larger cycloolefins gave polymeric imines (**182**) as illustrated with cyclohexene, **181**, in Scheme 36.

Schulz *et al.* (*382*) have obtained oxaziridines (**184**) by ozonolyses of alkenes in the presence of ammonia or primary amines and, in some cases, actually isolated or identified the α-aminoalkyl hydroperoxide (**183**) intermediate. An example is shown in Scheme 37.

Fields (*379*) designated the ozonolyses carried out in the presence of hydrogen cyanide as "cyanozonolyses." Although he did not actually isolate the α-cyanoalkyl hydroperoxide (e.g., **186**) intermediates, the final products, hydroxycarboxylic acids or derivatives thereof, must have arisen from such. Examples are given in Scheme 38, starting with cyclooctene (**185**) and indene (**179**), which were converted to **187** and **188**, respectively.

C. OZONIDE FORMATION IN PARTICIPATING SOLVENTS

Although, in general, hydroperoxides of type **160** or **162** (Scheme 34), or derivations thereof, are the peroxidic ozonolysis products when participating solvents are employed, this is not always the case. Ozonolysis of 1,2-diphenylindene (**189**) gave the corresponding ozonide, **190**, in high yield,

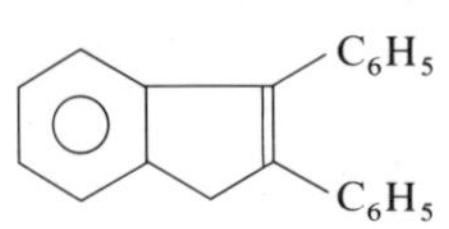

189

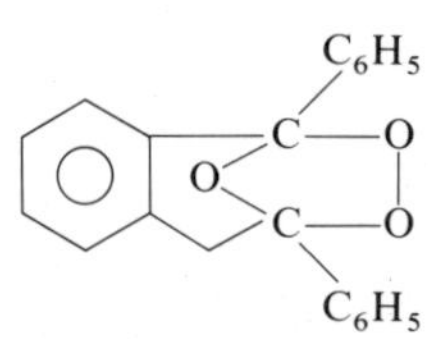

190

both with cyclohexane and with glacial acetic acid as solvents (*106*). Similarly, ozonolyses of certain tetra-, penta-, and hexasubstituted cyclobutenes (*130,198,207*), as well as 2,3-disubstituted indenones (*383*) and certain other five-membered cyclic olefinic types (*268,384*), produced ozonides even with methanol as the solvent. Perhaps in these special cyclic cases, either the hydroperoxides are produced, but immediately cyclize and eliminate the participating solvent to give the ozonide, or else the ozonide is produced by other than the Criegee mechanism, as suggested by Reinhardt *et al.* (*130*) (see Section I, C of this chapter).

D. INTRAMOLECULAR INTERACTIONS

A special class of α-oxyalkyl hydroperoxides are those produced by the intramolecular addition of a hydroxyl or carboxyl group to the Criegee zwitterion (*300–304,348*). These have already been illustrated in Scheme 8 (Chapter VI, Section I) and Scheme 30 (Chapter VI, Section VI, B). It is noteworthy that some of these ozonolyses were carried out in methanol solution, showing that the intramolecular alcohol–zwitterion reaction is more facile than the corresponding intermolecular reaction.

III. Direction of Cleavage of Primary Ozonides of Unsymmetrical Olefins

Certain olefins which are unsymmetrically substituted at the carbon–carbon double bond, such as methyl oleate (*229*), apparently give approximately equal amounts of the two possible pairs of Criegee moieties during ozonolysis [e.g., Eq. (14)]. In other cases, however, where greater differences in the electronic properties of the R and R′ groups [Eq. (14)] exist, one pair of moieties is produced preferentially, or predominantly, over the other. The question of whether or not such a preference exists with various compounds and, if so, why, has been of considerable interest (*62,81,108,117,119,120,123–125,188,271,337,338,353,385–393*), dating back even to the time of Harries (*85,88*).

$$\mathrm{RCH{=}CHR'} \xrightarrow{O_3} \mathrm{R{-}\underset{\underset{H}{|}}{C}{=}O} + \mathrm{R'{-}\underset{\underset{H}{|}}{C}{=}\overset{+}{O}{-}\overset{-}{O}} \text{ and } \mathrm{R{-}\underset{\underset{H}{|}}{C}{=}\overset{+}{O}{-}\overset{-}{O}} + \mathrm{R'{-}\underset{\underset{H}{|}}{C}{=}O} \qquad (14)$$

The recent interest in this problem began with its utilization as a method of studying the nature of the initial attack of ozone on a carbon–carbon double bond (Chapter IV) (*62,108,117,119,120,387*). For example, as seen in Scheme 39, ozonolysis of 1,2-dibenzoylpropene (**191**) in methanol produced phenylglyoxal (**195**) and methoxyhydroperoxide (**197**) (*117,120*). This could be rationalized on the basis of either (a) the addition of ozone to the carbon–carbon double bond of **191** according to Markownikoff's rule to give a four-membered ring adduct (**191** → **192** → **194**), which could cleave only to **195** and **196**; or (b) the preferential cleavage of a five-membered ring adduct in the direction of the inductive flow of electrons, as shown in **193** (*62,117,119,120*). In contrast, however, only the cleavage of the five-membered adduct (**199**), controlled by the inductive effect of the methoxyl group, would explain the results of ozonolysis of methoxydibenzoylethylene (**198**) to give ester **202** and hydroperoxide **205** (from **201**). The two-step addition, controlled by the resonance effect of the methoxyl group of **198**, should result in **200** as the preferred four-membered ring adduct and, consequently, produce **195** and **204** (from **203**) (*62,120*). Such reasoning led to the favoring of a five-membered ring for primary ozonides. This has since been proved correct, as discussed in Chapter IV, Section I. Therefore, the interest in the preferential formation of one particular pair of Criegee moieties over the other has shifted to the viewpoint of using this information to elucidate the mechanism of cleavage of primary ozonides.

It is difficult to determine with certainty which pair of Criegee moieties is formed predominantly during ozonolyses of unsymmetrical olefins. Of the older methods, the most reliable involved ozonolysis in methanol or ethanol and subsequent separation and characterization of the alkoxy hydroperoxide(s) formed by the trapping of the carbonyl oxide(s) produced. Some examples of this type are shown in Table XVIII. Another of the older methods involved ozonolysis in a nonparticipating solvent, as illustrated by examples 10 and 11 of Table XVIII. Because ozonides were obtained, the conclusion was reached that the predominant carbonyl oxide was the one shown; experience had taught that carbonyl oxides do not add to ketones that have no special activation (unless excess ketone is present; see Chapter V, Section I, and Section I, A of this chapter). Another method that appears to be reliable in most cases involves the determination under mild conditions

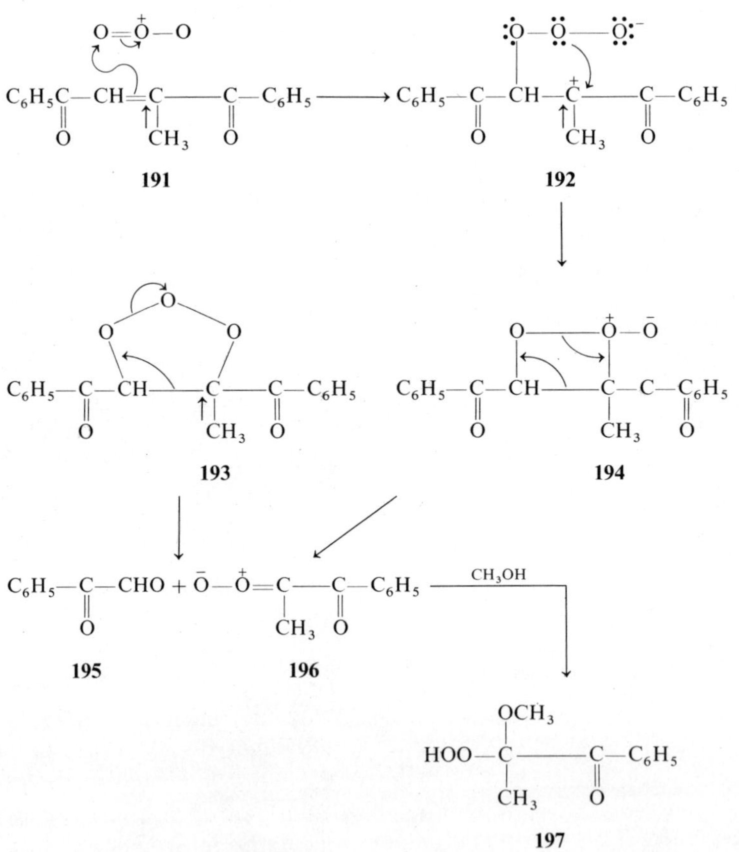

SCHEME 39. Ozonolyses of unsymmetrical olefins.

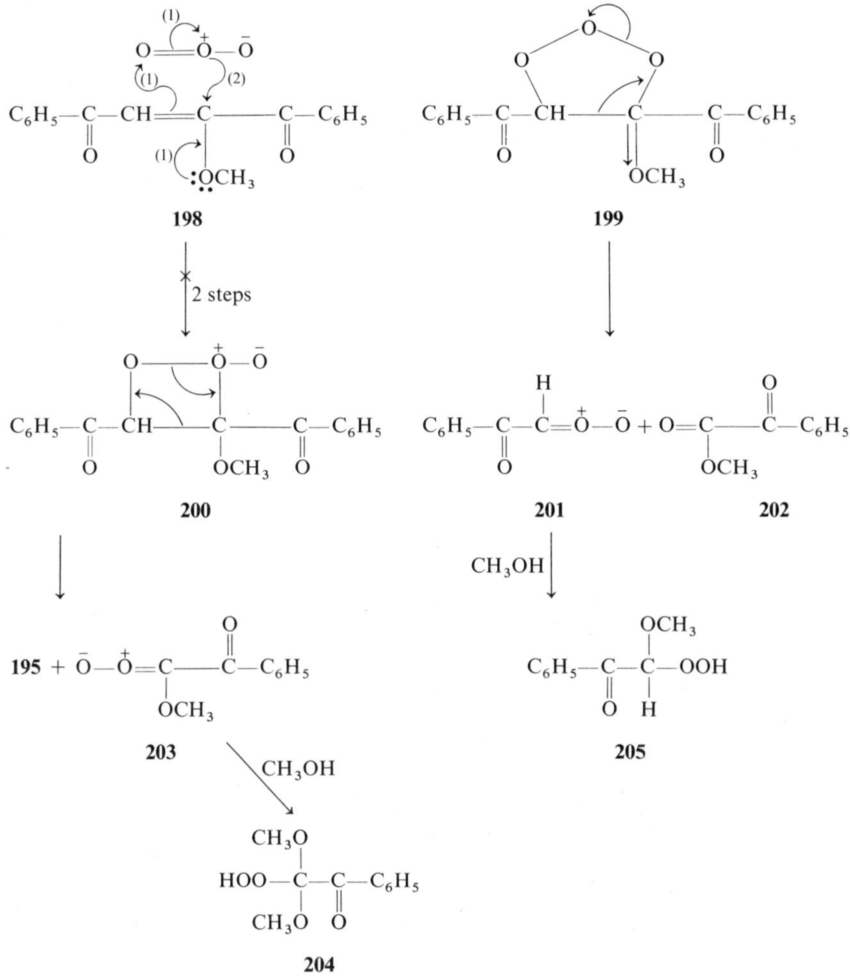

SCHEME 39 (*Continued*)

of the yield of aromatic aldehyde formed along with a methoxyalkyl hydroperoxide (from ozonolysis of ArCH$=$CR$_2$ in methanol), after showing that 1 mole of hydroperoxide is produced per mole of olefin ozonized (*271,353, 387–389*). A particularly useful method developed involves polarographic determination of the proportions of the two possible α-alkoxyalkyl hydroperoxides after determining the total hydroperoxide yield (*337,392,394*). Polarographic curves of reaction mixtures are compared with those of known α-alkoxyalkyl hydroperoxides. The curve is a function of the hydroperoxide used, and intensities are proportional to concentrations.

TABLE XVIII

SOME PREFERENTIAL CLEAVAGE PRODUCTS FROM OZONOLYSIS OF UNSYMMETRICAL OLEFINS

Example	Solvent	Compound ozonized	Predominating carbonyl oxide and/or carbonyl moieties	Major product(s)	Yield (%)	References
1	CH_3OH	$C_6H_5C(=O)CH{=}C(CH_3)C(=O)C_6H_5$ (trans)	$C_6H_5C(=O)C(CH_3){=}\overset{+}{O}{-}\overset{-}{O}$	$C_6H_5C(=O)C(OCH_3)(CH_3){-}OOH$	88	*117*, *366*
2	CH_3OH	Cis isomer of above	Same	Same	59	*120*
3	CH_3OH	$C_6H_5C(=O)CH{=}C(OCH_3)C(=O)C_6H_5$	$C_6H_5C(=O)CH{=}\overset{+}{O}{-}\overset{-}{O}$	$C_6H_5C(=O)CH(OCH_3)OOH$	84	*120*
				$C_6H_5C(=O)C(=O){-}OCH_3$	74	
4	CH_3OH	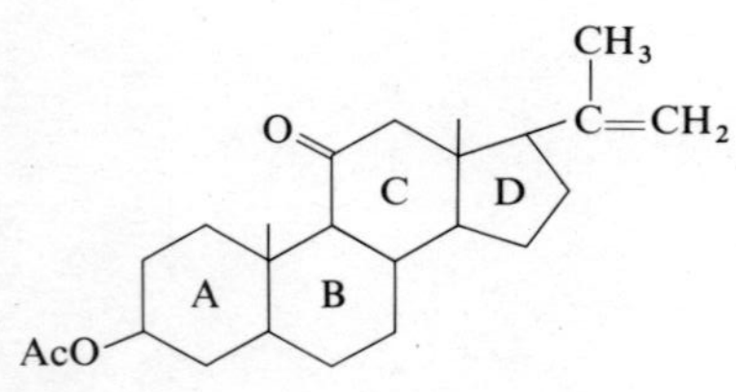	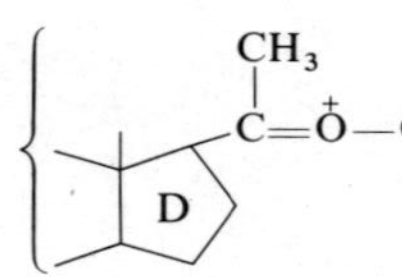	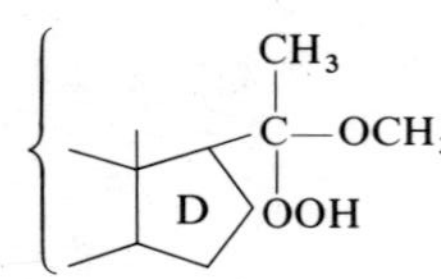	68	*266*

5	CH_3OH	like example 4 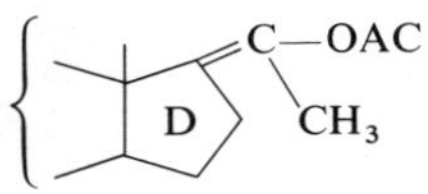	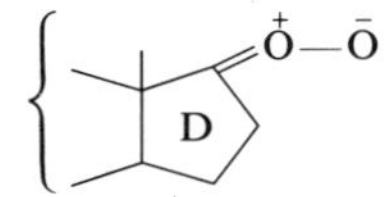	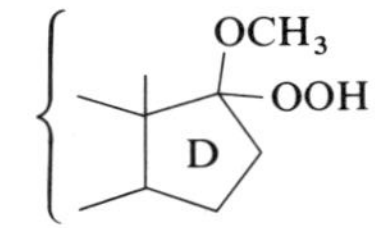	86	*266*
6	CH_3OH	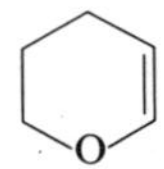	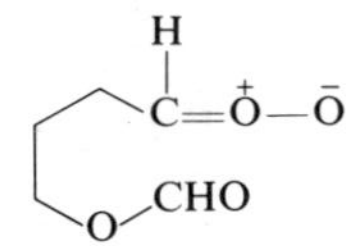	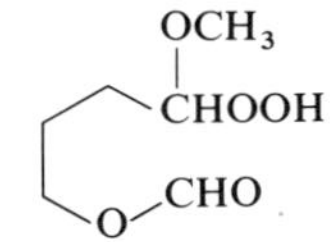	97	*269*
7	CH_3OH	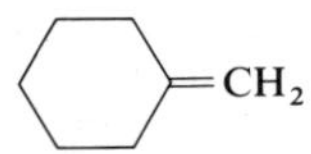	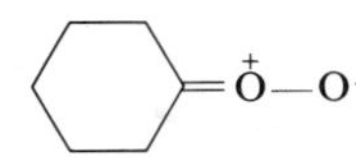	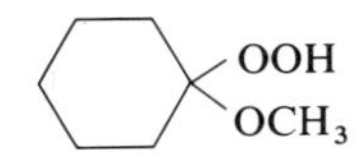	80	*264*
8	CH_3CH_2OH	$CH_3(CH_2)_5CH{=}CH_2$	$CH_3(CH_2)_5\overset{H}{C}{=}\overset{+}{O}{-}\bar{O}$	$CH_3(CH_2)_5CH(OCH_3)OOH$	92	*265*
9	CH_3OH	$H_2C{=}CH(CH_2)_3OH$	$HO(CH_2)_3\overset{H}{C}{=}\overset{+}{O}{-}\bar{O}$	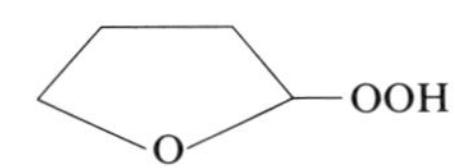	76	*303*
10	CH_3Cl	$(CH_3)_2C{=}CH_2$	$(CH_3)_2C{=}\overset{+}{O}{-}\bar{O}$	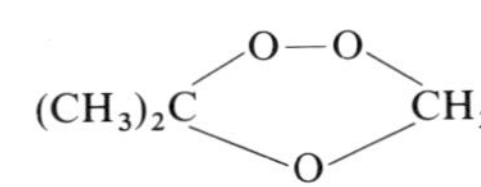	~	*62,105*
11	Pentane	$(CH_3)_2C{=}C(CH_3){-}C({=}O)OCH_3$	$(CH_3)_2C{=}\overset{+}{O}{-}\bar{O}$	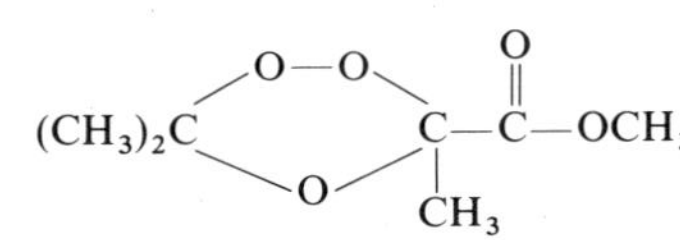	~	*62,189*

Some unreliable methods also are in the literature. In cases where it has been impossible to isolate the alkoxyalkyl hydroperoxides, the reaction mixtures have been heated or allowed to stand until peroxide decomposition was complete, after which the nonperoxidic ozonolysis products were determined and related to the presumed hydroperoxide precursor (*62,108,120, 390,391*). Kolsaker and Bailey (*395*), however, have shown that under these conditions the alkoxyalkyl hydroperoxide and aldehyde moieties often interact and the decomposition products of the adduct may be quite different from those of the alkoxyalkyl hydroperoxide itself; thus, erroneous conclusions are reached. In another series of studies, the decision as to which carbonyl oxide was predominantly formed was based on identification of rearrangement products, assuming that these involved rearrangement of the predominant carbonyl oxide moiety (*123–125,386*). An elaborate scheme was devised to explain the results obtained. However, it is now evident that the rearrangements involved an ozonide, or some other peroxidic ozonolysis product, rather than a carbonyl oxide (cf. *395–397*) (see also Chapter IX, Section IV). Thus, the authors' peculiar results and the elaborate mechanistic explanation (*123–125,386*) have little meaning. Their apparent success (*123,124*) in explaining examples 2 and 3 of Table XVIII appears to be purely fortuitous.

As indicated earlier, the examples of Table XVIII can be rationalized on the basis of the collapse of the primary ozonide in the direction of the inductive flow of electrons (see Scheme 39, **193** and **199**). This causes the carbonyl oxide to be formed predominantly at the carbon atom of the primary ozonide to which groups with the greatest electron-donating power are attached, as has been pointed out by several authors (*62,120,266,269, 271,385,387,388,390,391*).

Fliszár and co-workers (*337,338,387–389,392*) have made a very complete, detailed, and quantitative study of the direction of cleavage of primary ozonides and have correlated their results in regard to substituent effects by means of Hammett- and Taft-type relationships and conformational considerations. They found that the proportions of the two possible carbonyl oxides from various olefins were not affected by temperature or solvent (*337,389*). The transition states assumed are illustrated in Scheme 40. In general, the carbonyl oxide preferentially formed is the one with which the substituent is better able to stabilize the developing positive charge.

The first studies were carried out with *trans*-olefins, in one case an aromatic series, R—C_6H_4CH═CH—R′ (*389*) and in the other a purely aliphatic series, RCH═CHR′ (*392*). It is to be noted that with trans primary ozonides electronic rather than conformational effects are most important, since both substituents would be either diaxial or diequatorial (see Chapter VI, Section IV, A). In each case R was varied while R′ was kept constant and vice versa.

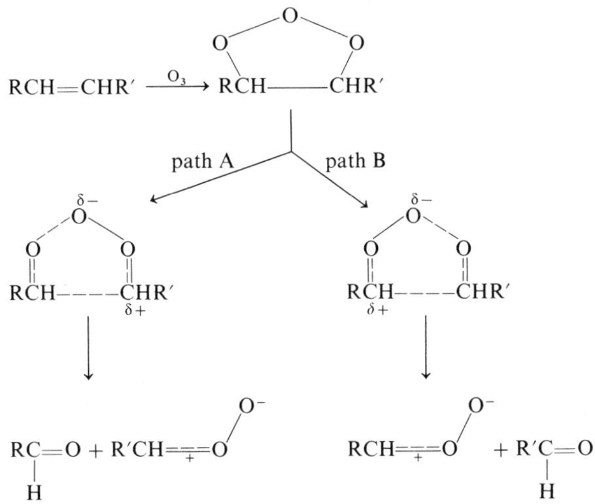

SCHEME 40. Cleavage of unsymmetrical primary ozonides

Some of the results are shown in Tables XIX and XX. It can be seen that, in general, the extent to which the carbonyl oxide is formed at a given carbon increases as the electron-attracting power of the substituent (R or R′) decreases, or its electron-donating power increases. In the aromatic series, there was a good correlation of R with the Hammett σ constants. Further, the R or R′ groups affect the carbonyl oxide formation in the same order

TABLE XIX

PROPORTIONS OF CLEAVAGE OF *trans*-R-$C_6H_4\overset{\alpha}{C}H{=}\overset{\beta}{C}HR'$ YIELDING $R'CH{=}\overset{+}{O}{-}\overset{-}{O}$[a,b]

	R′						
R	$COOCH_3$	CH_2OH	H	C_6H_5	COOH	CH_3	$COCH_3$
p-NO_2	—	—	0.86	0.91	—	—	—
m-NO_2	—	—	0.84	—	—	—	—
m-Cl	0.42	—	0.67	—	0.84	—	—
p-Cl	0.36	0.40	0.57	0.64	—	0.90	—
H	0.23	0.26	0.41	0.50	0.61	0.82	0.90
p-CH_3	—	—	0.29	0.39	0.46	—	—
3,4-CH_2O_2[c]	—	—	—	—	—	0.75	0.85
p-OCH_3	—	—	0.21	—	0.39	0.68	0.80

[a] Reprinted, with permission, from Fliszár and Granger (*389*). Copyright by he American Chemical Society.

[b] The average accuracy of these values is ± 0.02.

[c] For this entry, the general formula should be $RC_6H_3CH{=}CHR'$.

TABLE XX

PROPORTIONS OF CLEAVAGE OF trans-RCH=CHR′ TO GIVE $R'CH{=}\overset{+}{O}{-}\overset{-}{O}$[a,b]

R	R′	Proportion $R'CH\overset{+}{O}\overset{-}{O}$	
		Exptl.	Calcd.
CH_3	$COCH_3$	0.53	0.525
	COOH	0.45	0.457
	H	0.38	0.384
	CH_2OH	0.23	0.229
	$COOCH_3$	0.15	0.152
H	$COCH_3$	0.64	0.642
	COOH	0.57	0.575
	CH_2OH	0.33	0.322
	$COOCH_3$	0.22	0.223
CH_2OH	COOH	0.74	0.740
	$COOCH_3$	0.38	0.376
CH_2Br	CH_3	0.87	0.865
	H	0.80	0.80
	$COOCH_3$	0.54	0.534
CH_2Cl	CH_3	0.89	0.889
	H	0.83	0.833

[a] Reprinted in part, with permission, from Fliszár and Granger (*392*). Copyright by the American Chemical Society.

[b] The accuracy of these values is within ± 0.01.

regardless of the nature of the other substituent. For olefins of type RCH=CHR′, the authors developed an equation by which they could calculate the predicted percentage produced of a given carbonyl oxide, knowing the substituents R and R′ for a given olefin (*392*). As seen in the partial Table XX the agreement between calculated and experimental values is remarkable.

At first sight the decreasing order with which the R′ groups in Table XIX and the R and R′ groups in Table XX stabilize the developing carbonyl oxide (i.e., $COCH_3 > CH_3 > COOH > H > CH_2OH > COOCH_3 > CH_2Br > CH_2Cl$) might seem anomalous in regard to the positions of $COCH_3$ and COOH relative to those of other groups, especially $COOCH_3$. Fliszár and Granger (*389,392*) explained this on the basis that a carbonyl group tends to stabilize the positive charge on the carbonyl oxide through resonance as shown in Scheme 41 (**206**). This would be important where $X = CH_3$ or OH, but not where $X = OCH_3$, owing to the more important contribution from **207**. In addition, with the carboxyl group, hydrogen bonding could

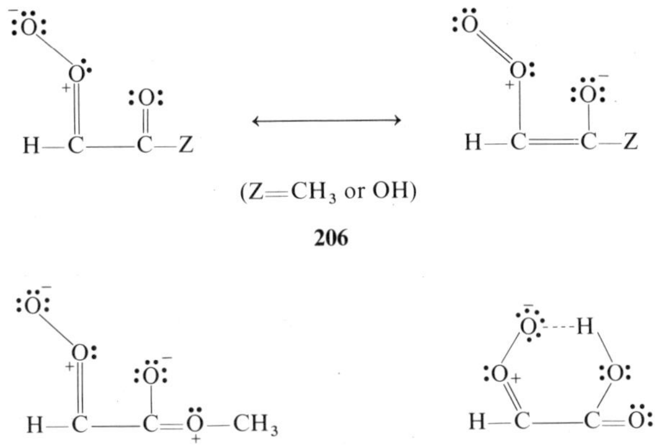

206 **207** **208**

SCHEME 41. Effects of carbonyl groups or carbonyl oxide stablilization

help stabilize the carbonyl oxide, as shown in **208**. This striking difference between carboxyl and carbomethoxyl groups in stabilizing carbonyl oxides has also been observed by Johnson and Bailey (*398*) during ozonation of cinnamic acids and esters.

Results with olefins having only alkyl substituents around the double bond were more complex (*337,338*). For 1-alkenes the results correlated with inductive effects (i.e., the order of decreasing electron withdrawal or increasing electron release, inductively), as shown in Table XXI, and, consequently, with Taft plots (*337*).

Results with various *cis*- and *trans*-1,2-dialkylethylenes are shown in Table XXII (*337*). It can be seen that with *trans*-alkenes the hyperconjugative order ($CH_3 > Et > i\text{-}Pr > t\text{-}Bu$) rather than the inductive order governs the direction of cleavage of the primary ozonide and the stabilization of the forming carbonyl oxides (*337*). This is also true, but to a lesser degree, with the cis series, except for the *tert*-butyl group.

These results can be at least rationalized, if not completely understood, on the basis of competitions between electronic and conformational effects. With primary ozonides of 1-alkenes all substituents are equatorial, according to Renard and Fliszár (*328*), and inductive stabilization of the developing carbonyl oxide zwitterion is best transmitted by equatorial substituents (*338*). The energy differences between the diequatorial and the diaxial conformations of primary ozonides of *trans*-olefins can be either very small or large, but in either case the diaxial conformation is predominant (*328*). Axial substituents appear to stabilize the developing carbonyl oxide better through hyperconjugation. In the case of cis primary ozonides, the substituents are equatorial–axial and, whereas a methyl group tends to be equatorial in

TABLE XXI

DIRECTION OF OZONE CLEAVAGE WITH MONOSUBSTITUTED ETHYLENES[a,b]

R in $R—CH{=}CH_2$	Proportion of $RCH{=}\overset{+}{O}—\overset{-}{O}$
$ClCH_2$—	0.17
$BrCH_2$—	0.20
ICH_2—	0.27
C_6H_5—	0.59
CH_3—	0.62
Et—	0.63
i-Pr—	0.64
t-Bu—	0.70
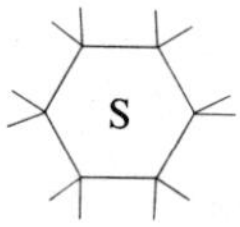	0.86

[a] Reproduced in part by permission of the National Research Council of Canada from *Canad. J. Chem.* **48**, 3002–3018, 1970 (*337*).
[b] The accuracy of these values is within ±0.01.

TABLE XXII

DIRECTION OF CLEAVAGE OF PRIMARY OZONIDES OF SELECTED *cis*- AND *trans*-1,2-DIALKYLETHYLENES ($RCH{=}CHR'$)[a,b]

		Proportion of $RCH{=}\overset{+}{O}—\overset{-}{O}$	
R	R′	*trans*-Olefin	*cis*-Olefin
Me	Et	0.56	0.52
Me	*n*-Pr	0.60	0.54
Me	*i*-Pr	0.71	0.57
Me	*t*-Bu	0.88	0.38
Et	*i*-Pr	0.66	—
Et	*t*-Bu	0.87	0.37
Me	C_6H_5	0.82	—
Me	*p*-MeO-C_6H_4	0.68	—

[a] Reproduced in part by permission of the National Research Council of Canada from *Canad. J. Chem.* **48**, 3002–3018, 1970 (*337*).
[b] The accuracy of these values averages ±0.01.

relation to propyl and isopropyl (and perhaps ethyl), the *tert*-butyl group tends to be equatorial in relation to smaller groups (*328,337*). The results concerning *cis*-olefins in Table XXII show that the equatorial substituent incorporates into the carbonyl oxide preferentially, as predicted by Bauld–Bailey Rule 2 (Chapter VI, Section IV, A).

For 1,1-disubstituted, trisubstituted, and tetrasubstituted ethylenes, the carbonyl oxide favored is the one stabilized by the more electron-releasing groups (*337*). With alkyl substituents, hyperconjugation continued to be more important than the inductive effect. With phenyl groups, resonance plays an important role, as seen by comparing the last two entries of Table XXII.

In the above discussions only electronic and conformational factors appeared to be important. One instance where a steric factor seems to play a role was mentioned in Chapter VI, Section VI, B. In Scheme 27, of the two possible carbonyl oxides from primary ozonide **116**, only **117** was produced, and the two geminal methyl groups were shown to be responsible for this preference (*347*).

IV. Summary of Routes to Peroxidic Products

At present, ozonolysis of most olefinic compounds appears to involve the following steps: (1) an initial concerted addition of ozone to the double bond, which is generally thought to be a 1,3-dipolar cycloaddition, to give an unstable 1,2,3-trioxacyclopentane primary ozonide; (2) decomposition of the primary ozonide by a process best described as the reverse of a 1,3-dipolar cycloaddition (a 1,3-dipolar reversion) to give a carbonyl compound and a carbonyl oxide; the carbonyl oxide is the key intermediate of ozonolysis and can exist, when unsymmetrically substituted, in syn and anti isomeric forms; (3) various fates of the carbonyl oxide, as follows: (a) stereoselective 1,3-dipolar cycloaddition with an aldehyde (or in some cases a ketone) to give *cis*- and *trans*-ozonides (1,2,4-trioxacylopentanes), (b) dimerization to aldehyde or ketone diperoxides (1,2,4,5-tetraoxacyclohexanes), (c) polymerization or copolymerization (with the aldehyde or ketone moiety) to give polymeric peroxides and ozonides, (d) reaction with a protic, nucleophilic solvent to give an oxyalkyl hydroperoxide or tautomers, etc., thereof.

Razumovskii (*182*) and Wadt and Goddard (*78*) have estimated various parameters of these steps in the ozonolysis reaction. Most are fast reactions with relatively low activation energies. Figure 10 shows the thermochemistry calculated by Wadt and Goddard (*78*) for some of the steps they envision in the ozonolysis of ethylene in a nonparticipating solvent to give ethylene ozonide.

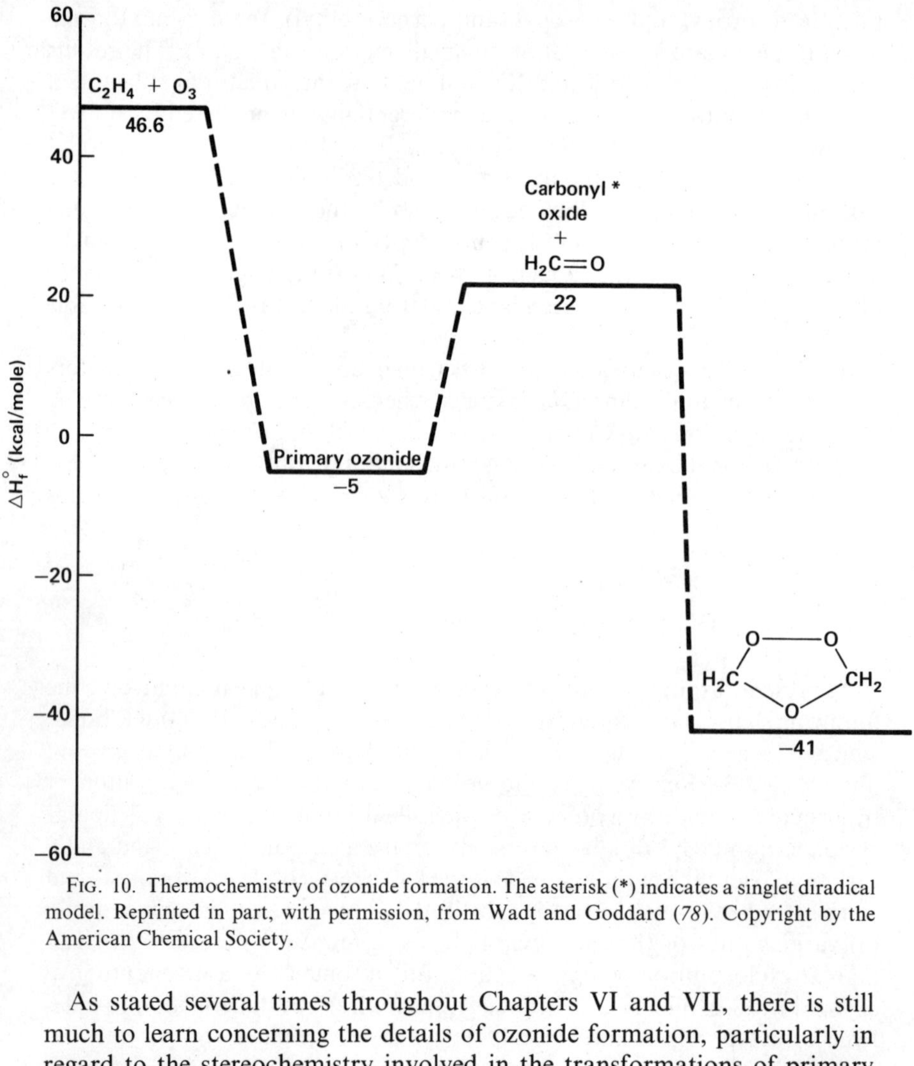

FIG. 10. Thermochemistry of ozonide formation. The asterisk (*) indicates a singlet diradical model. Reprinted in part, with permission, from Wadt and Goddard (*78*). Copyright by the American Chemical Society.

As stated several times throughout Chapters VI and VII, there is still much to learn concerning the details of ozonide formation, particularly in regard to the stereochemistry involved in the transformations of primary ozonides to final ozonides. For a brief review of a paper too recent to include here see the Epilogue at the end of this volume.

CHAPTER VIII

Ozonolysis of Olefins: Routes from Peroxidic to Nonperoxidic Products

Under certain circumstances nonperoxidic ozonolysis products are formed directly during an ozonolysis, without going through a peroxidic precursor other than a primary ozonide. Usually, however, this nonperoxidic ozonolysis product is an aldehyde or ketone, which is left behind when the accompanying carbonyl oxide dimerizes, polymerizes, or reacts with a participating solvent (Chapter V, Sections II, III, and IV; Chapter VII, Sections I, A and II). Thus, even in these cases there usually is (unless spontaneous decomposition occurs) at least one peroxidic ozonolysis product (aldehyde or ketone diperoxide, oligomeric peroxide, or α-oxyalkyl hydroperoxide, etc.). These, and ozonides, therefore, must be converted to nonperoxidic ozonolysis products. The conversion methods can be classified as oxidative (to carboxylic acids), reductive (to aldehydes and/or ketones), hydrolylic (to aldehydes and/or ketones), or thermal, photolytic, or catalytic decompositions to aldehydes, ketones, carboxylic acids, or esters, etc. (*26*,*62*) Sometimes the latter reactions are spontaneous and may involve rearrangements (Chapter IX). From a mechanistic viewpoint, these reactions are better classified as nucleophilic displacements at oxygen (usually resulting in reduction, see Section I of this chapter), acid- and base-catalyzed decompositions (Sections III and IV), and metal ion-catalyzed, thermal, or photolytic decompositions, some of which involve oxidative or reductive processes (Section V). All of these are general reactions of peroxides (*244*,*399*). An oxidative procedure, however, usually involves an oxidation of an aldehyde rather than of a peroxide.

It is beyond the scope of this book to go into the experimental details and techniques of these reactions. For such, the reader is referred to two other recent reviews (*115*,*116*).

I. Reductive. Nucleophilic Displacement at Oxygen

The majority of reagents which are used to reduce peroxidic ozonolysis products appear to make a nucleophilic attack on one of the oxygens of the peroxide linkage; this is analogous to reductions of other peroxides (*400–402a*). The generalized mechanism is illustrated in Eqs. (15) and (16). Examples are iodide anion (*403,404*), sulfides (*402,402a*, cf. *405*), phosphines (*231,232,297,298,342*), phosphites (*406*), lithium aluminum hydride (*292, 341,407*), methyllithium (*292,407*), and Grignard reagents (*402,402a*).

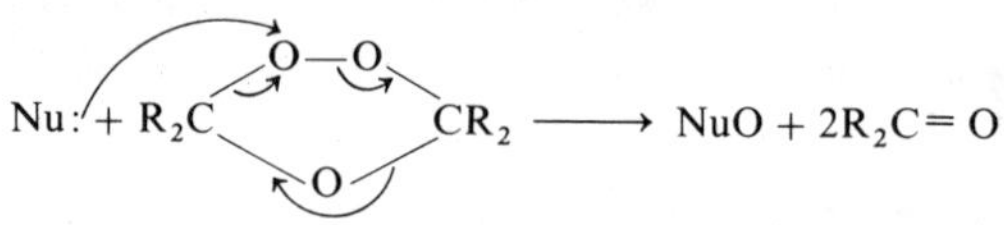

(15)

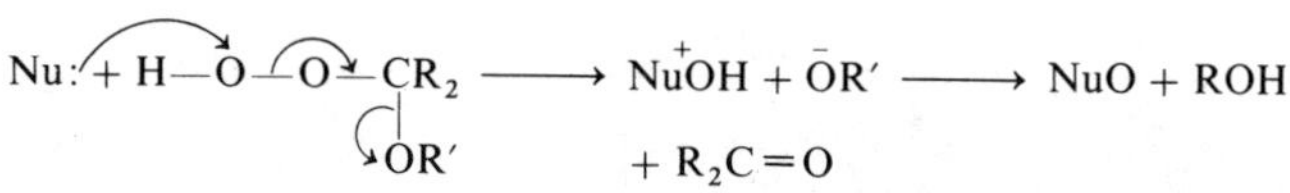

(16)

The oxidation of iodide ion to molecular iodine has long been used for both qualitative and quantitative determinations of peroxides, including peroxidic ozonolysis products (*243,244,399,408,409*). Reduction by iodide ion is also used to convert peroxidic ozonolysis products to aldehydes and ketones in good yield (*62,115,116*). The mechanism appears to be the type shown in Eqs. (15) and (16), followed by Eq. (17) (*403,404*).

$$O\bar{I} + I^- + 2H^+ \longrightarrow I_2 + H_2O \tag{17}$$

Sulfites and bisulfites are common reducing agents for peroxidic ozonolysis products (*62*). More important, ozonolysis in methanol, followed by dimethyl sufide reduction of the α-methoxyalkyl hydroperoxide product has become a superior method for the ozonolytic cleavage of olefinic double bonds to aldehydes and/or ketones (*410,411*); the sulfide is converted to sulfoxide. The mechanism most likely is that shown in Eq. (16) (cf. *402*).

Triphenylphosphine has been used extensively to reduce both ozonides (*231,232,297,298,342,412*) and α-alkoxyalkyl hydroperoxides (*413*); triphenyl or trimethyl phosphite has found similar usage (*406,414*). By these reactions, the phosphine is oxidized to the oxide, and phosphite to phosphate. For the mechanism of ozonide reduction by triphenyl phosphine, Eq. (15) was hypothesized by Lorenz and Parks (*231*) and established by Carles and Fliszár (*232,342*) and by Higley and Murray (*298*). Using ozonides labeled with ^{18}O, these authors showed that the triphenylphosphine attack occurred exclusively at a peroxidic oxygen. The results of Carles and Fliszár (*232,342*) are illustrated in Scheme 42.

Carles and Fliszár (*232,342*) found that the peroxidic oxygen attacked [(1) or (2) in **209**] depended on both electronic and steric factors. The results with **209a** and **b** indicated that the nucleophilic attack occurred predominantly at the oxygen next to the carbon bearing the more electron-attracting group, whereas the result with **209c** emphasized the importance of the steric

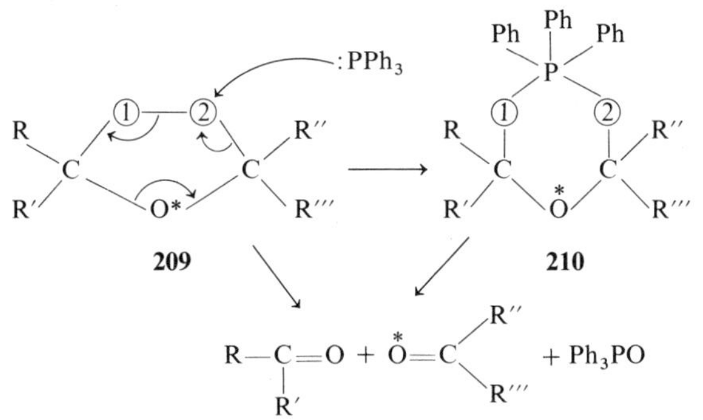

a: R=R′=R″=H, R‴=Ph   O*=^{18}O label
b: R=CH_3, R″=Ph, R′=R‴=H
c: R=R′=R″=Ph, R‴=H

SCHEME 42. Triphenyl phosphine reduction of ozonides.

factor. Further evidence for the nucleophilic attack involved unlabeled ozonides with which one substituent in **209** (R, R′, R″ or R‴) was a substituted phenyl group (Z—C_6H_4—) while the other substituents were hydrogen. The rates of the reactions decreased greatly as Z changed in the following order: p-NO_2 > m-Cl, p-Cl > H > p-CH_3 > p-OCH_3. Further studies led Carles and Fliszár (*232*) to the conclusion that an unstable intermediate, **210**, is first formed in the bimolecular reaction. A similar conclusion was also reached concerning phosphite reduction of ozonides (*406*).

Lithium aluminum hydride and sodium borohydride reductions of ozonides are well known and yield alcohols (*292,341,407,415,416*). Carles and Fliszár (*342*) suggested that the mechanism is similar to that of phosphine reductions. Story *et al.* (*292,341,407*) formulated the mechanism shown in Scheme 43. They also suggested that methyllithium reductions (*341,407*) occur by a similar mechanism. It should be pointed out, however, that although the nucleophilic attack on oxygen must be valid, some of the other details of Scheme 43 led to erroneous conclusions (*292,341,407*) when applied to determining the position of ^{18}O in labeled ozonides (cf. *298*).

Grignard reagent reductions of ozonides also have been performed (*417*). By analogy to mechanisms suggested for reduction of other peroxides by Grignard reagents (*402a,418*), the mechanism with ozonides should be similar to that shown in Scheme 43 for hydride reductions, at least in regard to the first step.

Pyridine has been employed as a unique reducing agent in ozonolyses (*62,419,420*); it was assumed that attack of pyridine on the Criegee carbonyl oxide occurred to give an aldehyde or ketone and pyridine oxide. Further work has cast doubt on this explanation (*421,422*) and at present the reaction appears to be much more complicated (*116*).

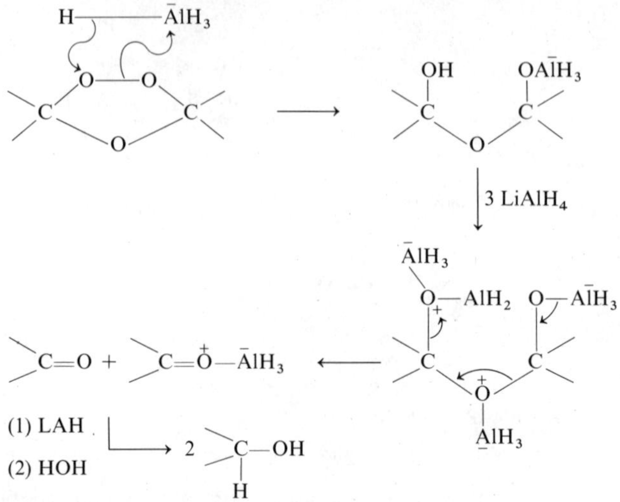

SCHEME 43. Hydride reduction of ozonides.

II. Other Reductive Methods

A. BORANE

Borane–amine and borane–tetrahydrofuran reductions of ozonides have been reported (*423*). Diaper and Strachan (*423*) suggested that the mechanism involves electrophilic attack of the borane on the ether oxygen as shown in Scheme 44. It also seems possible that the attack could occur at a peroxidic oxygen.

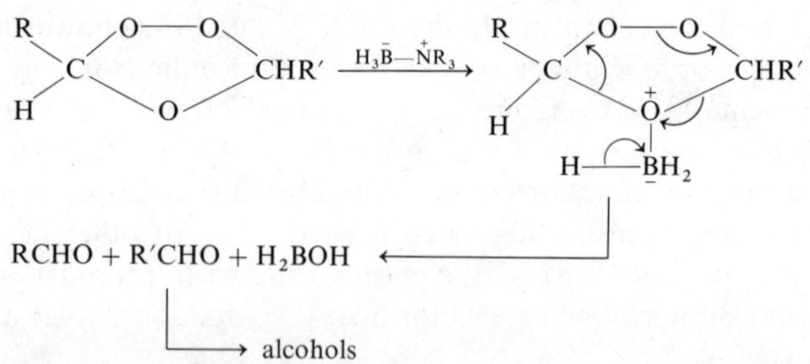

SCHEME 44. Borane reduction of ozonides.

B. GLYOXALIC ACID

Kolsaker, Bernatek *et al.* (*424*) have reported high-yield reductions by glyoxalic acid of peroxidic ozonolysis products formed in acetic acid. Acidic conditions are necessary. Ozonides and aldehyde or ketone diperoxides are not reduced. The reaction appears to involve a Baeyer–Villiger oxidation of the glyoxalic acid, as shown in Eq. (18).

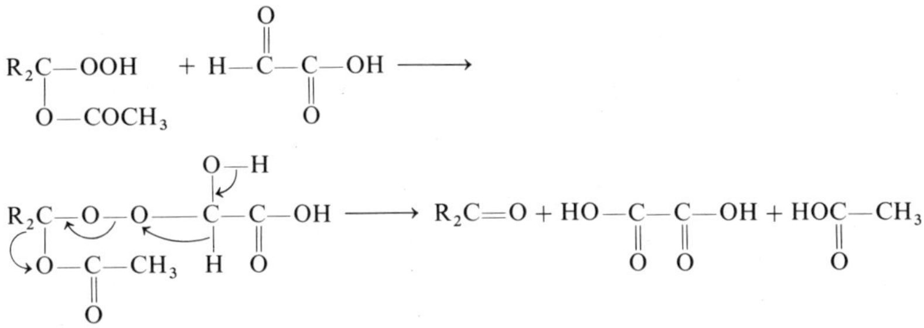

(18)

C. TETRACYANOETHYLENE

Ozonolysis of olefins in the presence of tetracyanoethylene generally results in high yields of aldehydes and/or ketones plus the epoxide of tetracyanoethylene (*149,361*). Because tetracyanoethylene will not reduce ozonides or other peroxides (*361*), it was suggested that the reduction involves either the primary ozonide (*361*) or the carbonyl oxide (*360,361,425*). The latter would consist of a 1,3-dipolar cycloaddition of the carbonyl oxide to the olefin, as outlined in Eq. (19). If this actually is the mechanism, it is an additional fate of the Criegee carbonyl oxide not outlined in Chapter III (Section I, C), Chapter V, or Chapter VI. Such possibilities were mentioned in Chapter VII, Section I. Intuitively, one would expect the adduct to be isolable if Eq. (19) were valid.

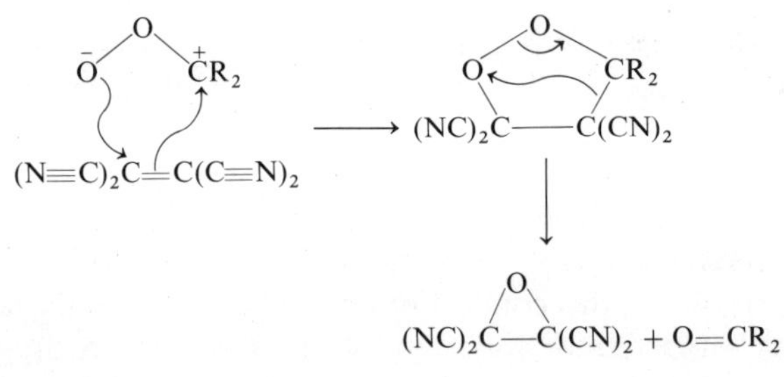

(19)

D. CATALYTIC

Catalytic hydrogenation is a common reductive method for peroxidic ozonolysis products (*26,62,115,116*). A radical-type mechanism would be expected (e.g., *415,426*).

III. Acid-Catalyzed Decompositions. Solvolysis

Harries suggested two decomposition routes for ozonides. One was hydrolysis to give aldehydes or ketones and hydrogen peroxide, and the other was decomposition to an aldehyde (or ketone) and a carboxylic acid (*23,24, 85*) (see Chapter III, Section I, A). For the latter to occur, Harries suggested that water is necessary as a catalyst. In order to account for these two pathways, Rieche (*53,54,257*) proposed a reaction scheme involving hydrolytic cleavage of the ether linkage of an ozonide to give **211**, presumably via nucleophilic substitution at carbon. This is followed by decomposition of **211** to an aldehyde or ketone plus **212**, which can either lose hydrogen peroxide and yield an aldehyde or ketone, or rearrange to a carboxylic acid plus water if one R of **212** is hydrogen (*26,116*). There is considerable disagreement, however, concerning this scheme (*116*). Some workers believe that water is only incidental and a purely thermal decomposition is involved (*427*). At present, it appears more likely that the two procedures referred to by Harries involve "acidolysis" (Section III, A) and base-catalyzed decompositions of ozonides and other peroxidic ozonolysis products (Section IV).

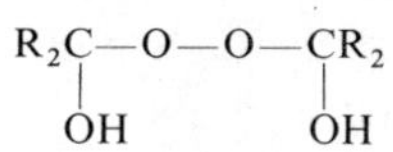

211

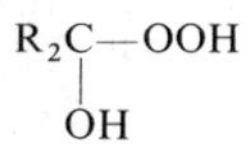

212

A. "ACIDOLYSIS"

Bernatek *et al.* (*428,429*) have found that decomposition of ozonides by means of formic acid, or acetic acid plus perchloric acid, yield performic or peracetic acid along with the expected carbonyl compounds. They suggested the mechanism shown in Scheme 45 (i.e., **213**). Overall, the reaction is a reduction of the ozonide by formic acid. The protonation and cleavage is shown at the more basic ether oxygen. It also could occur at a peroxidic oxygen.

Solvolytic ozonolyses were also carried out with maleic acid and other olefins in the presence of formic acid (*368,369*). In these cases the peroxidic ozonolysis product probably was **214**, with decomposition occurring as shown (Scheme 45). In this connection, it has been observed that ozonolysis of cyclohexene in methanol, followed by oxidation with hydrogen peroxide in formic acid results cleanly in adipic acid (*375*). It was suggested that this superior property of formic acid (in relation to other acids) was due to its reducing action followed by the oxidizing action of the resulting performic acid (*375*).

Yurév *et al.* (*430*) have studied the acid-catalyzed decomposition of 1-hexene and 5-decene ozonides and suggest a mechanism different from that of Bernatek *et al.* Their mechanism, also outlined in Scheme 45, suggests acid catalysis at a peroxidic oxygen of **215**, followed by cleavage to a carbonium ion intermediate (**218**), which in turn cleaves to an aldehyde (**217**), and a protonated Criegee zwitterion (**216**). The latter decomposes to a car-

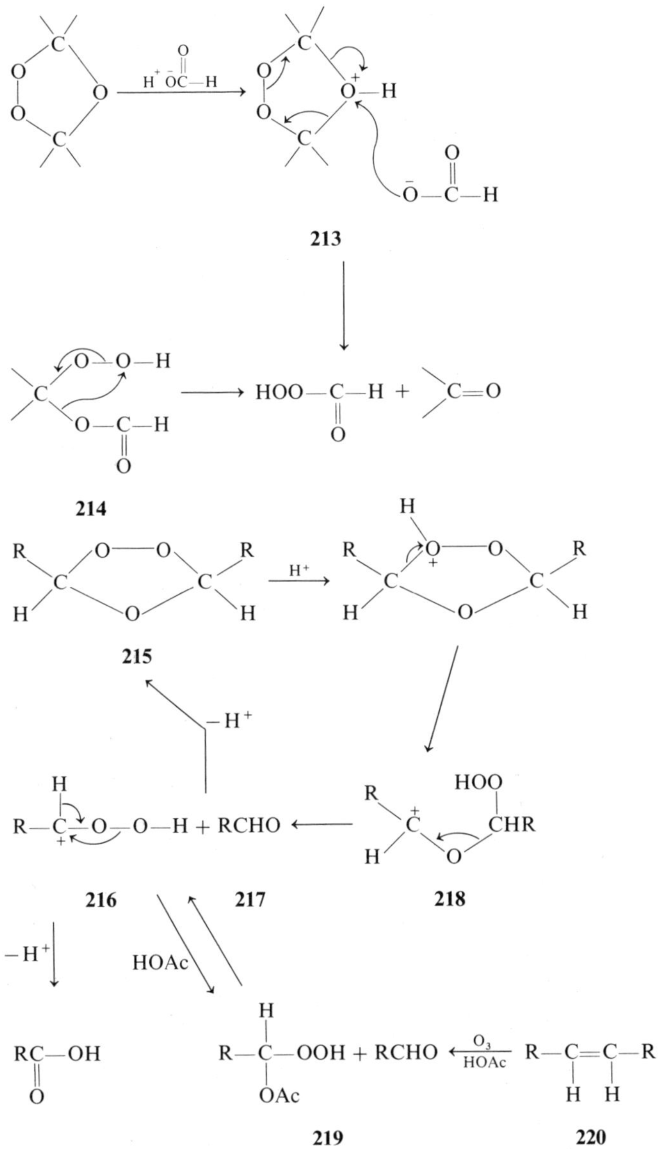

SCHEME 45. "Acidolysis."

boxylic acid. As evidence for this scheme, they cite that decomposition of 5-decene ozonide (**215**) in acetic acid plus a trace of *p*-toluenesulfonic acid gave a small amount of acyloxyalkyl hydroperoxide (**219**). Inversely, ozonolysis of *cis*-decene (**220**) in acetic acid followed by acid-catalyzed decomposition of the supposed product (**219**) gave a small amount of ozonide **215**.

Kinetics showed the decomposition rate constant for 1-hexene ozonide to be directly proportional to the concentration of the *p*-toluenesulfonic acid catalyst.

B. HYDROLYSIS OF α-OXYALKYL HYDROPEROXIDES

Although simple α-alkoxyalkyl hydroperoxides (**221a,x**) are either fairly stable to hydrolysis under neutral conditions or undergo other reactions (see Section IV) (*271,431,432*), aryl types (**221b,x**), (*431*) and dialkyl types (**221c,x** and **221c,z**) (*60,432*) appear to be readily hydrolyzed to the corresponding aldehyde or ketone and hydrogen peroxide. For example, Subluskey *et al.* (*431*) obtained high yields of aromatic aldehydes by steam distillation of methanolic reaction mixtures containing **221b,x**; the intermediate probably was **221b,y**. Similar results were obtained by Nebel (*432*) at room temperature using tetrasubstituted or 1,1-disubstituted ethylenes.

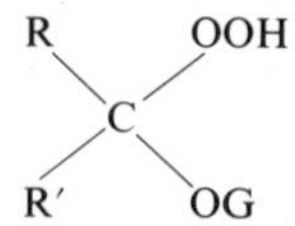

221a: R=alkyl, R′=H
221b: R=aryl, R′=H (**x:** G=CH_3; **y:** G=H; **z:** G=acyl)
221c: R=R′=alkyl

Ozonolyses carried out in water should give peroxides of type **221a,y**, **221b,y**, or **221c,y** directly. In several cases aldehydes have been obtained by decomposition of the hydroxy hydroperoxide (*433–435*). The method appears to be especially good for aromatic aldehydes (*434,435*) (this will be treated also in Part II).

IV. Base-Catalyzed Decompositions

The decomposition of ozonides and/or other peroxidic ozonolysis products to an aldehyde or ketone plus a carboxylic acid has been recognized since the time of Harries (*85,436*). Harries (*24,85*) and Staudinger (*51*) suggested that acids were produced by rearrangement of a species that is equivalent to the Criegee carbonyl oxide (see Chapter III, Section I, A and B).

Criegee and Korber (*437*) studied the decomposition of various ozonides, to a carboxylic acid and an aldehyde or ketone, in various solvents and found the rate of decomposition to be faster in nucleophilic than in nonnucleophilic solvents; the highest rate was with alcohols, especially methanol. With ozonides of type **222** (Scheme 46, R′ = H), the carboxylic acid involved the R group almost exclusively; the more electron-attracting the R group, the faster the reaction. In all cases there was a definite isotope effect when the ring hydrogens were replaced by deuterium. Because of these findings Criegee and Korber (*437*) proposed a mechanism analogous to the Kornblum–De La Mare (*438*) mechanism for base-induced decomposition of dialkyl peroxides. This is illustrated in Scheme 46, structure **222**. Similar results and mechanisms were suggested by Ellam and Padbury (*439*) for decompositions of ozonides with dimethyl sulfoxide, dimethylformamide, and amines.

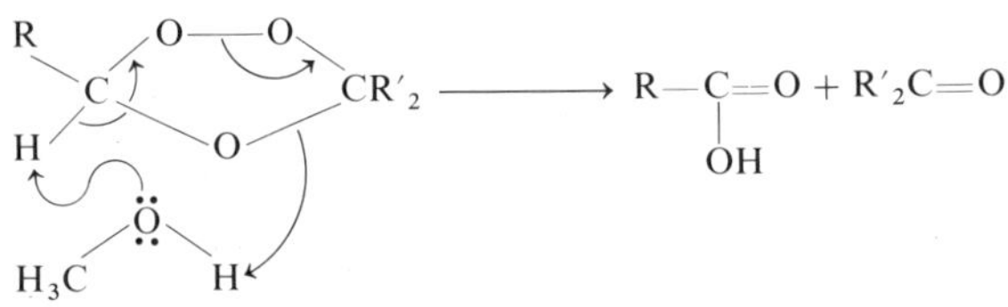

222

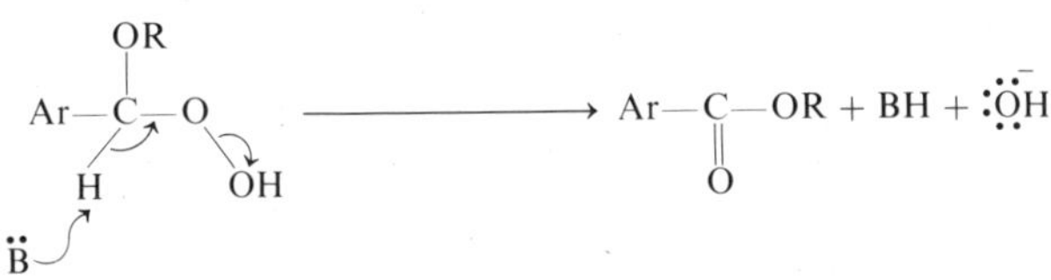

223

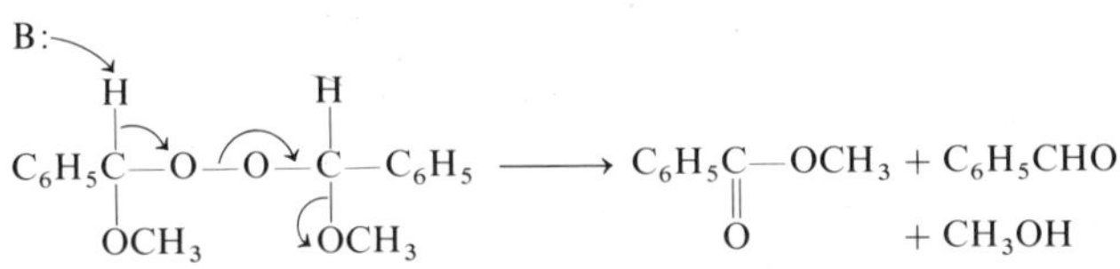

224

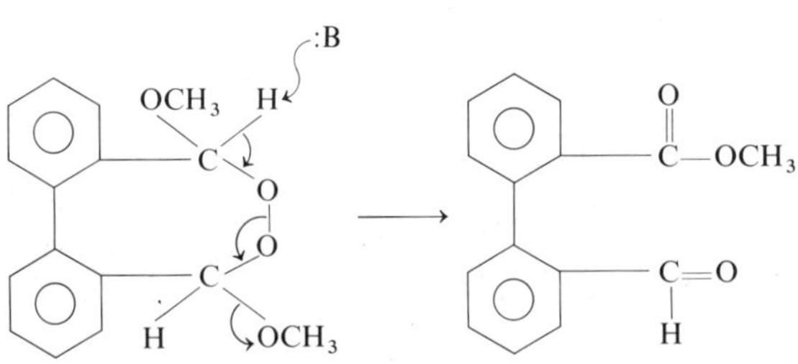

225

SCHEME 46. Base-catalyzed decompositions.

Ellam and Padbury (*440*) also have reported the quantitative decomposition of α-alkoxybenzyl hydroperoxides into the corresponding alkyl benzoates by means of dimethyl sulfoxide or amines. The reaction is illustrated by **223** (Scheme 46).

Similar reactions have been reported for various adducts, etc., of α-oxyalkyl hydroperoxides (*271,281,282,371,373,395,441*), as illustrated in Scheme 46 (**224** and **225**).

In addition to these strong base-catalyzed decompositions, it seems reasonable to assume that decompositions of α-alkoxyalkyl hydroperoxides in nucleophilic solvents (e.g., *267,269,442*) occur by such a mechanism; it is well known that hydroxyalkyl hydroperoxides decompose in this way (*433*) as well as by decomposition to an aldehyde or ketone and hydrogen peroxide as described in the preceding subsection.

Tertiary hydroperoxides are known to decompose in the presence of base to give molecular oxygen (*244,399,408,443*). This is probably the explanation for the reported ozonolysis of **226** in ethanol, followed by decomposition of the product in the presence of sodium hydroxide, to give **228** and molecular oxygen (*444*). The peroxidic ozonolysis product probably was **227**.

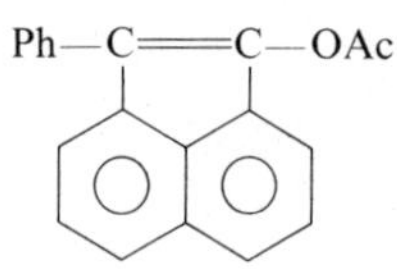

226

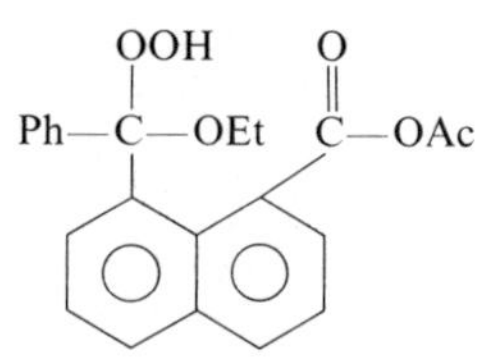

227

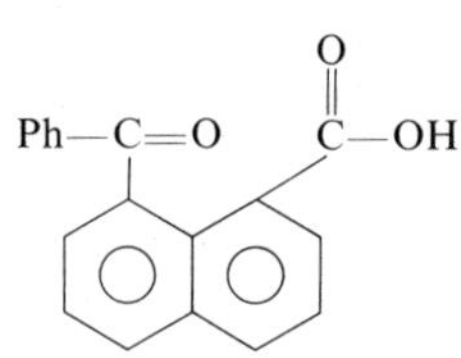

228

V. Pyrolysis and Photolysis

Pyrolytic cleavage of ozonides has been recognized and utilized by various investigators (*116*). Sometimes anomalous (Chapter IX, Section VII) as well as expected products are obtainable. Criegee *et al.* (*109*) suggested that the slow decomposition of 2,3,3-trimethyl-1-butene ozonide to give the corresponding ketone diperoxide, *tert*-butyl acetate, and formaldehyde polymers (*62*) involves breakdown to the carbonyl oxide and aldehyde from which it was derived (see, however, Chapter IX, Sections I and VII). Razumovskii and co-workers (*234,358,427,445–447*) also suggested heterolytic cleavage and a zwitterion intermediate in rationalizing thermal decompositions of ozonides. The products were acids, aldehydes, ketones, and anomalous products (Chapter IX, Section VII). Some of the decompositions were carried out in nucleophilic solvents including water. Razumovskii *et al.* (*427*) considered the process to be thermal rather than solvolytic, even in the presence of water.

Most of the published work concerning thermal decompositions of ozonides and other peroxidic ozonolysis products, however, is best explained by homolytic cleavage and radical intermediates.

Bernatek and Hvatum (*448*) studied the thermal decomposition of certain benzofuran ozonides in hydrocarbon solvents at temperatures up to 100°C and found the reactions to be strictly first order with no induced decomposition, at least during the first half of the reaction while the concentration of decomposition products was low. They considered the rate-controlling step to be homolytic cleavage of the O—O bond. Activation energies of 28–29 kcal/mole were determined. Privett and Nickell (*311,449*) also suggested such a homolytic peroxide cleavage to explain the formation of aldehydes, carboxylic acids, and anomalous products (Chapter IX, Section VII) from pyrolysis of ozonides of oleates and other fatty acid esters.

The most instructive work concerning the mechanism of pyrolysis (and photolysis) of ozonides and ketone diperoxides is that of Story *et al.* (*450–454*). Pyrolysis and photolysis of a mixture of the *cis*- and *trans*-1,2-diisopropylethylene ozonides in pentane or heptane afforded the same products, but in different ratios, as shown in Table XXIII (*450,451*). The authors account

TABLE XXIII

THERMAL AND PHOTOCHEMICAL DECOMPOSITIONS OF 1,2-DIISOPROPYLETHYLENE OZONIDE[a]

Product	Yield (mole %) 100°C, 12 hours	Yield (mole %) $h\nu$
$CH_3CH_2CH_3$	24[b]	35[b]
$(CH_3)_2CHCH(CH_3)_2$	3	33
$(CH_3)_2CHCHO$	77	14
$(CH_3)_2CHCOOH$	62	7
CO_2	0	14
CO	5	41
$HCOOH$	22	43

[a] Reprinted, with permission of publisher, from Story *et al.* (*451*). Copyright by Pergamon Press, Ltd.

[b] Assuming 2 moles of propane per mole of ozonide.

for these results by the mechanism shown in Scheme 47 (*451*). For photolysis, pathway C is major and pathway A is minor: just the opposite is true with pyrolysis. Heicklen *et al.* (*455*) suggested that in pathway A of pyrolysis the hydrogen atom transfer is synchronous with the homolytic cleavage of the O—O bond, as shown in **230**. They based this on their determination of the activation energies for pyrolyses of several ozonides (ethylene, propene, isobutene, and 2-butene) in the vapor phase and finding that the values

(24–27 kcal/mole) were considerably lower than the calculated value for —O—O-bond cleavage (32–38 kcal/mole). They also found the decompositions to be unimolecular and first order (*455*).

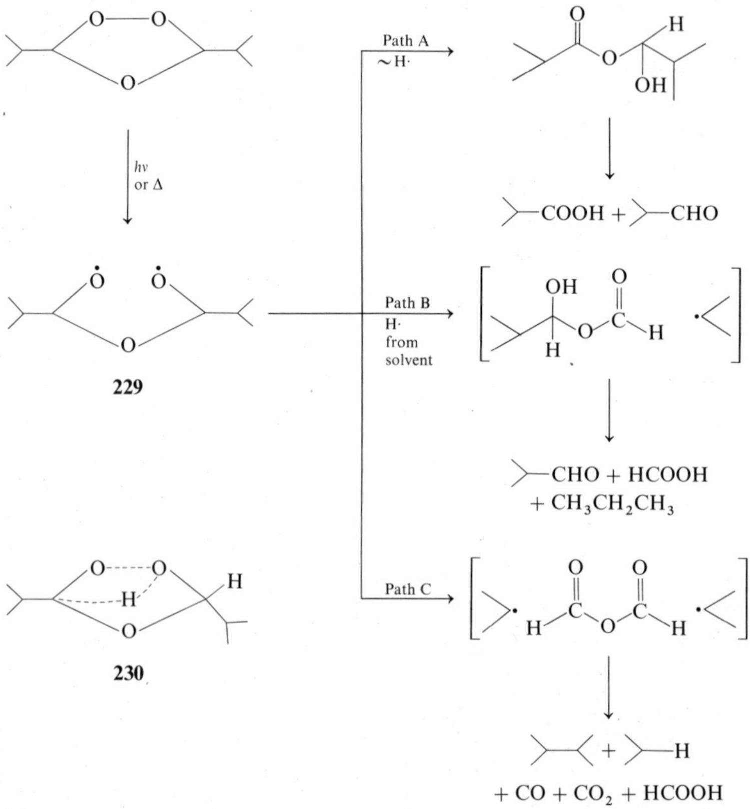

SCHEME 47. Thermal and photolytic decompositions of diisopropylethylene ozonide.

In contrast to the above, photolysis and pyrolysis of cyclopentene ozonide gave similar ratios of products, especially when the pyrolysis was conducted at 180°C (Table XXIV) (*450,451*). The rationalization is shown in Scheme 48. In this case, pathway A of Scheme 47 does not occur because of the restraints of the cyclic system and, thus, the results from photolysis and pyrolysis are similar. In favor of the mechanisms of Scheme 48 is the fact that the activation energy for the pyrolysis of cyclopentene ozonide (ca. 36 kcal/mole) is equal to the calculated value based on O—O bond cleavage (*455*).

The utilization of the photolysis of ozonides was continued both by Story *et al.* (*452*) and by Criegee *et al.* (*113,195,202,203,211*). Some of the novel

TABLE XXIV

THERMAL AND PHOTOCHEMICAL DECOMPOSITIONS OF CYCLOPENTENE OZONIDE[a]

Product	Yield (mole %) 180°C, 15 min	*hν*
Cyclopropane	37	35
$CH_3(CH_2)_2CHO$	9	10
$CH_3(CH_2)_2COOH$	14	14
Glutaric anhydride	3	18
CO	43	50
HCOOH	44	36

[a] Reprinted by permission of the publisher from Story *et al.* (*451*). Copyright by Pergamon Press, Ltd.

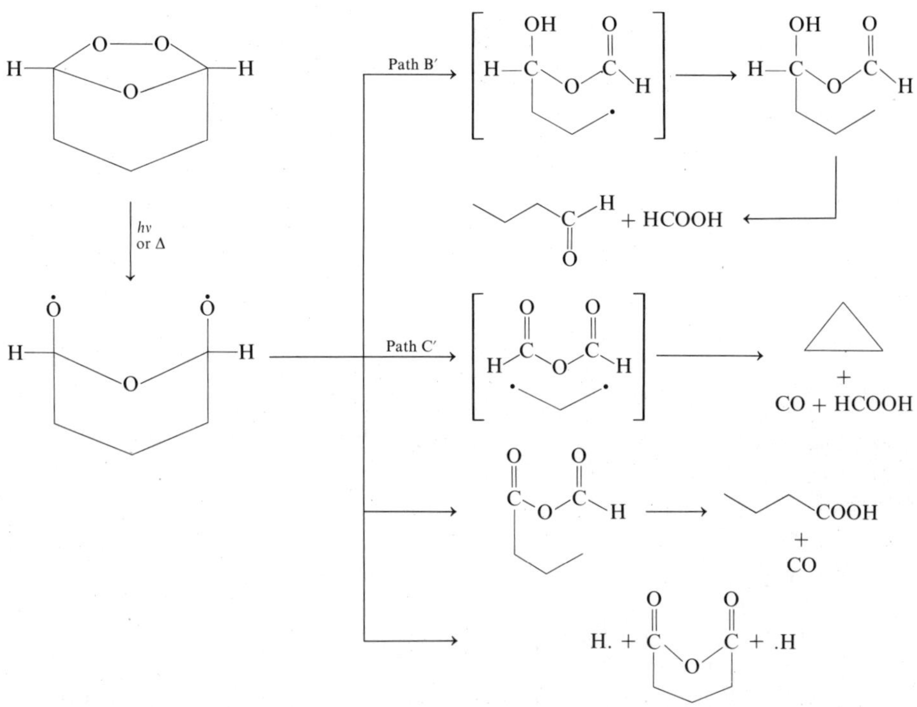

SCHEME 48. Pyrolysis and photolysis of cyclopentene ozonide.

results are outlined in Scheme 49. The synthesis of unstable cyclobutadiene derivatives (e.g., **231**) was evidenced by the isolation of dimers **232** and **233**, the former arising from the dimerization of two different cyclobutadienes formed in the same reaction mixture (*195*). Another interesting example is that of Kobayashi *et al.* (*215*), illustrated by the transformations **234** → **235**, etc. In some cases (*211*,*452*) peroxidic rearrangements occurred (see Chapter IX, Section VII).

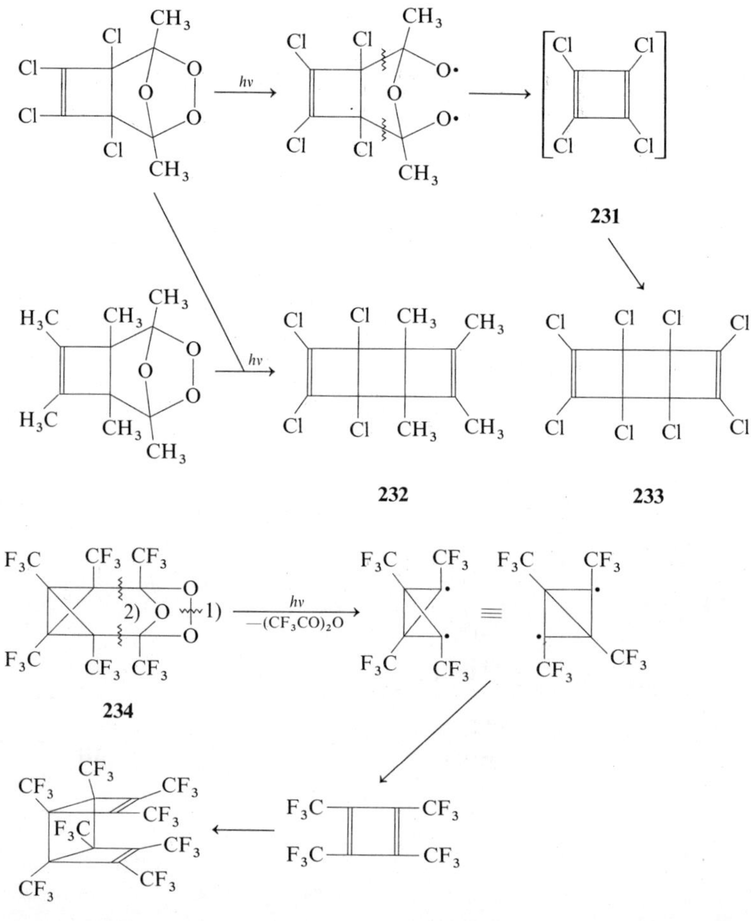

SCHEME 49. Syntheses via photolysis of ozonides.

Story and co-workers (*453*,*454*) also have pyrolyzed and photolyzed ketone diperoxides and have, thereby, synthesized macrocyclic hydrocarbons. An example is shown in Scheme 50. Although ketone diperoxides can be obtained by ozonolysis (Chapter V, Section II, Chapter VII, Section I, A), Story *et al.* (*454*) employed other synthetic methods. From diperoxide **236**, products **237**, **238**, and **239** were obtained in 20%, 10%, and 14% yields, respectively.

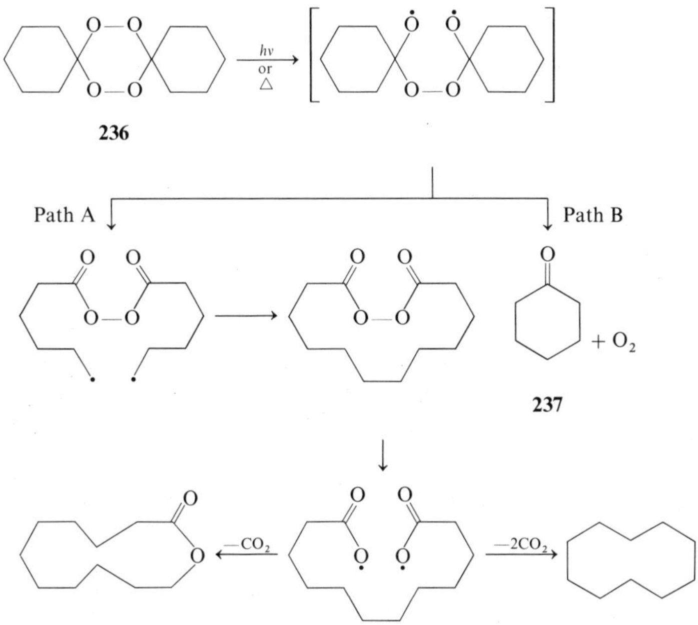

238 **239**

SCHEME 50. Pyrolysis and photolysis of ketone diperoxides.

By a pathway analogous to B of Scheme 50, Cadogan *et al.* (*456*) has obtained benzophenone in high yield from pyrolysis of benzophenone diperoxide.

Metal ion-catalyzed decomposition of ozonides and α-alkoxyalkyl hydroperoxides also appear to involve homolytic peroxide linkage cleavages (*265,269,457,458*). An example is shown in Scheme 51.

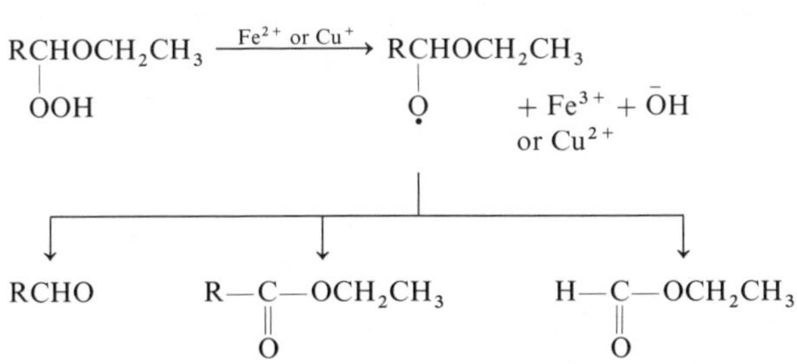

SCHEME 51. Metal ion-catalyzed decompositions.

For further discussions of decompositions of these types, see Chapter IX, Section VII).

VI. Miscellaneous

Fliszár and co-workers (*343,394,459–462*) have studied the mass spectral and polarographic decompositions of ozonides, α-alkoxyalkyl hydroperoxides, and ketone diperoxides and have proposed mechanisms for the reactions involved.

The lead tetraacetate test for hydroperoxides (with O_2 evolution) (*243,244,274,399,463,464*) is completely applicable to α-alkoxyalkyl hydroperoxides (e.g., *268*). The products are probably esters, originating via pathways similar to those proposed for other hydroperoxides (*244,399,463,464*).

Oxidative decompositions of peroxidic ozonolysis products are widely used (*115,116*). Since most of these are carried out in the presence of water, the actual oxidation process probably involves aldehydes rather than peroxidic ozonolysis products.

Special degradative ozonolyses, such as amozonolysis (*378*), cyanozonolysis (*379*), and ozonolyses in the presence of hydrogen peroxide (*370*), were discussed in Chapter VII, Section II, B.

CHAPTER **IX**

"Anomalous" Ozonolysis of Olefins

I. Introduction and General Theory

So-called "abnormal" or "anomalous" ozonolysis products were recognized early by Harries (*23,24,26,46,85,89,255,465–467*). Since then, many examples have been reported in the literature. Space will permit the inclusion of only a fraction of these in this review. All types, however, will be illustrated. With most "anomalous" ozonolyses, both the carbon–carbon double bond and one of the carbon–carbon single bonds next to it are cleaved. Thus, a product (or products) with one less carbon atom than expected is obtained. In general, but not exclusively, the systems that undergo "anomalous" ozonolysis as a major reaction are certain allylic compounds (Section II of this chapter), α,β-unsaturated carbonyl compounds (Sections III and IV), certain substituted cinnamic esters (Section V), and certain indoles and benzofurans, etc.(to be discussed under heterocyclic compounds in **Vol. II**). The reactions involved with these types are illustrated in Scheme 52 (*62,92,467*).

$$R_2C{=}CHCR'_2G \xrightarrow[(2)\,HOH]{(1)O_3} R_2C{=}O + HCOOH\ (\text{or CO}) + R'_2C{=}O + GH \qquad (a)$$

R and R′ = H, alkyl, aryl; G = OH, OR, NR_2

$$R_2C{=}\underset{R'}{\underset{|}{C}}{-}\underset{Z}{\underset{|}{C}}{=}O \xrightarrow[(2)\,HOH]{(1)\,O_3} R_2C{=}O + R'\underset{O}{\underset{\|}{C}}{-}OH + Z{-}\underset{O}{\underset{\|}{C}}{-}OH\ (\text{or } CO_2 + HZ) \qquad (b)$$

R and R′ as above, Z = H, R, OH, OR, etc.

$$Ar{-}CH{=}CH{-}\underset{O}{\underset{\|}{C}}{-}O{-}R'' \xrightarrow{O_3} ArOH + R''O{-}\underset{O}{\underset{\|}{C}}{-}\underset{O}{\underset{\|}{C}}{-}H + CO \qquad (c)$$

Ar = OH and OR substituted phenyl groups; R″ = alkyl or aryl

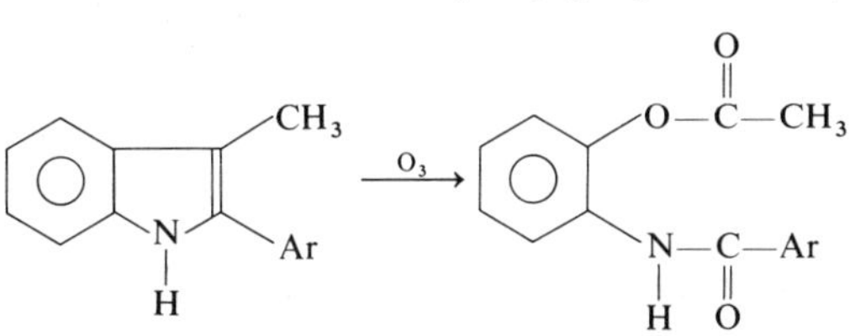

(d)

SCHEME 52. Types of "anomalous" ozonolyses.

In addition, other types often undergo "anomalous" ozonolysis as a minor side reaction (Sections, VI and VII). Since these reactions are normal for the systems and conditions involved, the terms "abnormal" and "anomalous" are, in principle, incorrect. They will be used only for convenience.

Sometimes anomalous ozonolyses are used for purposes of synthesis (*115*, *468–470*) or proof of structure (*471*). An example of the use in synthesis is shown in Scheme 53.

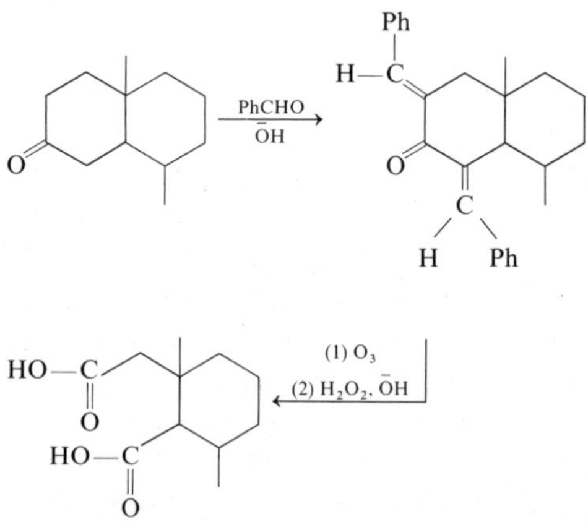

SCHEME 53. Example of anomalous ozonolysis in synthesis (*468*).

The importance of being able to recognize structures that might give so-called anomalous ozonolysis products is well illustrated by the surprise of Vlad and Souček (*472*) when they obtained levulinic acid (**241**), rather than the expected product (**242**), upon ozonolysis of linalool (**240**), and by the dilemma of Wibaut and Beyerman (*473*), who obtained nicotinic acid (**245**)

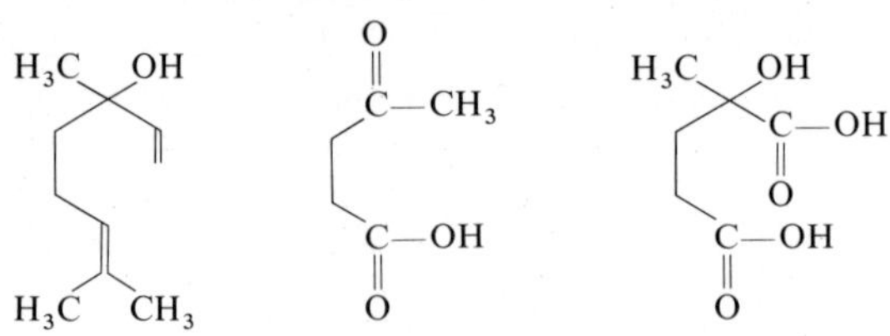

240 **241** **242**

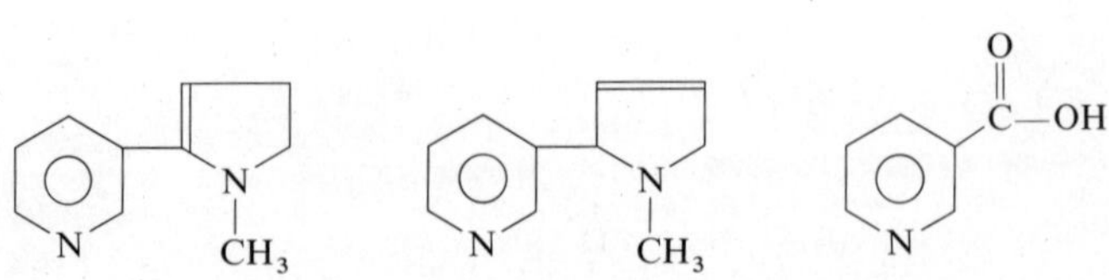

243 **244** **245**

upon ozonolysis of a dihydronicotyrine, but were uncertain whether it was a "normal" ozonolysis product of **243** or an "abnormal" ozonolysis product of **244**. Of even greater importance is the knowledge that anomalous ozonolysis products can be avoided by careful ozonolysis conditions and work-up, as described later in this section.

Harries (*24,59,85,89,466*) rationalized anomalous ozonolysis by means of rearrangement of a peroxidic species that was the forerunner of the Criegee carbonyl oxide (Chapter III, Section I, A). Such a rearrangement was also included in Staudinger's mechanism (*51*) (Chapter III, Section I, B) and, in modernized form, was the explanation offered by the Criegee mechanism (*62,110–112*) (Chapter III, Section I, C). Young (*474*) and Leffler (*92*) proposed rearrangements of a Staudinger molozonide (Chapter III, Section I, B) (*62,467*).

Following these proposals, it was shown that several compounds that gave anomalous products under usual work-up conditions afforded little or none of these products when the ozonolyses were carried out at temperatures of $-10°$ to $-70°C$, overozonation was avoided, and the cold ozonolysis mixtures were immediately reduced (*95,108,395,396*; see also *361,410*). Thus, abnormal ozonolyses can usually be prevented by a careful choice of reaction conditions. This, additionally, indicates that the anomalous products do not result from rearrangement of a primary ozonide or of a carbonyl oxide; these reactive intermediates should have disappeared before the reductions were carried out. Instead, anomalous ozonolysis products must arise from rearrangement of the peroxidic ozonolysis product, as illustrated in Scheme 54 [Eqs. (a), (b), and (c)], from oxidation of a methylene, etc., group followed by rearrangement, etc. (to be covered in **Vol. II**), or from oxidation of normal ozonolysis products as a result of overozonation or of an oxidative work-up procedure. For example, many ozonolysis reaction mixtures are treated with hydrogen peroxide (e.g., *468,469,475–477*), (see Scheme 53); the anomalous product could then result from a Baeyer–Villiger oxidation and rearrangement (*478*), as illustrated in Scheme 54, Eq. (d). The other rearrangements in Scheme 54 are also mechanistically similar to a Baeyer–Villiger reaction in the sense that all involve migrations to positive oxygen.

Another possible, but less likely, source of abnormal ozonolysis products was suggested by Story and Burgess (*397*). It involved the generation of a peroxy acid by ozonation of an aldehyde (to be discussed in **Vol. II**), followed by a Baeyer–Villiger reaction between the peroxy acid and an aldehyde. Since ozone attacks aldehydes much more slowly than olefins, (**Vol. II**), this should not occur except in cases of overozonation and could not, in most cases, account for anomalous ozonolysis as a major reaction. Although Story and Burgess (*397*) claimed to have shown that aldehydes can be oxidized to peracids in the presence of olefins, a careful analysis of their data,

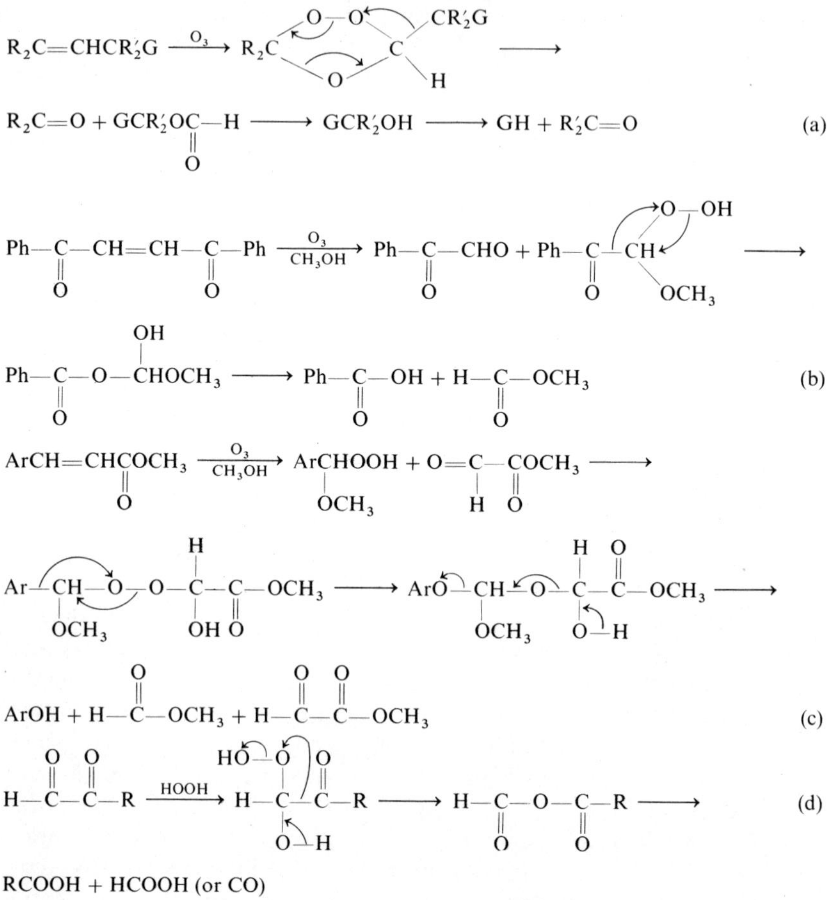

SCHEME 54. Rearrangements of peroxidic ozonolysis products in anomalous ozonolyses.

which again involves oxidation of cyclohexanone to 6-hexanolactone, indicates that their conclusion is not warranted (cf. error in this regard in Chapter VI, Section V). It is quite possible, however, that with ozonolyses carried out in a participating solvent, a Baeyer–Villiger reaction could occur between the resulting α-oxyalkyl hydroperoxide (Chapter V, Section IV; Chapter VII, Section II) and an aldehyde or ketone, as pointed out by Kolsaker and Bailey (*395*) and illustrated in Scheme 54, Eq. (c) (see also Section V of this chapter).

Further discussions of rationale for anomalous ozonolysis can be found in the following subsections, where particular types of olefinic compounds are discussed.

II. Allylic Compounds

The earliest example of an abnormal ozonolysis of an allylic compound appears to involve farnesol (**246**), which upon complete ozonolysis gave, along with the expected products, the anomalous product formic acid (*465*).

$$(CH_3)_2C{=}CHCH_2CH_2\underset{\displaystyle CH_3}{\underset{|}{C}}{=}CHCH_2CH_2\underset{\displaystyle CH_3}{\underset{|}{C}}{=}CHCH_2OH$$

246

TABLE XXV

ABNORMAL PRODUCTS FROM OZONOLYSIS OF ALLYLIC COMPOUNDS[a]

Compound	Abnormal product[b]	Yield (%)
$CH_3CH{=}CHCH_2Cl$[c]	HCOOH	O[c]
$CH_3CH{=}CHCH_2Br$[c]	HCOOH	O[c]
$CH_3CH{=}CHCH_2OCOCH_3$	HCOOH	O
$CH_3CH{=}CHCH_2OC_6H_4NO_2$-*p*	HCOOH	8
$CH_3CH{=}CHCH_2OC_6H_5$	HCOOH	15
$CH_3CH{=}CHCH_2OH$	HCOOH	23
	$H_2C{=}O$[d]	17
$CH_3CH{=}CHCH_2OC_2H_5$	HCOOH	25
$CH_3CH{=}CHCH_2N(C_2H_5)_2$	HCOOH	54
$C_6H_5CH{=}CHCH_2OH$	HCOOH	60
$CH_3CHOHCH{=}CH_2$	CH_3COOH	25
$CH_3CH(OC_2H_5)CH{=}CH_2$	CH_3COOH	35
$CH_3CH_2CHOHCH{=}CHCH_3$	CH_3CH_2COOH	15
	$H_2C{=}O$[d]	20
$CH_3(CH_2)_2CHOHCH{=}CHCH_3$	$CH_3(CH_2)_2COOH$	50
	$CH_3(CH_2)_2CHO$[d]	35
$CH_3(CH_2)_3CHOHCH{=}CH_2$	$CH_3(CH_2)_3COOH$	38
	$CH_3(CH_2)_3CHO$[d,e]	50

[a] Reprinted, with permission, from Young *et al.* (*474*). Copyright by the American Chemical Society.

[b] Unless otherwise specified, this is the percentage of the indicated acid in a mixture of volatile acids obtained in 30–80% yield after oxidation with Ag_2O and NaOH. α-Substituted acids were not determined.

[c] The mixture of isomeric allyl halides (primary and secondary) was ozonized, and the mixture of acetic and formic acids obtained fit the theoretical values based on the composition of the mixture of isomeric halides.

[d] Actual yield obtained after hydrogenation of the ozonolysis reaction mixture.

[e] In addition, a 35% yield of α-hydroxycaproic acid was obtained.

Young *et al.* (*62,92,467,474*) have made an extensive study of anomalous ozonolyses of allylic compounds of type RCH=CHCHR′G in nonparticipating solvents. Some of the results are shown in Table XXV. Although the authors used an oxidative work-up procedure (hydrolysis followed by Ag_2O and NaOH), they reported spontaneous decomposition of the ozonolysis reaction mixtures; they also obtained, in all cases tried, comparable yields of anomalous products by a reductive work-up procedure. Both Milas and Nolan (*95*) and Raulins and Sibert (*396*) obtained normal results in several of these same cases by a reductive work-up procedure, but the reductions were performed immediately on the cold ozonolysis solution. Thus, it appears that in Young's cases, the anomalous results were due to rearrangements of ozonides before the work-up, as illustrated in Scheme 54, Eq. (a), rather than oxidation of "normal" nonperoxidic ozonolysis products (more definitive work is needed, however).

The order in which group G [in Eq. (a), Scheme 54] influences the formation of anomalous products is: —Cl, —Br, or —OAc < —OC_6H_4–NO_2–*p* < —OC_6H_5 < —OH < —OEt < —NEt_2 (see Table XXV). This is the order in which the migratory aptitutudes for —CH_2G should increase [in

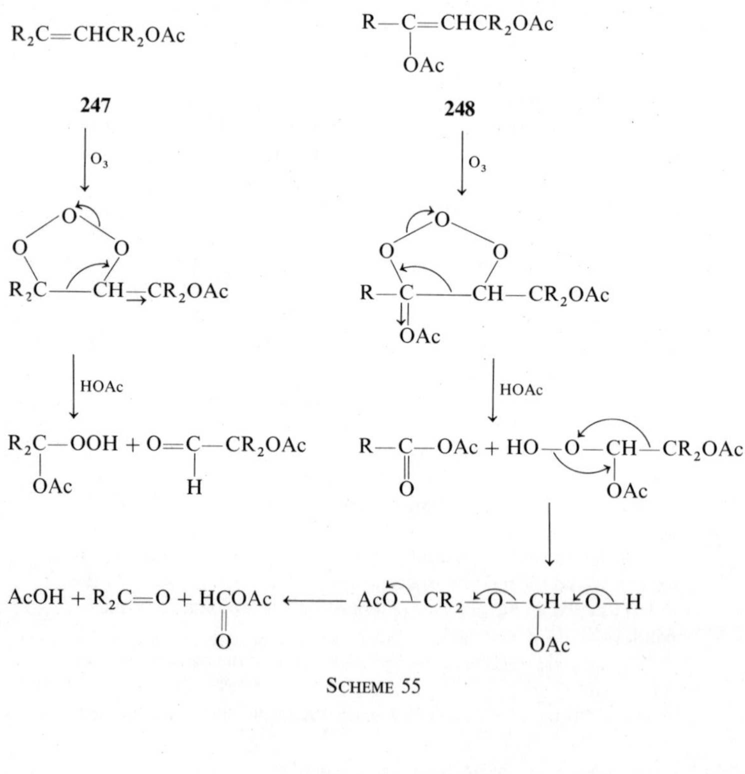

SCHEME 55

the ozonide of Eq. (a), Scheme 54], since this is the order of decreasing electron attraction by group G.

In regard to the findings of Young as outlined above, it has been reported that, although simple allyl acetates (e.g., **247**) do not afford anomalous products (*474*), those with a second acetate group do, provided the second acetate group is attached to the carbon of the double bond farther removed from the allylic acetate group (e.g., **248**) (*479*). The reactions, however, were carried out in the presence of acetic acid, and the explanation probably depends on which carbon of the double bond forms the zwitterion (Chapter VII, Section III), as illustrated in Scheme 55.

The principle illustrated in Scheme 55 is further exemplified by example 1 of Table XXVI. When this reaction was carried out in acetic acid solution, the product shown was obtained in 52% yield, whereas in methylene chloride solution, the anomalous product resulted in only 5% yield (*480*). The presence of the carbomethoxy group causes the zwitterion to be produced on the side necessary to allow a rearrangement of the types illustrated in Scheme 55. This explanation appears much more reasonable than that given by the authors (*480*).

Many more anomalous ozonolyses of allylic systems have been reported in the literature, but space permits reference to only a few, none of which appear to have used an oxidative work-up, although some were worked up hydrolytically (*396,480–491*). Some of these are illustrated in Table XXVI.

III. α,β-Unsaturated Carbonyl Compounds

The literature contains numerous examples of anomalous ozonolyses of α,β-unsaturated aldehydes and ketones, carboxylic acids, esters, etc. An early example is described in the review by Long (*26*). Some typical examples are given in Table XXVII. Others can be found in earlier reviews (*62,92*). None of the examples in Table XXVII were given a hydrogen peroxide work-up. It is not certain, however, how much of the anomalous result was caused by overozonation or a hydrolytic process, which conceivably, could produce hydrogen peroxide and anomalous results via a Baeyer–Villiger oxidation. For example, Proctor (*492*), Caunt *et al.* (*493*), and Vogel *et al.* (*494*) were able to obtain normal ozonolysis products by use of a minimum of ozone and a reductive work-up procedure, whereas excess ozone or an oxidative work-up gave rise to anomalous products (*492,493*). Nevertheless, most of the examples of Table XXVII (*117,120,124,366,383,495–513*) do not appear to have involved overozonation, and the anomalous results seem to have arisen from the peroxidic ozonolysis products themselves.

TABLE XXVI

ANOMALOUS OZONOLYSIS PRODUCTS FROM VARIOUS ALLYLIC COMPOUNDS

Example	Compound	Anomalous product(s)	Yield (%)	References
1	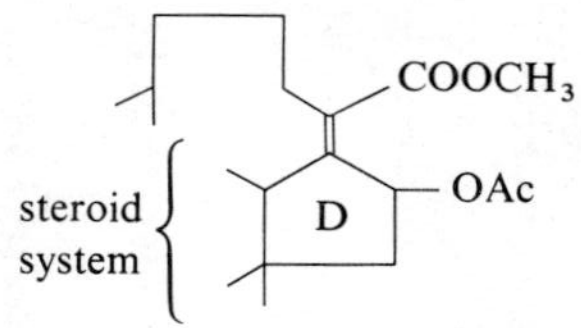	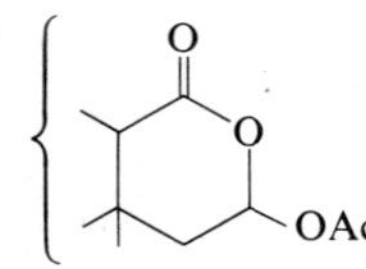	52	*480*
2	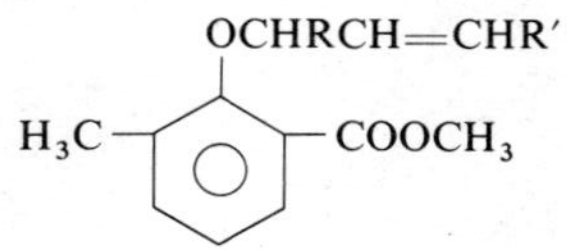	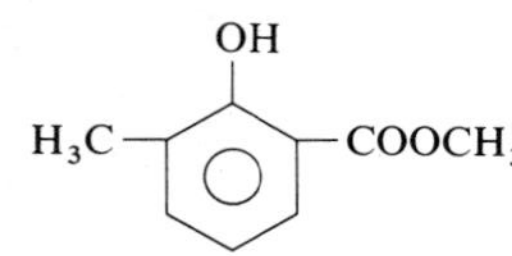 [a]	8–30	*396,484*
3	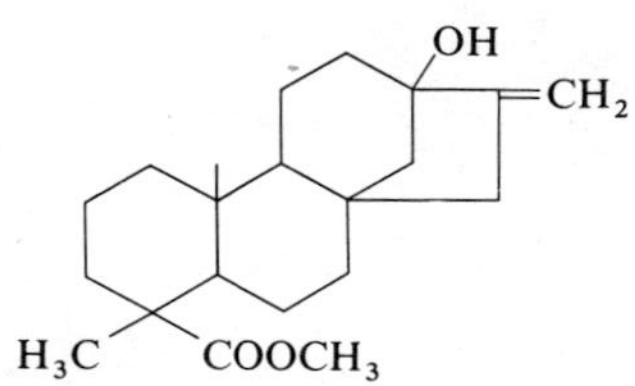	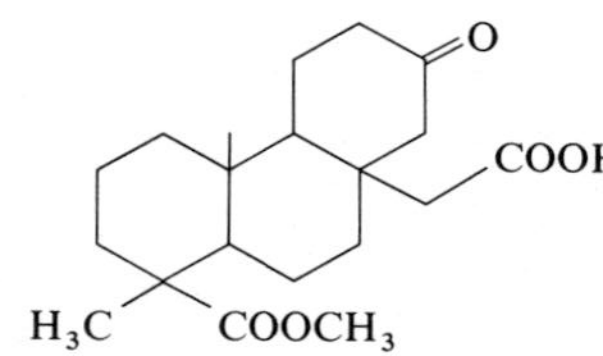 [a]	46	*489*
4	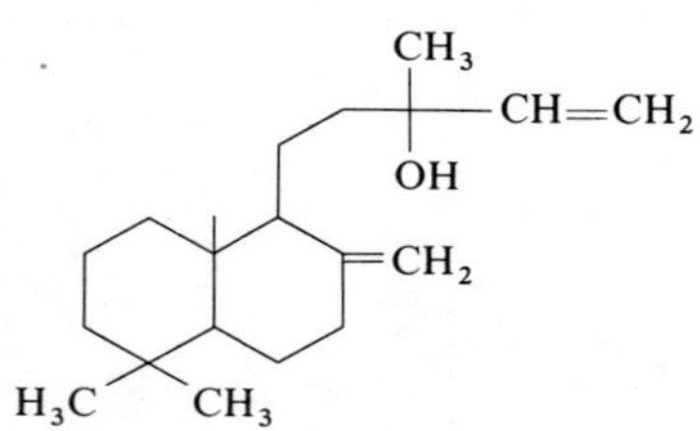	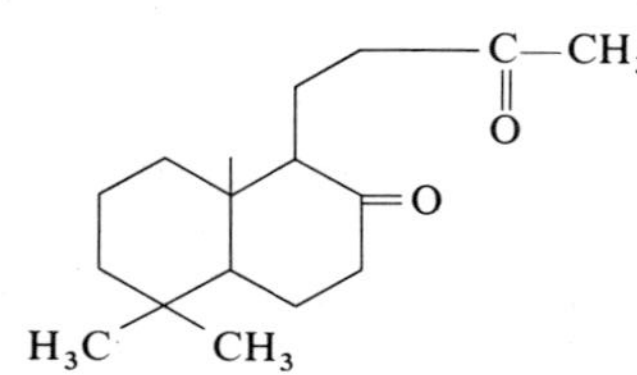 [a]	35–55	*491*

5	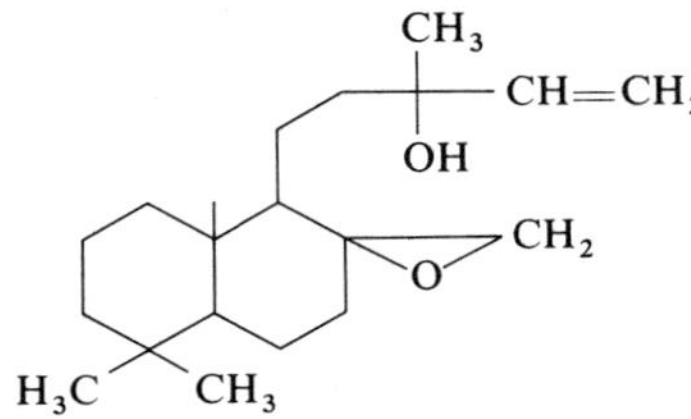	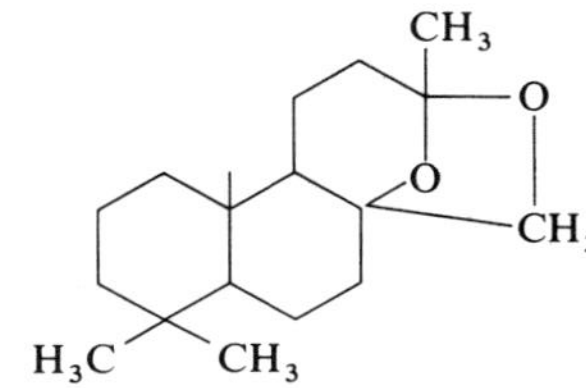 [a]	24	*490*
6	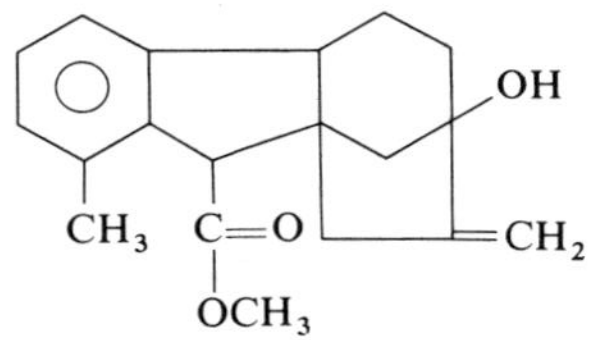	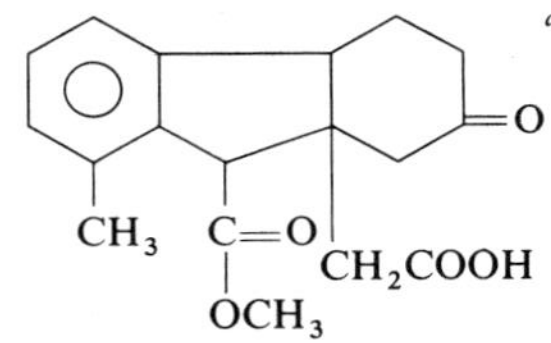 [a]	10–20	*486*
7	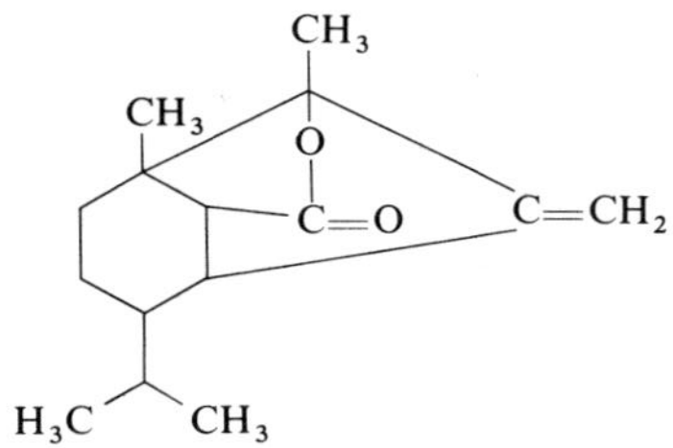	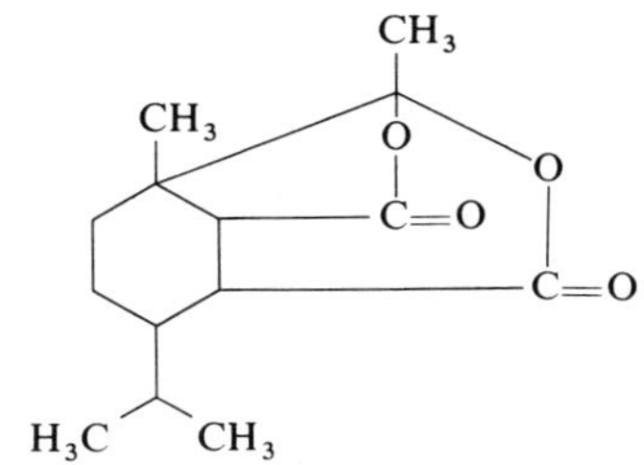	—[b]	*488*
8	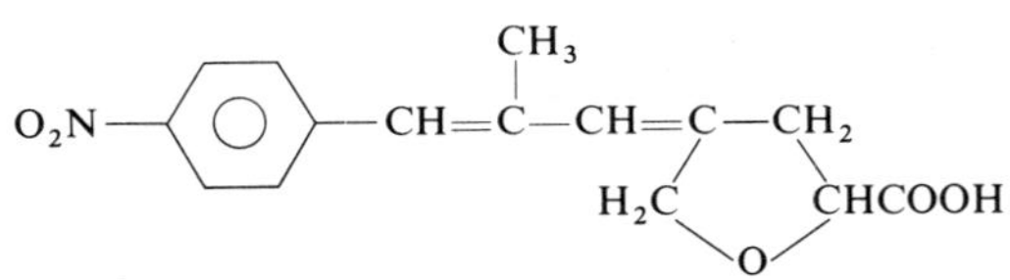	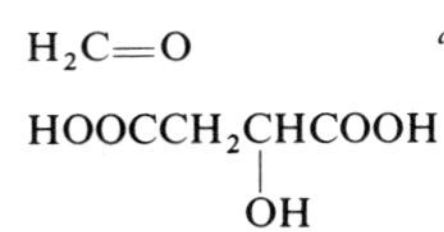 [a]	49	*487*

[a] Hydrolytic work-up.
[b] Yield not given.

TABLE XXVII

ANOMALOUS OZONOLYSIS PRODUCTS FROM α,β-UNSATURATED CARBONYL COMPOUNDS

Example	Compound	Anomalous product(s)	Yield (%)	References
1	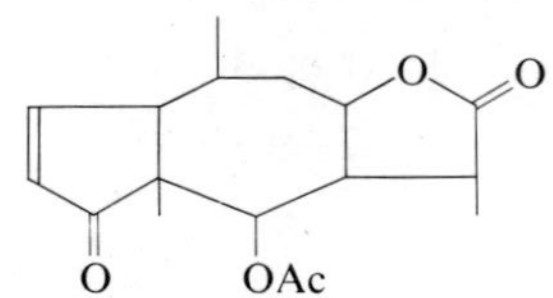	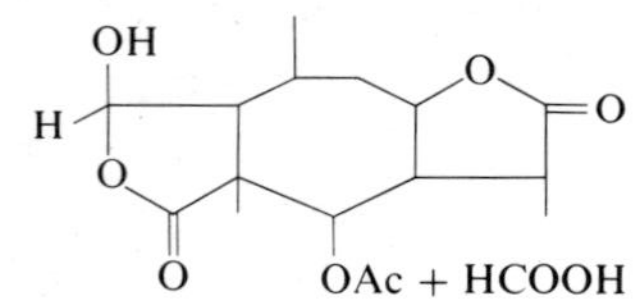	24	*495,496*
2	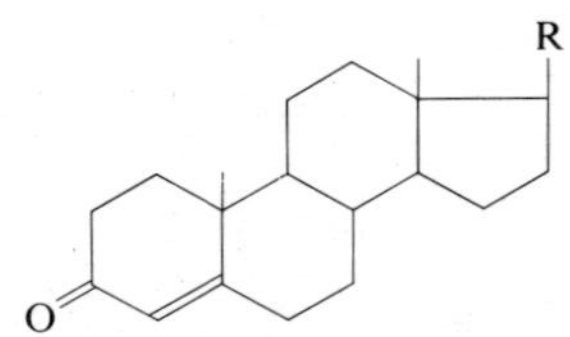 cholestenone, etc.	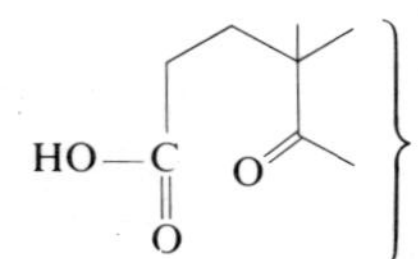	45–70	*497,498*
3	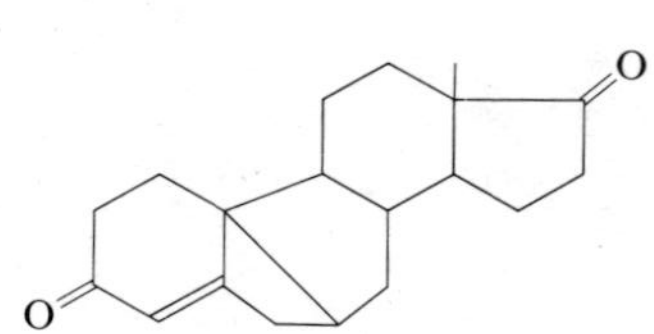	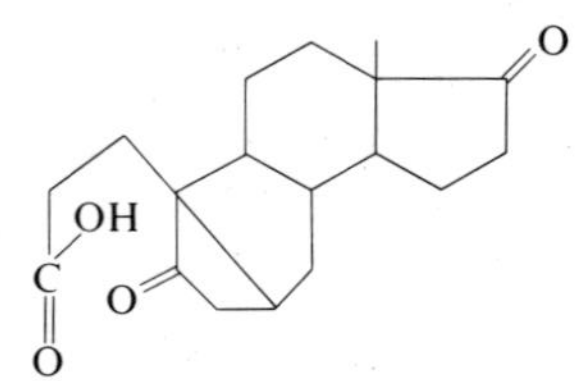	High	*499*
4	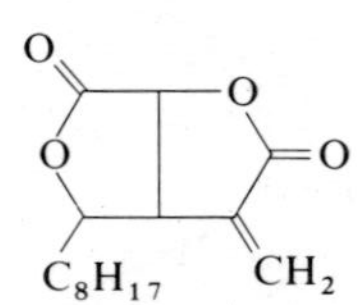	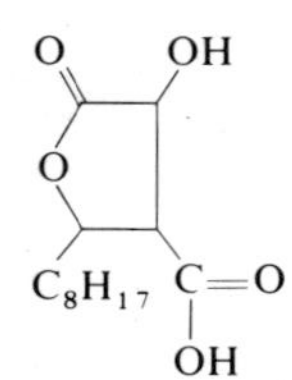	55	*500*

5	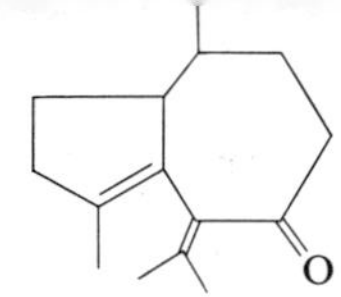	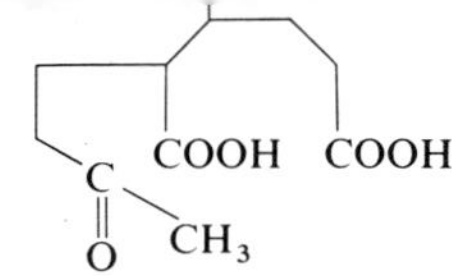	—	*501*
6	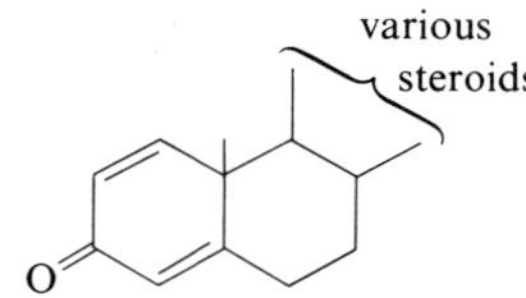	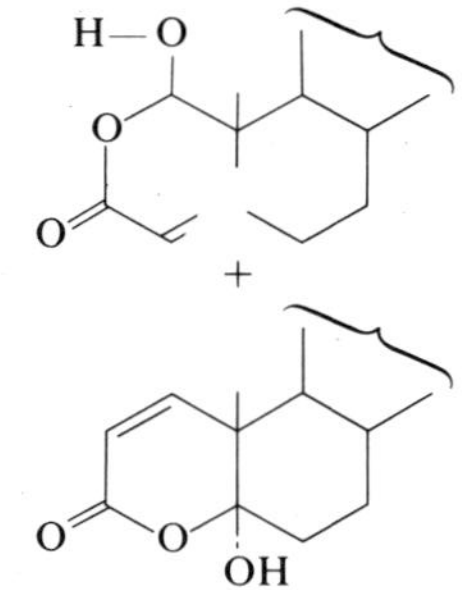	~ 10	*502–505*
7	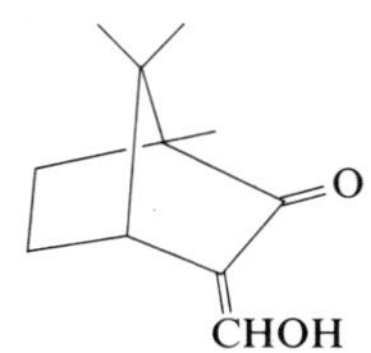	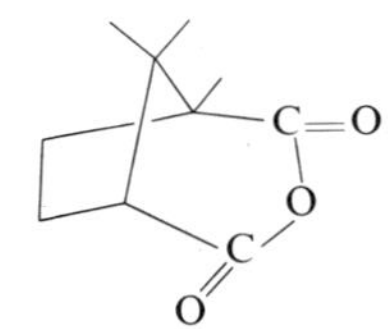	56	*506*
8	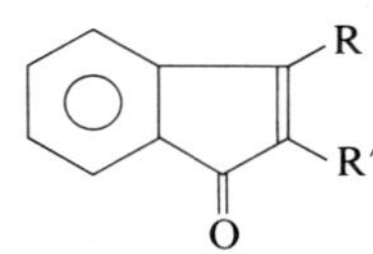	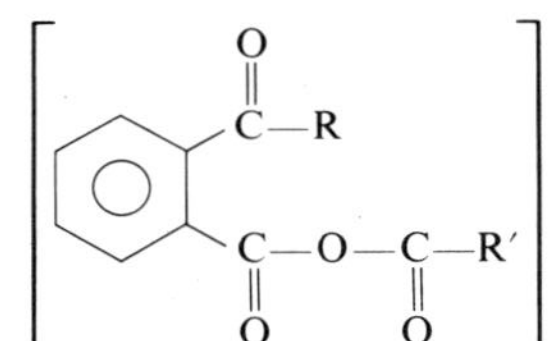	—	*383*
9	$Ph{-}\underset{\|}{\overset{}{C}}(=O){-}CH{=}C(CH_3){-}C(=O){-}Ph$	$Ph{-}C(=O){-}OH + CH_3C(=O){-}OCH_3$[a]	—	*117,120,366*

TABLE XXVII (*Continued*)

Example	Compound	Anomalous product	Yield (%)	References
10	$HOC(=O)-CH=C(CH_3)-C(=O)OH$	$CH_3C(=O)OH + CO_2 + H-C(=O)-OH$[b]	98	*507*
11	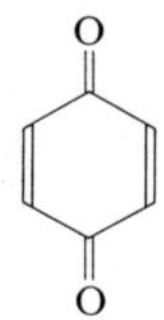	HCOOH, CO_2, $H-C(=O)-C(=O)-OH$, $HO-C(=O)-C(=O)-OH$	—	*508,509*
12	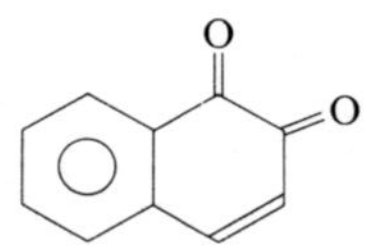	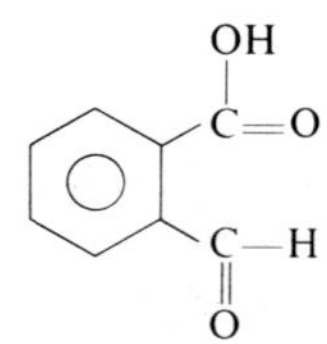	77	*510*
13	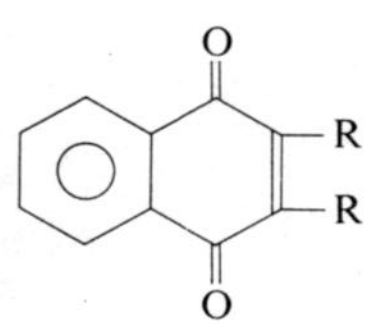	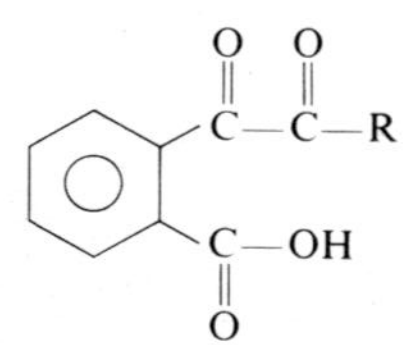 + RCOOH or CO (when one R=H)	*ca*, 100	*124,511–513*

[a] Ozonolysis and decomposition in methanol solution.
[b] Ozonolysis in aqueous solution.

Barton and Seoane (*514*) have proposed an ozonide fragmentation mechanism to explain the anomalous results from ozonolyses of α,β-unsaturated carbonyl systems, as illustrated in Scheme 56, Eqs. (a) and (b). For an α,β-unsaturated aldehyde or ketone, the mechanism requires the formation of a hydrate (**249**, Scheme 56). Although such reactions are reasonable, rearrangements of peroxidic ozonolysis products, as illustrated in Scheme 54, Eq. (b) or in Scheme 56, Eq. (c), seem more likely. These mechanisms are analogous to those of other peroxidic rearrangements, including the Baeyer–Villiger reaction, in which common migrating groups are acyl, carboxyl, carbomethoxy, etc. (*478*).

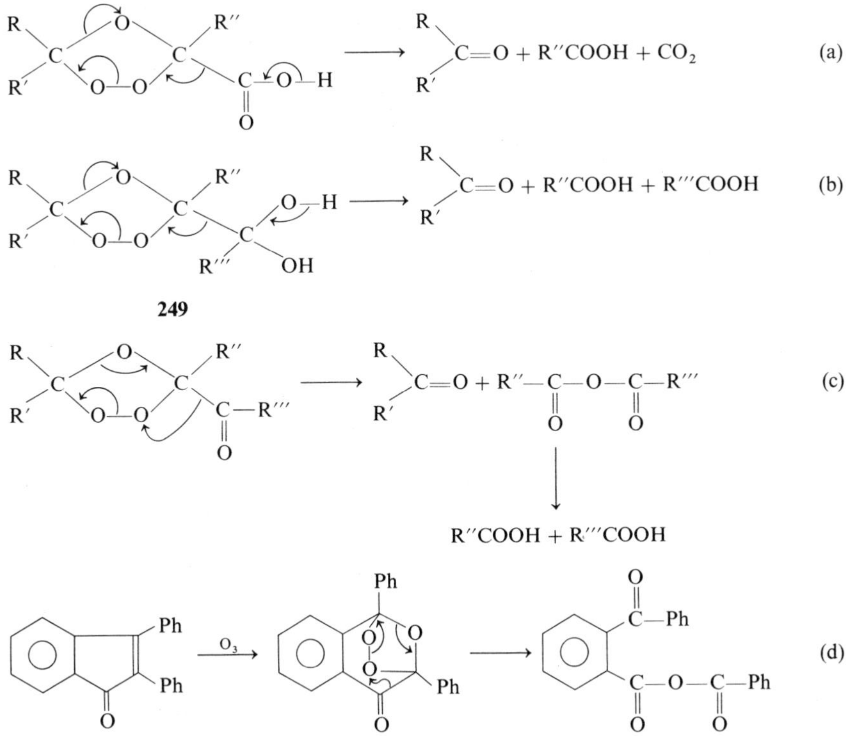

SCHEME 56. Anomalous ozonolyses of α,β-unsaturated carbonyl types

Only in a few instance have the peroxidic ozonolysis products involved as intermediates in anomalous ozonolyses been isolated and their rearrangements directly studied. Criegee *et al.* (*383*) characterized ozonides of certain 2,3-disubstituted indenones [example 8, Table XXVII; Scheme 56, Eq. (d)] and showed that they decomposed either in solution or on standing to give the anomalous product type shown. Although they proposed that

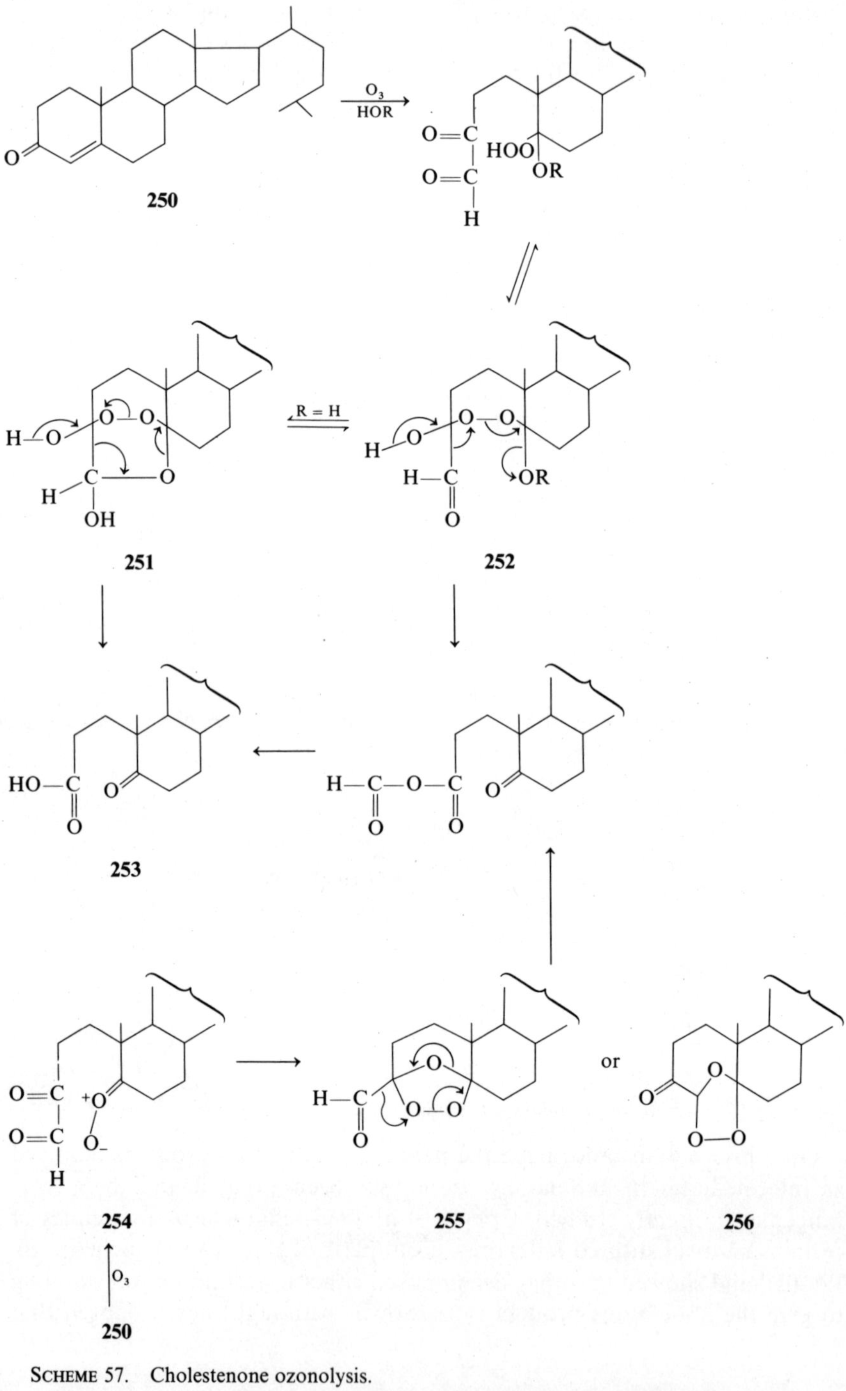

SCHEME 57. Cholestenone ozonolysis.

the rearrangement involved a carbonyl oxide zwitterion (*383*), the rearrangement of the ozonide itself, as shown in Scheme 56, Eq. (d), seems more likely, at least for the decompositions in solution (see also Section VII of this chapter).

Bailey *et al.* (*117,120,366*) have ozonized various 1,2-dibenzoylethylene derivatives in methanol (example 9, Table XXVII and have isolated the corresponding methoxyalkyl hydroperoxides [e.g., Scheme 54, Eq. (b)] in a few cases. These were shown to undergo rearrangement and decomposition readily, presumably by the mechanism shown in Scheme 54, Eq. (b). By analogy, the rearrangement in example 10 probably involved the corresponding hydroxyalkyl hydroperoxide, since the ozonolysis was performed in aqueous solution (*507*). Similar results were obtained with phenylmaleic and chloromaleic acids (*507*).

Dauben *et al.* (*497*) and Lefebvre *et al.* (*498*) have obtained crystalline peroxidic ozonolysis products from cholestenone (example 2, Table XXVII; and **250**, Scheme 57) using acetic acid and aqueous solvents. These peroxides, characterized as **251** and **252** (R = Ac), readily decomposed to anomalous product **253**, presumably as shown in **252** or, perhaps, **251** when the solvent is aqueous (*497*).

An oily, unstable peroxide also was obtained from ozonolysis of cholestenone in a nonparticipating solvent (*498*). If it were an ozonide, it probably had structure **255** rather than **256**, since carbonyl oxide **254** could cyclize either way (see Chapter V, Section I; Chapter VI, Section VI, B; and Chapter VII, Section I). Decomposition would be as shown in **255**.

A similar situation exists with examples 6 and 11–13 of Table XXVII. In fact, the product of example 12, **260**, can arise only from ozonide **258a** (or **258b**, but not **257**). Ozonide **257** should yield **259**, and **258b** could yield either **260** or **259**, as shown in Scheme 58. A minor product, phthalic anhydride (*510*), also could result from ozonide **258a** by a different mechanism (see Section VII of this chapter).

The solvent for example 3, Table XXVII, was acetic acid (*499*), and the peroxidic ozonolysis product should be analogous to **252**. Example 7 of Table XXVII is especially interesting. The high yield of anhydride (**263**) was produced in acetic acid solution (*506*). Much more of the "normal" product, **261**, was obtained in methanol or methylene chloride solution. The mechanism shown in Scheme 59 would explain the results. Moieties **261** and **262** should be preferentially produced (see Chapter VII, Section III). Acetic acid solvent could cause the Baeyer–Villiger reaction shown to be favored through acid catalysis. Perhaps, in methanol, **264** is produced but decomposes or reacts no further; and in methylene chloride, the unusual final ozonide decomposes via the hydroxyl group to **261**.

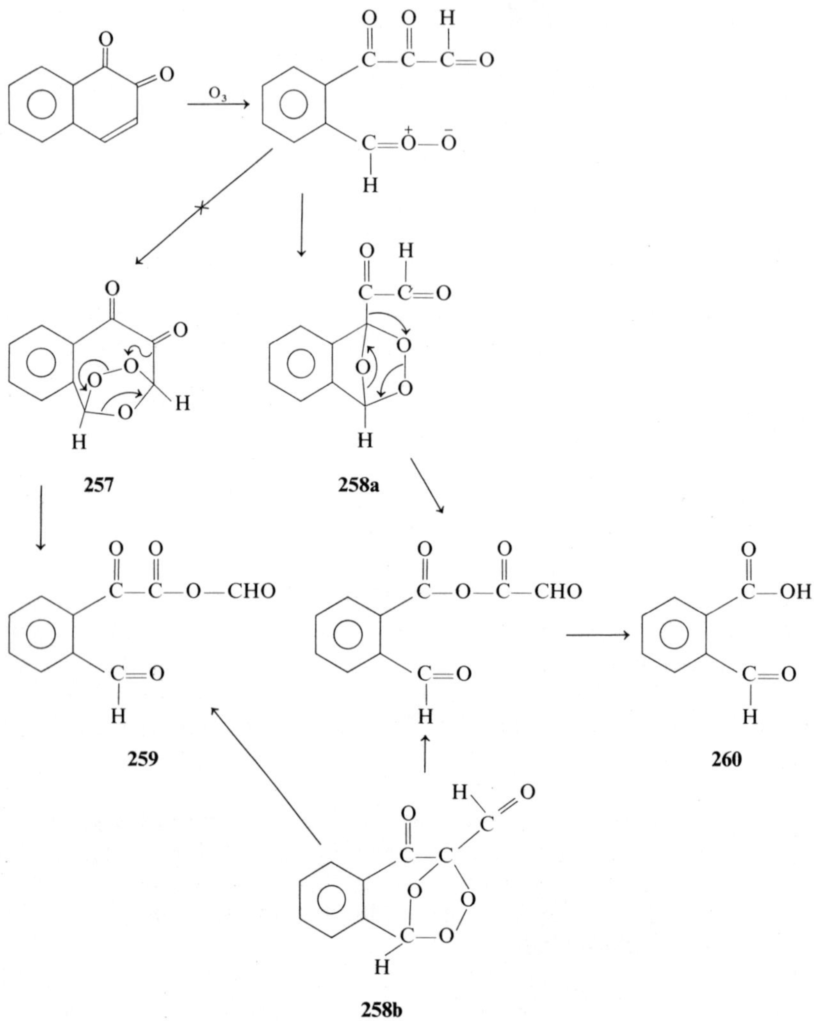

SCHEME 58. *o*-Naphthoquinone ozonolysis.

IV. 1,4-Naphthoquinones

Ozonolyses of 1,4-naphthoquinones (e.g., example 13, Table XXVII) require separate consideration because of the theoretical implications placed on them by Bernatek and co-workers (*123,124*) in regard to the directive influence of substituents in the cleavage of unsymmetrical olefins (Chapter VII, Section III). Example 13 of Table XXVII shows the type of anomalous product obtained in nearly quantitative yield from ozonolysis of 1,4-

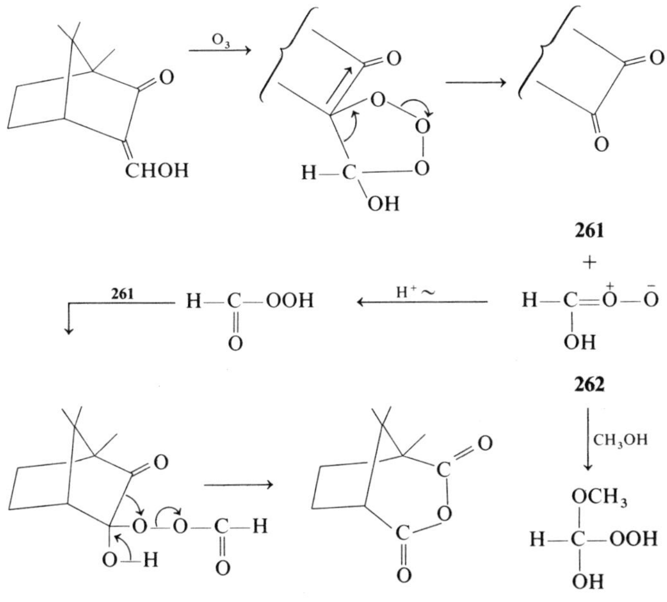

263 264

SCHEME 59

naphthoquinones in chloroform. With unsymmetrically substituted 1,4-naphthoquinones, the R and/or R′ groups exert a very strong directive influence as shown in Table XXVIII.

Bernatek (*511,512*) found that ozonolysis of 1,4-naphthoquinone, itself, resulted in approximately a 70% yield of carbon monoxide, during the ozonolysis, plus a 30% yield of a peroxidic ozonolysis product which remained in the reaction mixture and slowly decomposed to carbon monoxide and *o*-carboxyphenylglyoxal (**267**). On the other hand, treatment of the ozonolysis reaction mixture with water gave a nearly quantitative yield of *o*-carboxyphenylglyoxal (**267**) plus a 30% yield of formic acid. The peroxidic ozonolysis product could be isolated either as an oil or as a crystalline solid under certain conditions, but decomposed before it could be characterized (*511,512*). Similar, but less exact, results were obtained with 2-methyl- and 2,3-dimethyl-1,4-naphthoquinones (*512*). Because of these results, Bernatek and co-workers (*123,124,512,513*) concluded that the anomalous ozonolysis products, **267**, CO, etc., resulted from two competing mechanisms: rearrangement of a carbonyl oxide zwitterion (**265**), as suggested earlier by Criegee (e.g., *110*) and by the present author (*62*), and rearrangement of an ozonide (**266**), presumably as illustrated in Scheme 60. They considered the zwitterion rearrangement to be predominant (60–70%).

TABLE XXVIII

OZONOLYSIS OF 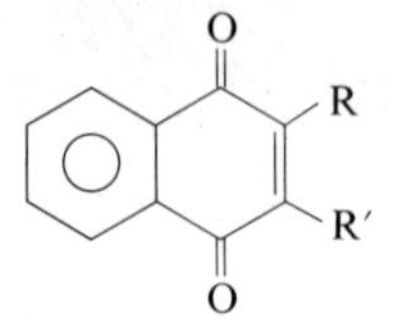

Example	R	R′	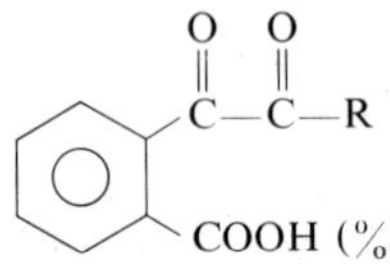 COOH (%)	References
1	CH_3	H	97	*124,512*
2	OCH_3	H	96	*124*
3	NH_2	H	95	*124*
4	O^-	H	90	*124*
5	NHAc	H	86	*124*
6	NMe_2	H	79	*124*
7	OAc	H	29	*124*
8	Ph	H	28	*124*
9	Cl	H	22	*124*
10	OH	H	10	*124*
11	CH_3	Cl	65	*513*
12	CH_3	OH	67	*513*
13	NH_2	Cl	95	*513*

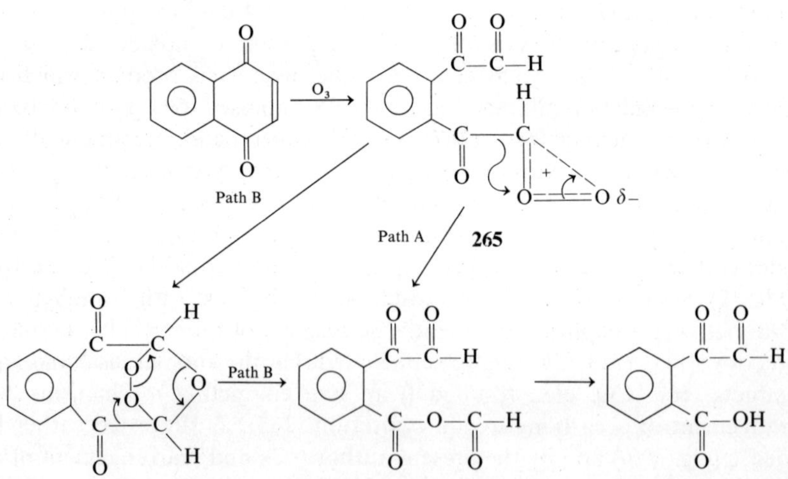

266 **267**

SCHEME 60

Moreover, since the unsymmetrical naphthoquinone results (Table XXVIII) showed that the substituent R, in **268**, exerted a strong influence in regard to which product, **270** or **272**, was predominent, Bernatek and co-workers (*123,124*) assumed that this was determined largely by which carbonyl oxide, **269** or **271**, was preferentially produced and subsequently rearranged, as illustrated in Scheme 61. On this basis, they developed a complicated theory to explain their results (*124*). However, their conclusions concerning the directive influence of substituents are so different from those discussed in Chapter VII, Section III, that their assumption as to the major source of the anomalous products cannot be accepted. For instance, with example 1, Table XXVIII, the predominant carbonyl oxide from **268** (R = CH_3) would have to be **269** (R = CH_3), which is completely contrary to previous findings with 1,2-dibenzoylpropene (*117,120*) and other alkyl-substituted ethylenes (see Chapter VII, Section III).

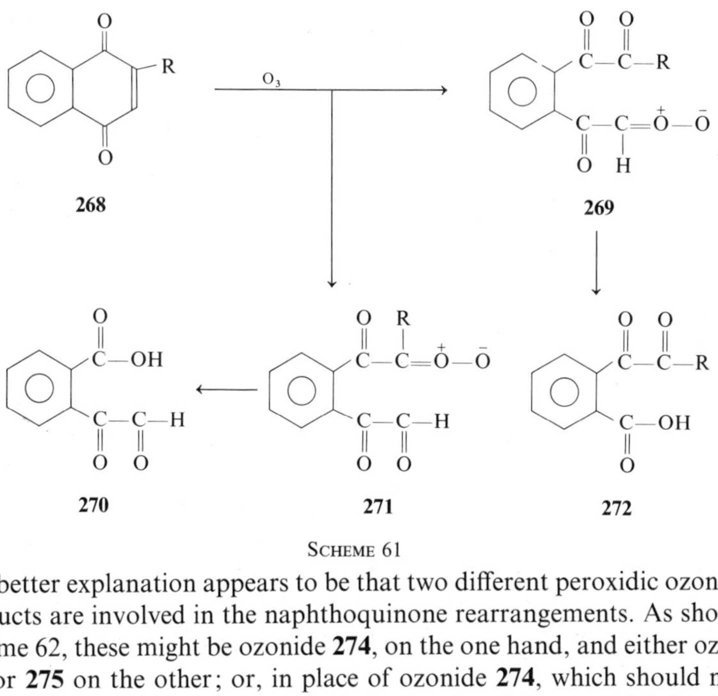

268 **269** **270** **271** **272**

SCHEME 61

A better explanation appears to be that two different peroxidic ozonolysis products are involved in the naphthoquinone rearrangements. As shown in Scheme 62, these might be ozonide **274**, on the one hand, and either ozonide **273** or **275** on the other; or, in place of ozonide **274**, which should not be formed preferentially (see Chapter V, Section I and Chapter VII, Section I), an oligomer of **269** and/or **271** might be the major product, which decomposes spontaneously. In any event, the substituent R not only could play a role in determining which carbonyl oxide is produced preferentially, but also in influencing the direction of ozonide rearrangement (i.e., which group migrates). At this stage the problem is too complicated to warrant further speculation.

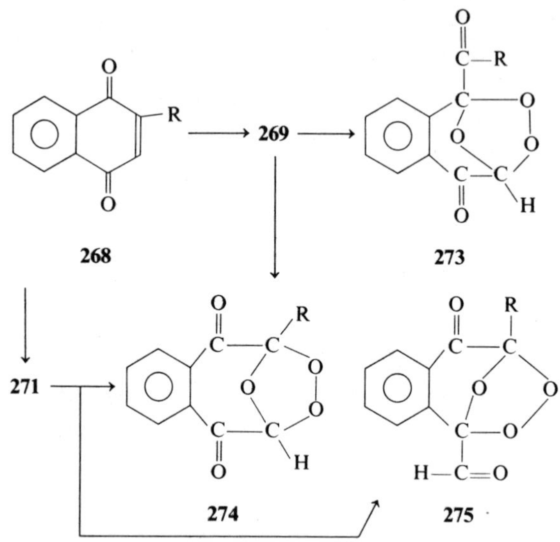

SCHEME 62

V. Cinnamic Esters

Although cinnamic esters can be classified as α,β-unsaturated carbonyl systems, they are treated separately because the rearrangement that occurs involves the aryl group rather than the carbonyl group of the system, and the anomalous product is a phenol (*62,92,395,467,515,516*).

Späth *et al.* (*515,516*) were the first to observe anomalous ozonolysis products from cinnamic esters. In all cases, the aryl group was highly substituted with methoxy and alkyl groups, as illustrated in Table XXIX. Thus, the aryl groups should have high migratory aptitudes, suggesting the mechanism illustrated in Scheme 63, Eq. (a). Although it was shown that the aldehydes obtained as normal ozonolysis products also react with ozone to give phenols, the yields in most cases were much lower than those obtained from the cinnamates (*516,517*); thus, aldehyde oxidation was not a major source of the anomalous results with the cinnamates.

Kolsaker and Bailey (*395*) have made a detailed study of ozonolysis of cinnamic esters and similar compounds in methanol. Some of the results are shown in Table XXX.

A comparison of the phenol yields in examples 1 and 3 of Table XXX gives a clue concerning the mechanistic course of the reactions. If the phenol yield arises solely from rearrangement of hydroperoxide **276** [Scheme 63, Eqs. (b) and (c)], the yield in example 3 should not be greater than that in example 1, since example 1 should represent the maximum extent to which

TABLE XXIX

ANOMALOUS PRODUCTS FROM OZONOLYSES OF CINNAMATES (516)

Example	Ar in ArCH=CHCOOCH$_3$[a]	Yield of ArOH (%)
1	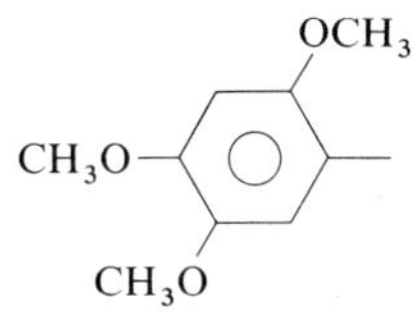	48
2	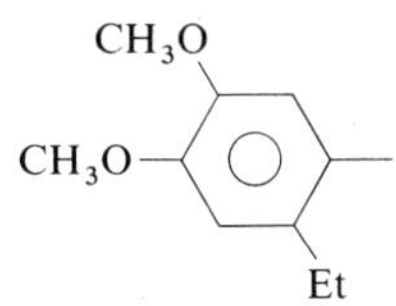	46
3	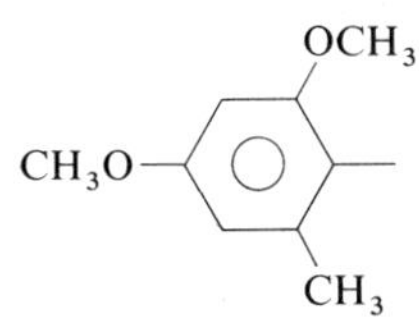	27
4	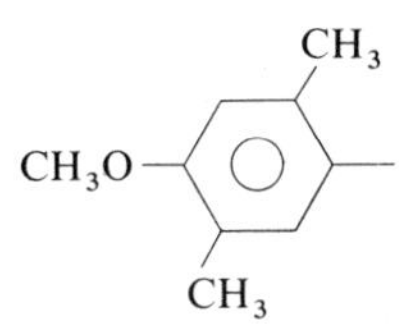	22
5	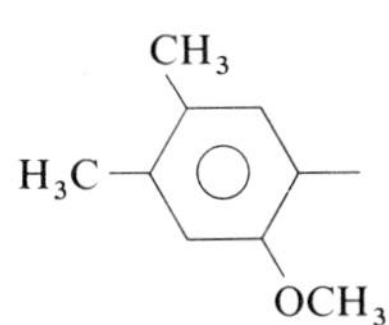	16
6	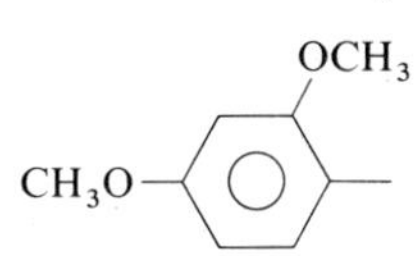	15
7	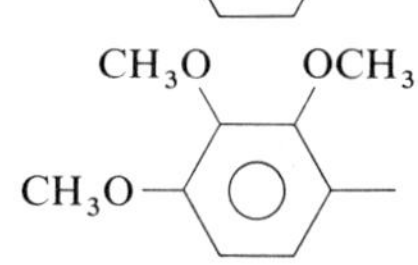	14
8	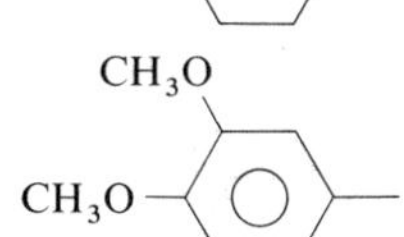	Trace

[a] Ozonized in CHCl$_3$.

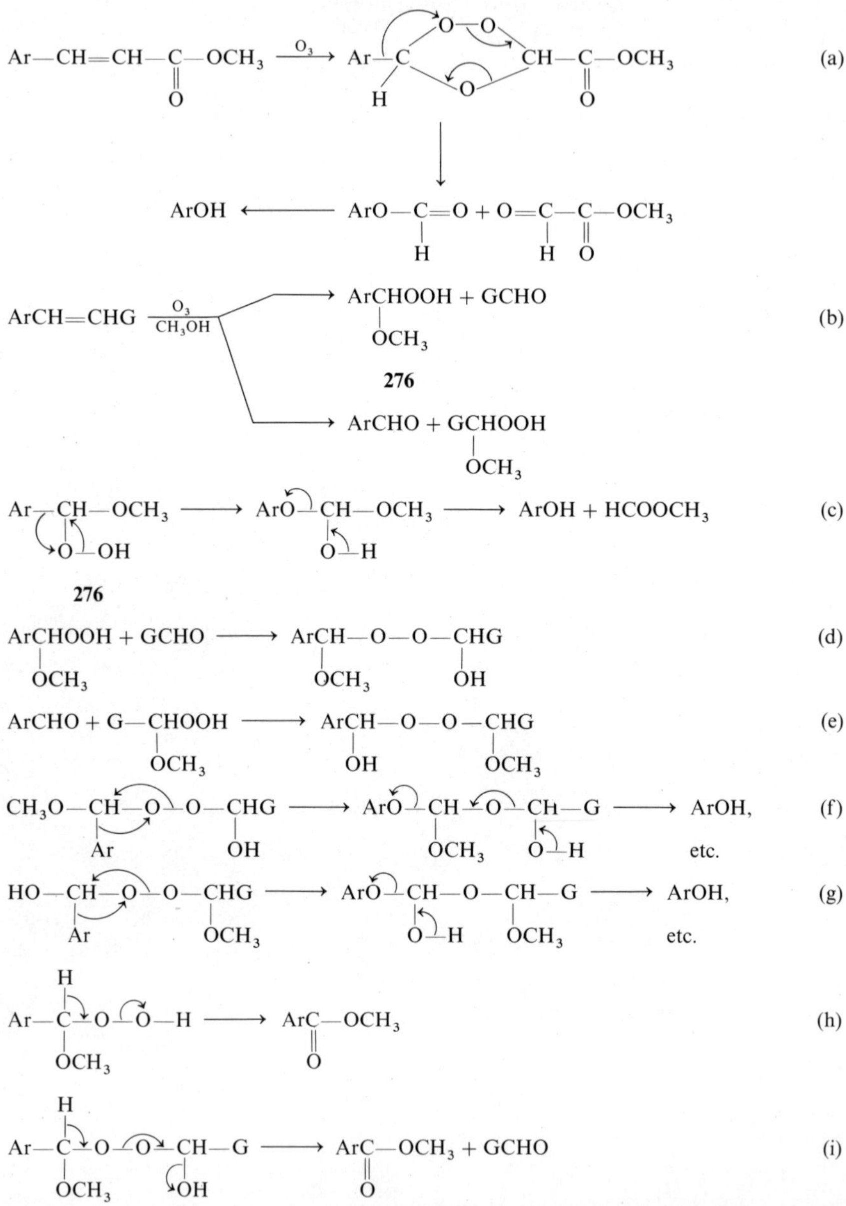

SCHEME 63

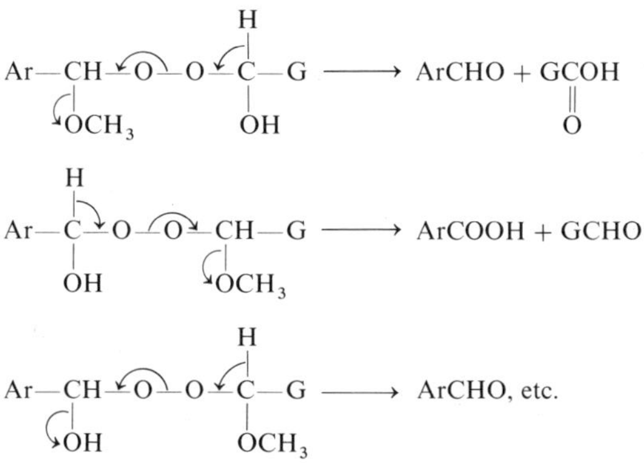

(j)

(k)

(l)

SCHEME 63 (*Continued*)

TABLE XXX

OZONOLYSES OF ArCH=CHG[a] IN METHANOL (*395*)

Example	Ar	G	Products (% yield)[b] ArOH	ArCHO	ArCOOH / $ArCOOCH_3$	$COOH(CH_3)$ / $COOH(CH_3)$
1	*p*-MeO–C_6H_4—	*p*-MeO–C_6H_4—	53	111	34	—
2	*p*-MeO–C_6H_4—	$COOCH_3$	40	25	29	51
3	*p*-MeO–C_6H_4—	CH_3	70	15	6	c)
4	2,4-diMeOC_6H_3—	$COOCH_3$	60	22	7	c)

[a] Work-up by letting the neutral reaction mixture decompose over a period of several days, followed by base extraction and gas–liquid chromatography.
[b] Total yield should equal 200.
[c] Products from G side not determined.

276 rearranges. Therefore, reactions such as those represented by Scheme 63, Eqs. (d)–(g) must also be occurring. These, along with Eqs. (h)–(l) of Scheme 63 explain all the data obtained (*395*). Equations (f) and (g) are typical peroxide rearrangements involving 1,2-shifts to positive oxygen. The higher yield of the phenolic product in example 4 than in example 2 of Table XXX is indicative of the greater migratory aptitude of the more highly methoxylated aryl group.

Acid and base decompositions of the ozonolysis mixtures were also performed (*395*). The yield of phenol was greater under acidic conditions, as would be expected from an acid-catalyzed Baeyer–Villiger-type rearrangement. With base decompositions, no phenol was obtained, indicating that reactions such as those represented by Scheme 63, Eqs. (h)–(l) became predominant, as would be expected, under base-catalyzed conditions.

VI. Miscellaneous Types

A. CAMPHENE AND OTHER CYCLIC SYSTEMS

Other types besides those mentioned in the preceding sections give major or minor amounts of anomalous ozonolysis products. One of the best-known examples is camphene (**277**), which under usual ozonolysis conditions gives largely lactone **279**, hydroxy acid **281b**, and an unsaturated acid (probably

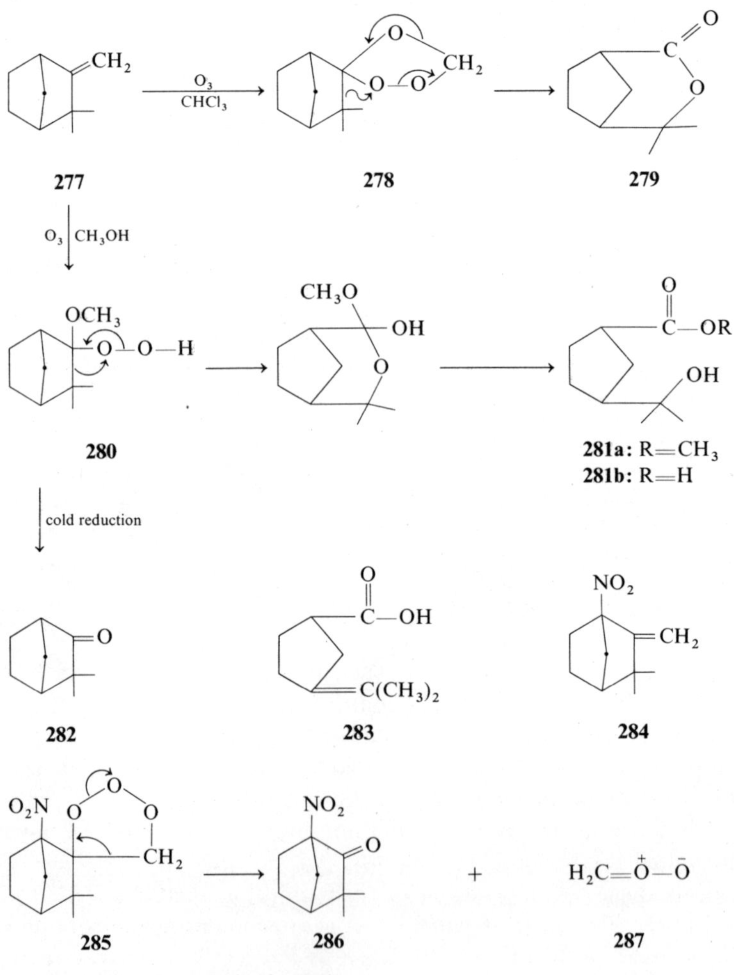

SCHEME 64. Ozonolysis of camphene.

283), when ozonized in chloroform (*518*), hydrocarbons (*108*), or acetic acid (*89*), and a hydroxy ester (**281a**), when ozonized in methanol (*62,108*). These rearrangements undoubtedly involve the highly unstable (*108*) ozonide (**278**) and the methoxyalkyl hydroperoxide (**280**), as illustrated in Scheme 64.

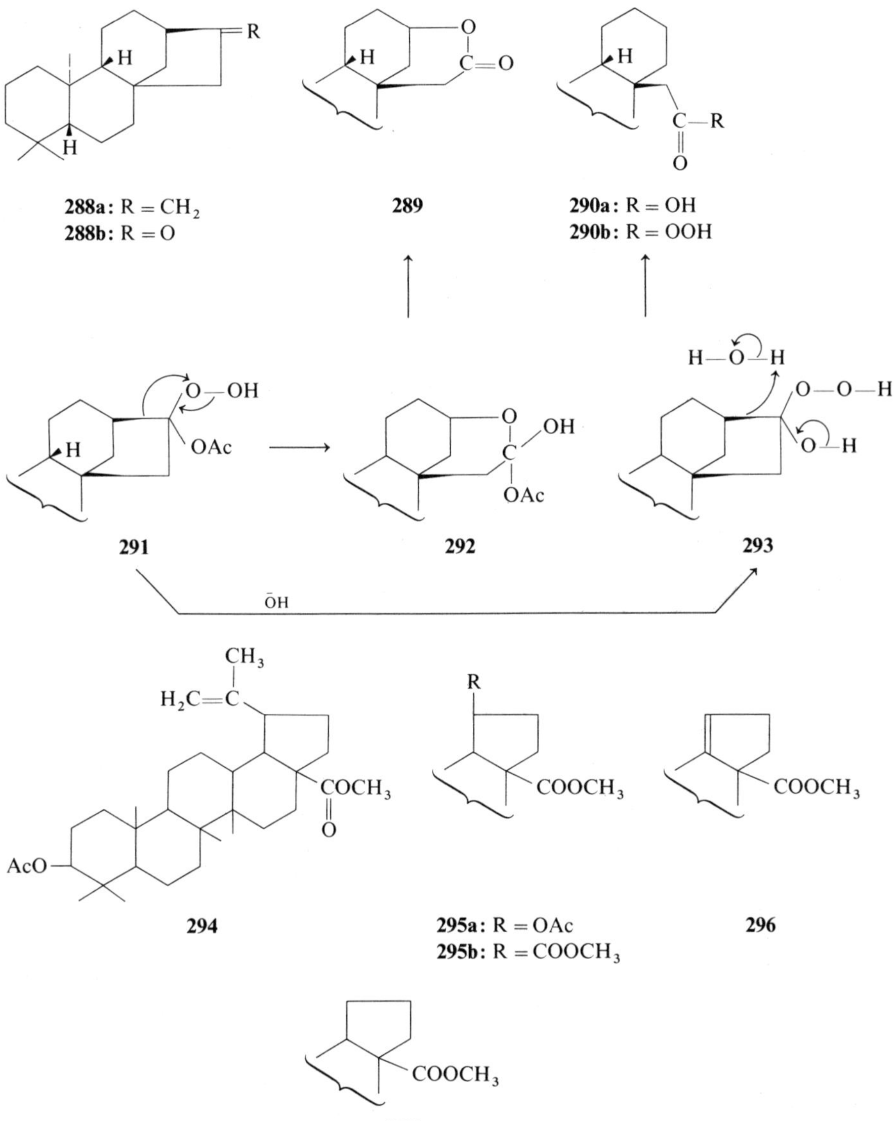

SCHEME 65

Reduction of the cold methanolic ozonolysis solution containing **280** provided the "normal" ozonolysis products, camphenilone (**282**), in 70–75% yield (*108*,*519*). The ease with which these rearrangements occur must be due to relief of strain energy. It is interesting that 1-nitrocamphene (**284**) is reported to give the corresponding camphenilone (**286**) in 91% yield under conditions that afford only "anomalous" products from camphene (*520*). This difference probably results from the effect of the electron-withdrawing nitro group in the primary ozonide (**285**), causing it to decompose to the camphenilone (**286**) and carbonyl oxide **287**. These moieties should not reunite to afford the ozonide (see Chapter V, Section I, A). Obviously, just the opposite pair is produced from camphene itself (*108*).

Another interesting and somewhat similar example is (−)-kuarene (**288a**), which upon ozonolysis in acetic acid gave, along with the major and expected product, **288b**, 5–6% yields of compounds **289** and **290a** (Scheme 65), after a hydrolytic work-up (*521*). Lactone **289** must have been produced as shown by **291** → **292** → **289**; this is analogous to one of the rearrangements in Scheme 64. The formation of **290a**, however, is novel. A possible explanation, similar to one suggested by the author (*521*), is hydrolysis of **291** to **293** under the basic conditions employed, followed by decomposition of **293** to **290b** and of **290b** to **290a**. Another possibility is a radical decomposition of **291**, analogous to that in Scheme 51. A similar example has also been reported in the steroid field, where ozonolysis of **294** in the presence of methanol gave, along with the expected ketone, 4–6% yields, each, of **295a**, **295b**, **296**, and **297** (*522*). The reaction leading to **295a** and its elimination product, **296**, is analogous to that yielding **289** from **288a**; product **295b** also is a rearrangement product. The production of compound **297** is analogous to that of compound **290a** from **288a**.

Ozonolyses of some other pinene-type terpenes also have resulted in similar minor anomalous products, but it is quite possible that these arose during the oxidative work-up (*523–525*).

In regard to the lactones just described from ozonolyses of camphene and (−)-kuarene, i.e., **279** and **289**, it is noteworthy that a minor product from ozonolysis of (−)-pin-2(10)-ene (**298**), previously reported to be lactone **299** (*526*), has recently been reformulated as bis(6,6-dimethylnorpinane-2,2-diyl) diperoxide (**300**) (*527*).

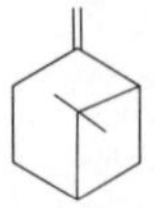

298

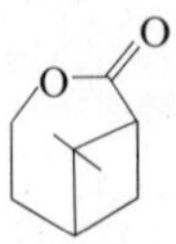

299

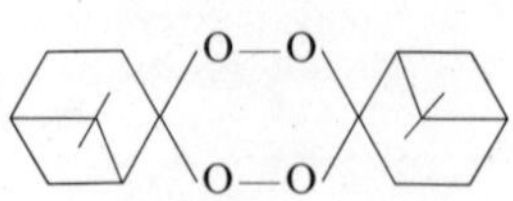

300

One abnormal ozonolysis product became so generally accepted that it was considered normal and characteristic. This is geronic acid (**302**) from ozonolysis of carotenoids with a β-ionone type ring, e.g., **301** (*528,529*). Recently, however, it has been shown that upon careful ozonolysis with 1 mole-equivalent of ozone followed by a reductive work-up, compound **303** is produced in good yield (*529*). Geronic acid (**302**), therefore, is an anomalous ozonolysis product of an α,β-unsaturated aldehyde, **303**. Surprising about this reaction is that the double bond closest to the carbonyl group of **301** is attacked in preference to the ring double bond (cf. Chapter IV, Section II, A). Geronic acid also has been obtained as a minor anomalous ozonolysis product from carotenoid types with an unconjugated diene system, e.g., **304** (*530*). This can arise as shown in **305**, followed by hydrolysis. 2,2-Dimethylpentanedioic acid (**306**), also has been observed as an anomalous ozonolysis product of β-ionone-type carotenoids, e.g., **301**, but this probably results from overozonation and/or an oxidative work-up (*528*).

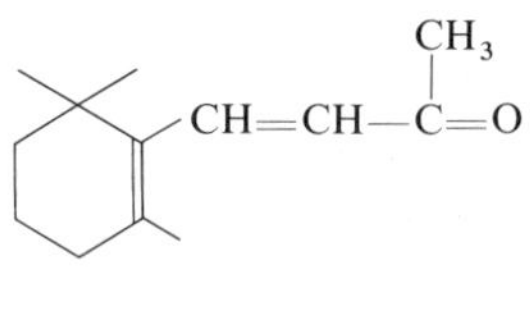

301

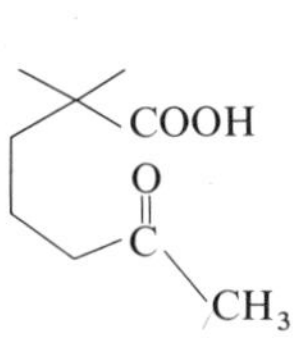

302

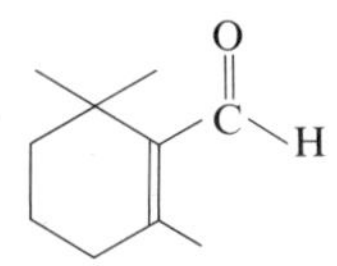

303

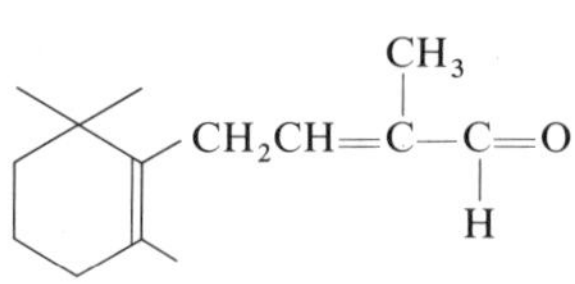

304

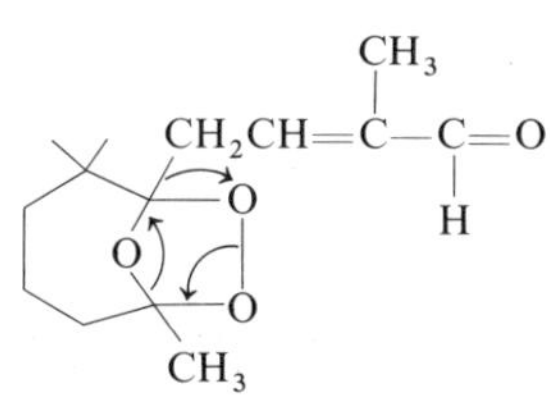

305

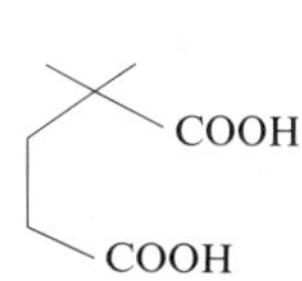

306

A number of other interesting anomalous ozonolyses of terpenoid compounds can be found in the literature (e.g., *467,531*), but these could well involve Baeyer–Villiger oxidations of normal ozonolysis products.

An unusual, but understandable, anomalous ozonolysis has been reported for bicyclo[2.2.2.]octadiene-2,5 (**307**) (*532*). The product, after a silver oxide work-up, was succinic acid. Silver oxide has been shown by Asinger (*533*) not to cause abnormal results with simple olefinic systems. The probable mechanism is outlined in Scheme 66.

B. SIMPLE OLEFINS

Anomalous ozonolysis products also arise from simple olefins in minor amounts. For example, careful ozonolyses of *trans*-1,2-di-*tert*-butylethylene

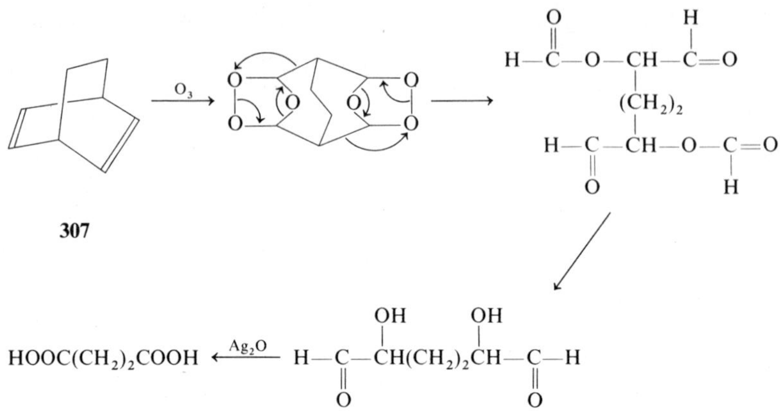

307

SCHEME 66

and of *trans*-2,2-dimethyl-3-hexene afforded small amounts of *tert*-butyl formate, as determined by gas chromatography (*128,397*). This could have arisen by rearrangement of the ozonide, as discussed in other cases and as shown in Eq. (20). Also a radical-type mechanism could be involved owing to thermal decomposition in the gas chromatograph (see next subsection).

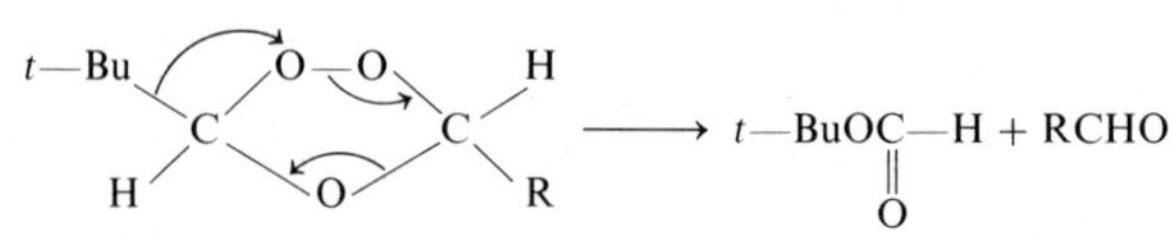

(20)

Ziegler *et al.* (*534*) have obtained various amounts of anomalous ozonolysis products by careful ozonolyses followed by the oxidative work-up with silver oxide developed by Asinger (*533*). The abnormal products included 2% pentanoic acid and 1% butanoic acid from 6-dodecene (**308**), and 13% of a mixture of 5-decanol (**310**) and 5-decanone, along with 1% acids of 4–6 carbon atoms, from 3-butyl-1-octene (**309**). Other abnormal ozonolyses reported by Ziegler (*534*) and by Marvel *et al.* (*535*) with cyclohexene and 4-vinylcyclohexene, described in the earlier review (*62*), involved a hydrogen peroxide-acetic acid work-up; thus, the anomalous products could have arisen from Baeyer–Villiger oxidations at that stage. The same is true in later work with 4-vinylcyclohexene (*536*). Hydrogen peroxide–acetic acid work-up was shown to increase the yields of abnormal products from 6-dodecene and 3-butyl-1-octene (*534*). The origin of pentanoic acid from ozonolysis of 6-dodecene, however, probably involves the same mechanistic route discussed throughout this subsection, since it was shown that the normal product, hexanal, was oxidized cleanly to hexanoic acid by silver oxide (*534*). The 5-decanol from 3-butyl-1-octene (**309**) could have arisen either by the ozonolytic rearrangement under discussion or, perhaps, by

$CH_3(CH_2)_4CH{=}CH(CH_2)_4CH_3$

308

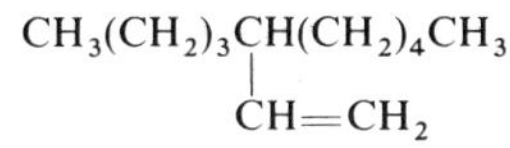

309

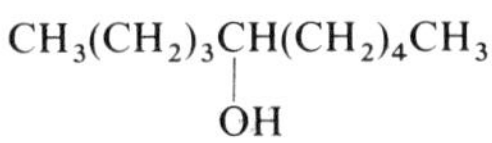

310

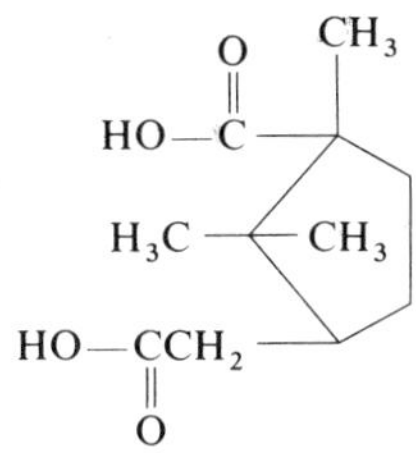

311

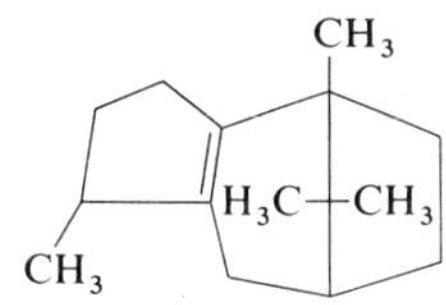

312

silver oxide oxidation of the normal product (*534*). The 5-decanol probably was oxidized to 5-decanone either by the silver oxide or by ozone.

Although most of the anomalous ozonolysis products above can be explained either by the rearrangement of peroxidic ozonolysis products, illustrated repeatedly in this section, or by a Baeyer–Villiger reaction during an oxidative work-up, certain other examples of anomalous ozonolyses are not of these types. Ziegler's isolation of small amounts of butanoic acid from ozonolysis of 6-dodecene and of low molecular weight acids from 3-butyl-1-octene ozonolysis (*534*) fall into this category and must have involved a competing ozone attack, or other oxidation at a secondary or tertiary C—H bond (to be discussed in **Vol. II**). Additional reported instances that also must have involved sp^3 carbon–hydrogen oxidation, assuming that the starting materials were pure, are small amounts of formaldehyde from 5-undecene (*537*) and compound **311** from **312** (*538*). Other cases of anomalous formaldehyde formation (*539,540*) may also have resulted by this pathway, or may have been due to impure starting materials.

Hydrogen, alkanes, alkenes, and similar compounds have been reported as anomalous ozonolysis products from various olefinic compounds (*53,62, 116,311,358,427,433,446,449,541–546*). For example, Greiner and coworkers (*544,545*) obtained some decane, decene, and eicosane from ozonolysis of 1-dodecene (**313**) after reductive, hydrolytic, oxidative, or thermal decomposition of the ozonide; and Stoll and Rouvé (*546*) obtained a small amount of **315** from ozonolyses of brassidyl and erucyl acetates (**314**). Although the conditions were other than thermal, the explanation must involve a competing thermal decomposition during the work-up. By analogy to the mechanisms discussed in Chapter VIII, Section V, these reactions can be rationalized as shown in Scheme 67 (cf. Scheme 47). Another reaction involving a competing thermal decomposition was the direct formation of phthalic anhydride in 19% yield from ozonolysis of 1,2-naphthoquinone

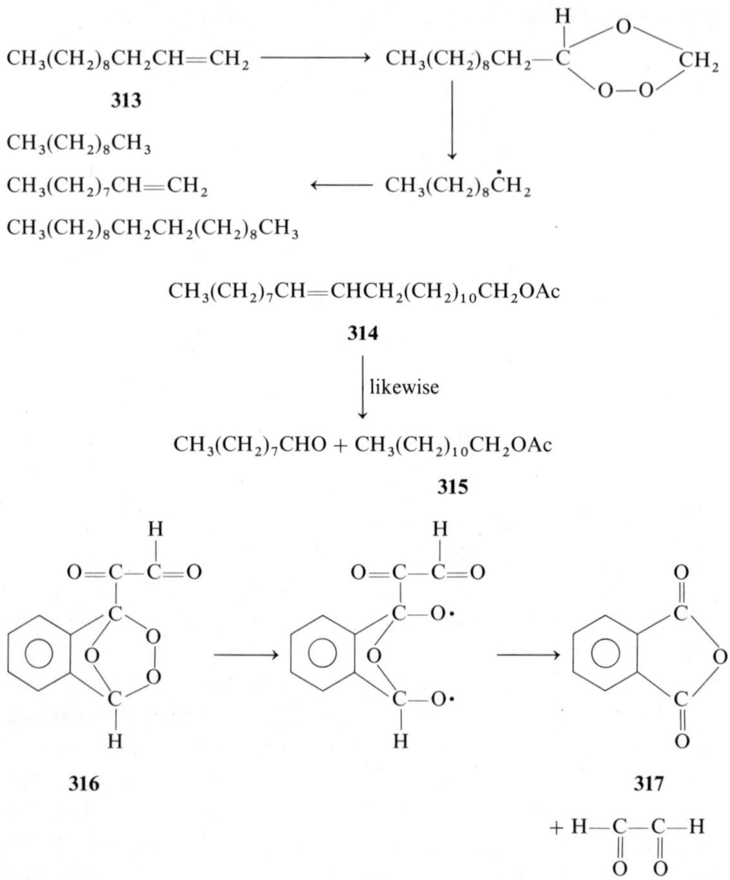

SCHEME 67. Minor ozonide decompositions.

(*510*) (see also Scheme 58 and Table XXVII, Section III of this chapter). Assuming that the labile peroxidic ozonolysis product was ozonide **316**, (**258a** in Scheme 58) the formation of phthalic acid (**317**) probably occurred as shown in Scheme 67 (cf. Scheme 47).

It is noteworthy that Privett and Nickell (*311*,*449*) obtained hydrocarbons and other anomalous products from ordinary thermal decompositions of ozonides, but that when the decompositions were carried out in the presence of hydrogen and a catalyst, no anomalous products were obtained. Presumably, intermediates such as **229** (Scheme 47) were reduced.

C. MISCELLANEOUS

Sometimes surprising, and therefore anomalous, ozonolysis products are obtained as the result of further unexpected reactions occurring with the expected products. Three examples will be given. The first deals with the isolation of considerable amounts of acetaldehyde along with acetone from ozonolyses of humalone (**318a**) and cohumalone (**318b**) (*547,548*). This, perhaps, is understandable if one considers that ozonolysis of the enol structures (**319**) of these compounds should yield **320**, which might yield some acetaldehyde via decarboxylation and oxidation, as shown in Scheme 68.

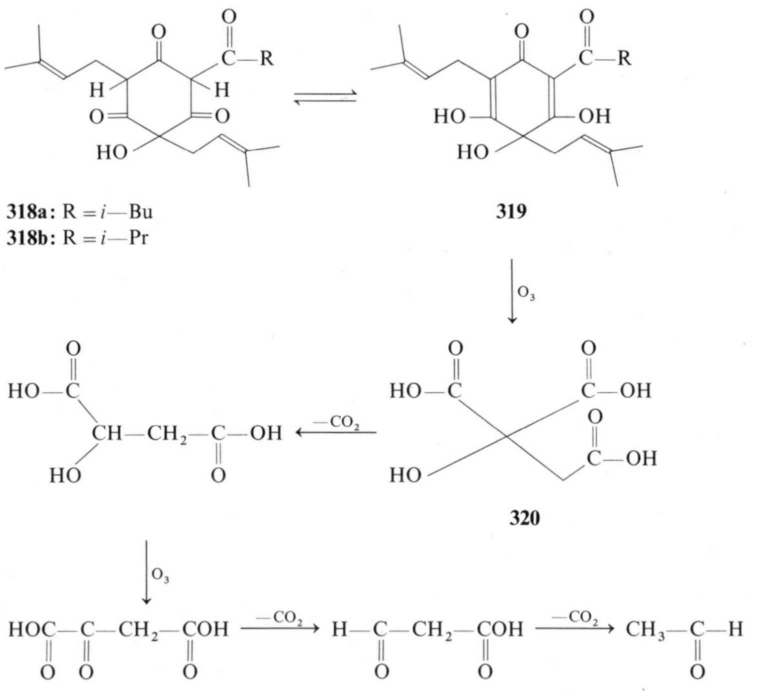

SCHEME 68

Both acetaldehyde and acetone also were, surprisingly, obtained from ozonolysis of compounds **321** and **322** (*549*). Further study showed that the anomalous product resulted as shown for **321** in Scheme 69.

The third of these curious results involved the hydrolysis of stable ozonide **323** to give benzoic acid and **325** (*550*). As shown in Scheme 70, rearrangement of **323** should give **324** instead of **325**. However, a base extraction was employed in the work-up, and this must have brought about a Cannizarro reaction with **324** to give **325**.

D. FEIST'S ACID

The structures of Feist's acid and its diethyl ester (*551,552*) and a rationalization of the results of ozonation of the diester has been a problem for many years (*62*). These substances were first assigned structure **326** because ozonation of the diethyl ester gave no formaldehyde, but rather a product

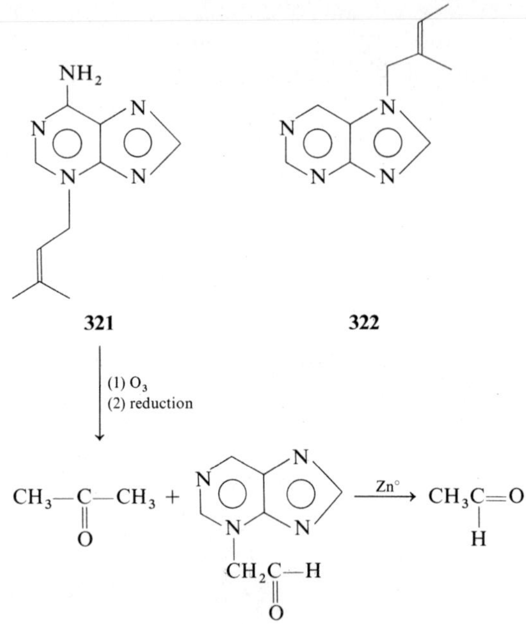

321 322

SCHEME 69

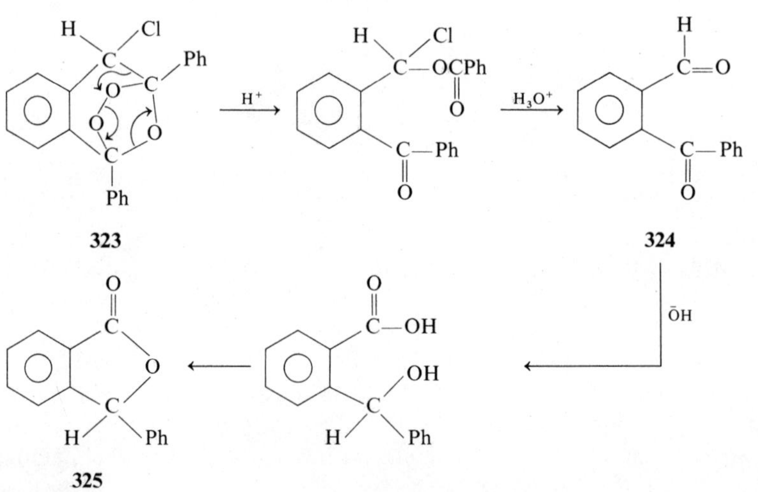

323 324 325

SCHEME 70

formulated as **327b** (*553*). Later, it was established that the structures of Feist's acid and its ester were as shown by **328** (*554–557*); the problem then changed to an effort to account for the strange ozonolysis result. A long, involved mechanistic rationalization was attempted in 1956 (*557*). Ten years later, Erickson (*558*) showed that the ozonolysis product of **328c** was not **327c**, but rather a mixture of products **329**, **331**, and **332**. Erickson and Mercer (*559*) proposed the mechanisms shown in Scheme 71 to rationalize the formation of these products of ozonation of **328c**. The epoxide, **332**,

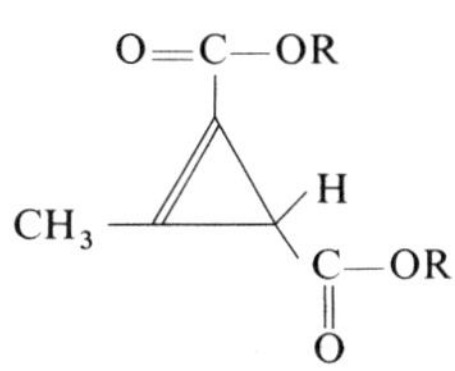

326a: R = H
326b: R = Et

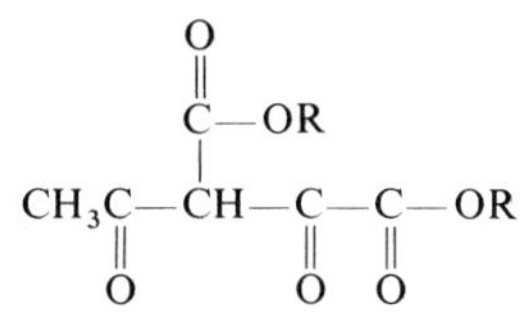

327b: R = Et
327c: R = Me

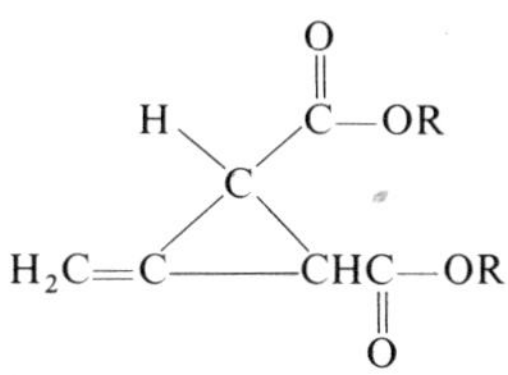

328a: R = H
328b: R = Et
328c: R = Me

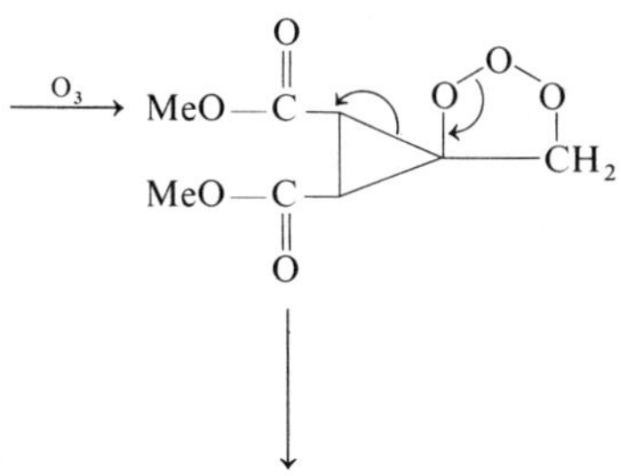

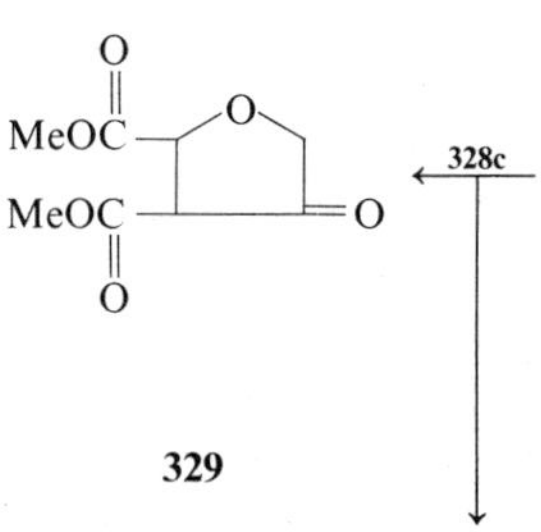

329

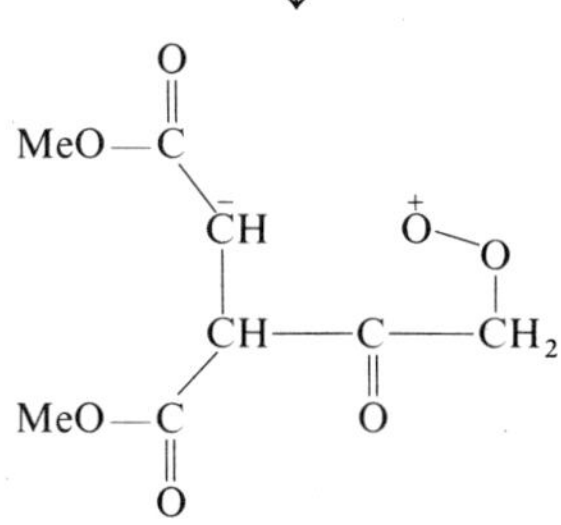

330

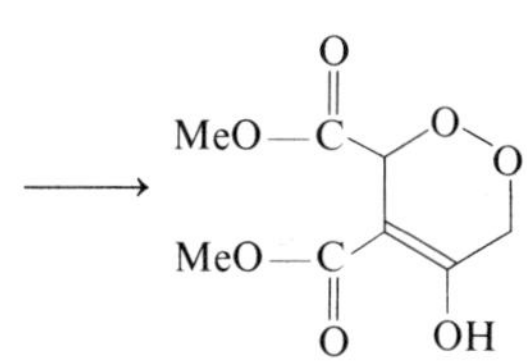

331

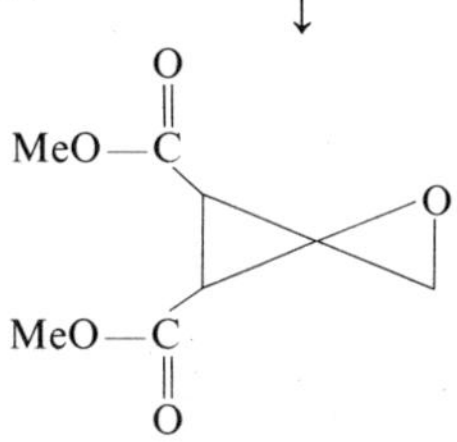

332

SCHEME 71. Feist's ester ozonation products.

was shown to be a secondary product, from epoxidation of **328c** by some species such as **330**.

Another anomalous ozonolysis product which is difficult to explain is an ester, such as **334**, as a minor but significant product from ozonolysis of **333** (*560–565*). The origin has been ascribed to a complicated conversion of carbonyl oxide zwitterions to peracids, and on to a diacyl peroxide, which loses carbon dioxide to produce the ester (*563–565*). If this actually is the mechanism, it must involve thermal decomposition of a peroxide (Chapter VIII, Section V; this chapter, Section VII).

$$R(CH_2)_nCH{=}CH(CH_2)_nR' \qquad\qquad R(CH_2)_n\underset{\overset{\|}{O}}{C}{-}O(CH_2)_nR'$$

333 **334**

Finally, it should be noted that in the past certain reports of so-called abnormal products have been due to impurities rather than to rearrangements or other causes (*566*). This even led one investigator to believe that an ozonide linkage could move from one position to another (*567*) and caused others to cast doubt on the reliability of ozonolysis for structural determination (*568*). Also, the dissatisfaction of certain other workers (*569*) with the ozonolysis method could have been avoided had they used more suitable reaction conditions (*570*), as suggested at the beginning of this chapter.

VII. Thermal and Photochemical

In Chapter VIII, Section V, thermal and photochemical decompositions of peroxidic ozonolysis products were discussed. Since what should be considered "normal" and "abnormal" in these reactions is somewhat nebulous, this section is, to some degree, only an extension of that section.

In some cases peroxidic rearrangements occur that resemble, in products formed, the ionic Baeyer–Villiger-type rearrangements discussed in the preceding sections. Examples are shown in Scheme 72.

On standing, the simple ozonide **335** decomposed to *tert*-butyl acetate (**337**), formaldehyde, and pinacolone diperoxide (*109*) (Chapter V, Section II). There is no evidence from which to decide whether the rearrangement to **337** is ionic, as shown in **335** and discussed in the preceding sections, or of a radical nature, as shown by **335** → **336** → **338** → **337** and, thus, analogous to the reactions of Chapter VIII, Section V. The formation of the diperoxide involved the Criegee carbonyl oxide, either as a zwitterion or as a diradical. Rearrangements similar to **335** → **337** were also reported by

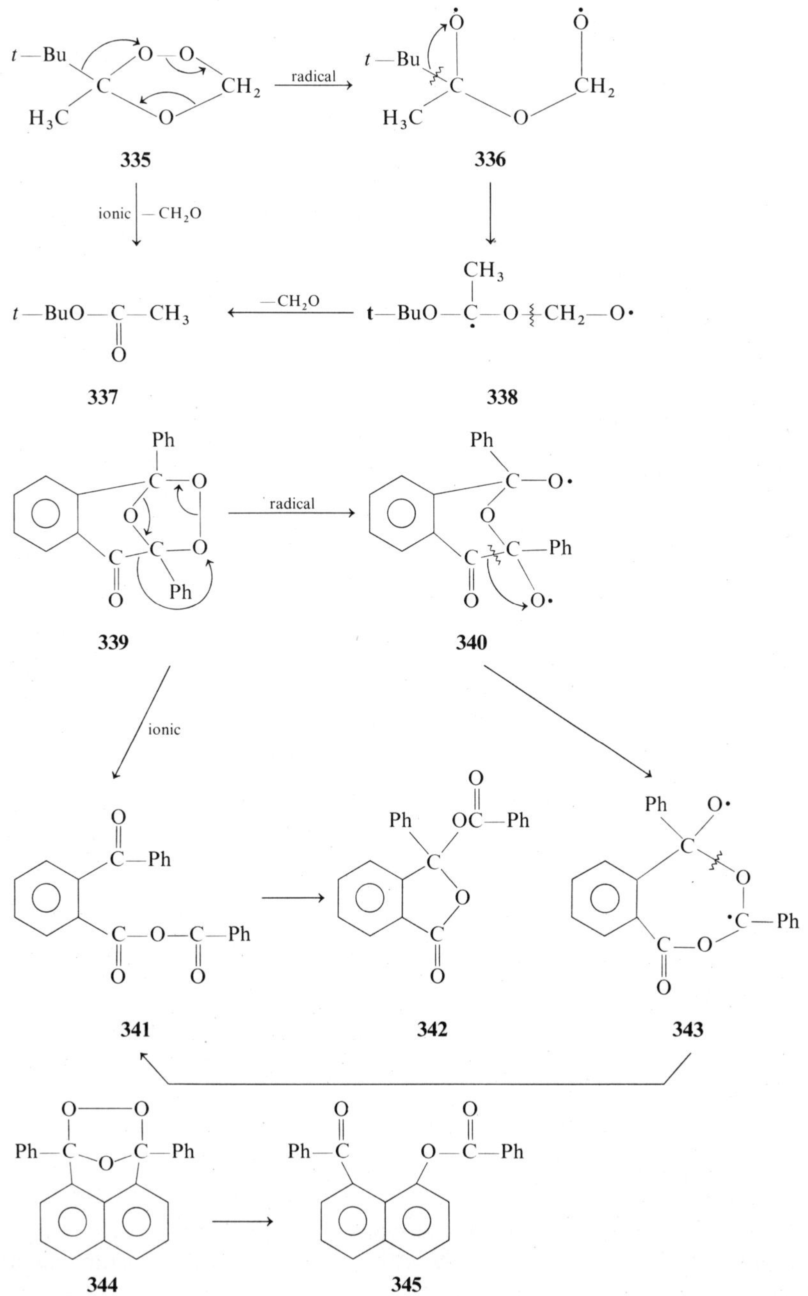

SCHEME 72. Thermal and photolytic ozonide rearrangements.

Razumovski *et al.* (*427,446*), who assumed that the decompositions involved the zwitterion.

The indenone ozonide rearrangement (e.g., **339** to **341** and/or **342**) has occurred in solution (*383*), on standing or heating (*383*), or under photolytic conditions (*571*). Again, one cannot be certain whether the thermal decomposition is of the Baeyer–Villiger type (**339** → **341**; see also Scheme 56) or of the radical type (**339** → **340** → **343** → **341**). The photorearrangement (*571*) would be expected to be of the latter type, however, as also suggested for the photorearrangement of **344** to **345** (*452*). The rearrangement of **341** to **342** is thermal (*571*) and might be either ionic or radical.

The examples that follow are sufficiently different from those shown in Scheme 72 and from the examples in Chapter VIII, Section V to wait until now for discussion.

Photolysis of hexamethyl–Dewar–benzene ozonide (**346**) not only gave the dimer of tetramethylcyclobutadiene via di-β cleavage of diradical **347** (cf. Scheme 49, Chapter VIII, Section V), but also gave **348** via mono-β cleavage of **347**, as shown in Scheme 73 (*452*). Similarly, ozonide **349**

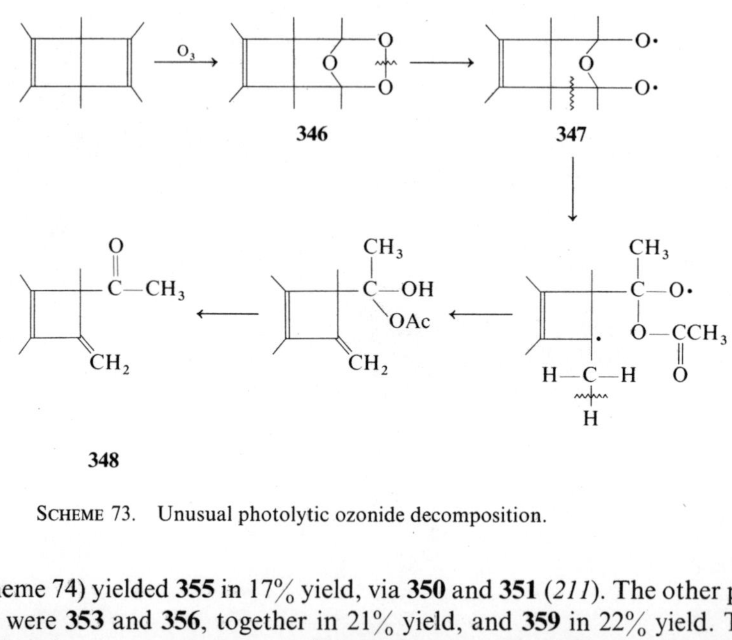

SCHEME 73. Unusual photolytic ozonide decomposition.

(Scheme 74) yielded **355** in 17% yield, via **350** and **351** (*211*). The other products were **353** and **356**, together in 21% yield, and **359** in 22% yield. These resulted from rearrangements of the types shown in Scheme 72, involving **352** and **354** (*211*). Compound **359** is the result of further rearrangements of **353**, via **357** and **358**.

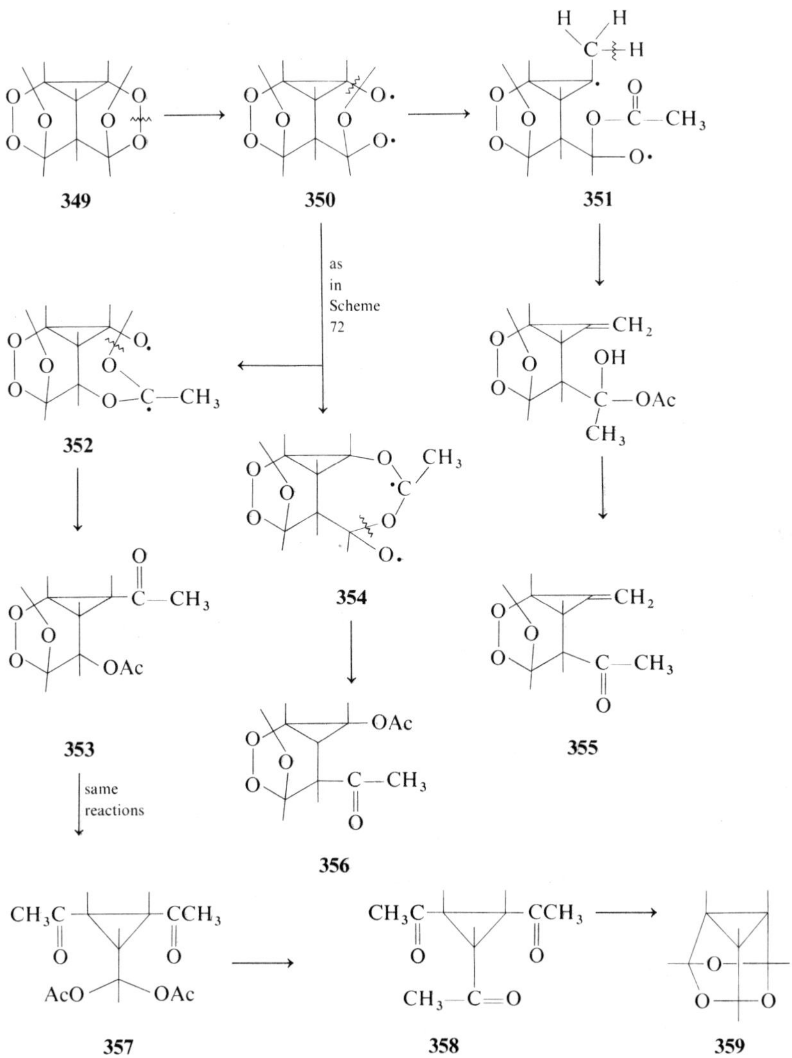

SCHEME 74. Photolysis of octamethylsemibullvalene diozonide.

Various products similar to those discussed in this section and in Chapter VIII, Section V, have also been obtained from thermal decompositions of α-methoxy- or α-hydroxy-alkyl hydroperoxides (*269,433*). These, too, appear to involve homolytic cleavage of the peroxide linkage.

CHAPTER X

"Special" Liquid-Phase Ozonolyses

Certain olefinic compounds, which are not of the "anomalous" type or types considered in Chapter IX, give novel results upon ozonolysis. Included in this category are haloethylenes, oxyphosphoranes, allenes, cycloheptatrienes, cyclooctatetraene, and benzvalene.

I. Haloethylenes

As discussed in Chapter IV, Section II, A, halogens directly attached to a double bond greatly decelerate the rate of ozone attack (e.g., *157*). In competitions between halogenated and nonhalogenated double bonds, ozonolysis appears to occur exclusively at the latter (*195,203,260,262*).

Many reports appear in the literature concerning the production of carboxylic acids by ozonolysis, followed by a hydrolytic work-up, of olefins with a halogen on the double bond (e.g., *169a,572–575*). Subsequent to these reports, Griesbaum and Keul (*576*) have shown that olefins of type **360** yield 2 moles of methyl esters of carboxylic acids when ozonized in methanol. Their mechanistic formulations are shown in Eq. (21). One ester arose from the acyl halide and the other from the methoxyalkyl hydroperoxide (*576*).

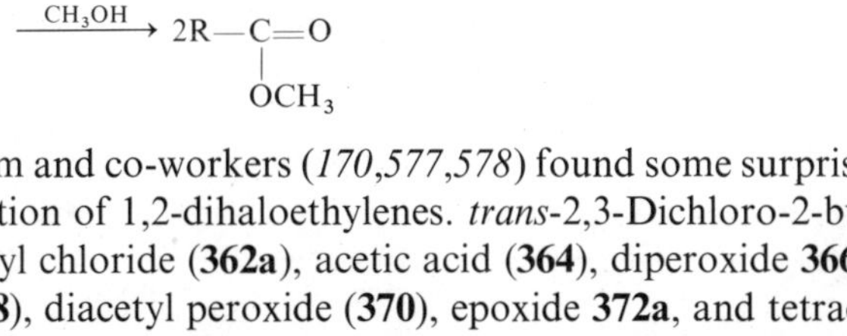

360 (21)

Griesbaum and co-workers (*170,577,578*) found some surprising products from ozonation of 1,2-dihaloethylenes. *trans*-2,3-Dichloro-2-butene (**361a**), yielded acetyl chloride (**362a**), acetic acid (**364**), diperoxide **366a**, acetic anhydride (**368**), diacetyl peroxide (**370**), epoxide **372a**, and tetrachloroethane (**367a**), as shown in Scheme 75 (*578*). From *trans*-2,3-dibromo-2-butene (**361b**), the same or analogous products were obtained, with the exceptions of **366b** and **372b** (*170,577*). Instead of epoxide **372b**, its rearrangement product (**373**) was isolated. In addition, **371** was produced during ozonation of **361b**. The mechanistic explanation outlined in Scheme 75 seems reasonable. Ozonide **365** was too unstable to isolate but indirect evidence for its presence was obtained (*578*). Compounds **367** and **371** were thought to arise via addition and substitution of halogen with **361**. Epoxide **372** could have resulted from direct ozonation of **361** (Chapter XI) or epoxidation of **361** by carbonyl oxide **363** or peroxides produced during the ozonation. Since less than 1 mole equivalent of ozone reacted, Griesbaum (*170,577*) favored the latter pathway. Although considerable starting material reacted with halogen, rather than with ozone, to produce **367** and **371**, some of the bromine had to come from ozone oxidation of hydrogen halide, as well as from **366**. Incidentally, compound **366a** was stable enough to isolate, but **366b** was not (*578*).

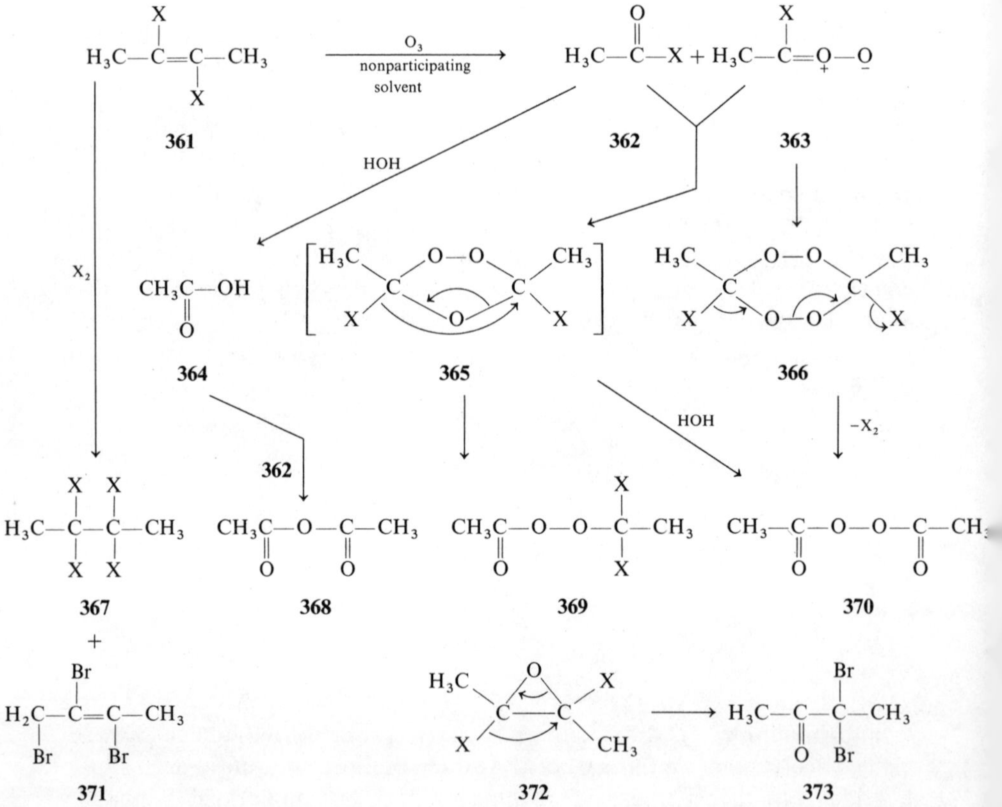

a: X = Cl **b**: X = Br

SCHEME 75. Ozonation of 2,3-dihalobutanes.

Epoxides also have resulted from ozonation of other haloethylenes (*579*, *580*), as will be discussed in Chapter XI. From tetrafluoroethylene, Gozzo and Camaggi (*579*) obtained not only difluoroformaldehyde and the epoxide of tetrafluoroethylene, but also hexafluorocyclopropane. A small amount of *cis*-1,2,3-trifluorocyclopropane also was obtained from ozonation of *cis*- and *trans*-1,2,-difluoroethylene, along with ozonide and epoxide (*580*). Both Gozzo and Camaggi (*579*) and Gillies (*580*) suggested mechanisms similar to that of Eq. (22) to explain the formation of cyclopropanes.

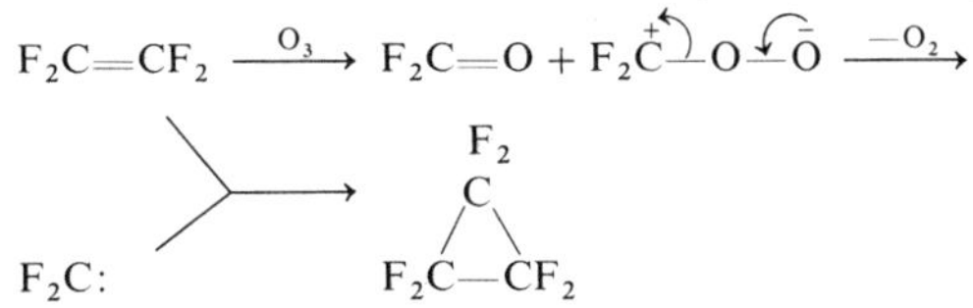

(22)

Gillies (*580*) obtained a *cis*:*trans* ozonide ratio of 10:90 from both *cis*- and *trans*-1,2-difluoroethylene. This is analogous to results described in Chapter VI, Section IV, for *cis*- and *trans*-2-butene.

II. Unsaturated Oxyphosphoranes

Ramirez *et al.* (*581–583*) have studied extensively the reactions of unsaturated oxyphosphoranes of types **374** and **385** with ozone. Only 0.5–0.6 mole of ozone per mole of **374** were required at $-75°C$ and the products were peroxide **379** (15–25%), anhydride **383** (15–20%), and diketone **382** (60–70%). With **385**, nearly 1 mole-equivalent of ozone reacted and the yields of products were 60% **386**, 5–10% **387**, and 25–30% **388**. Ramirez *et al.* (*583*) rationalized these results as shown in Scheme 76. In order to account for the low ratio of ozone to compounds **374**, they further suggested that intermediate **377** can oxidize 2 moles of **376**, a tautomeric form of the oxyphosphorane, to produce a total of 3 moles of diketone **382** plus **380** and/or **384**. They previously had shown that peroxide **379** does not ozidize the oxyphosphorane and that oxidation by molecular oxygen is much slower than ozone absorption.

To this reviewer many of the reactions of Scheme 76 appear unlikely. Carles and Fliszár (*406*) have suggested that the principal intermediate in the reactions is the normal ozonide (**391**) which should be quite readily formed since it is the ozonide of a five-membered unsaturated ring system (see Chapter V, Section I, A). They do not account, however, for the much less than 1 mole of ozone required per mole of **374**. Scheme 77 is a combination of the ideas of Carles and Fliszár (*406*) and of the present reviewer concerning the reactions of Ramirez.

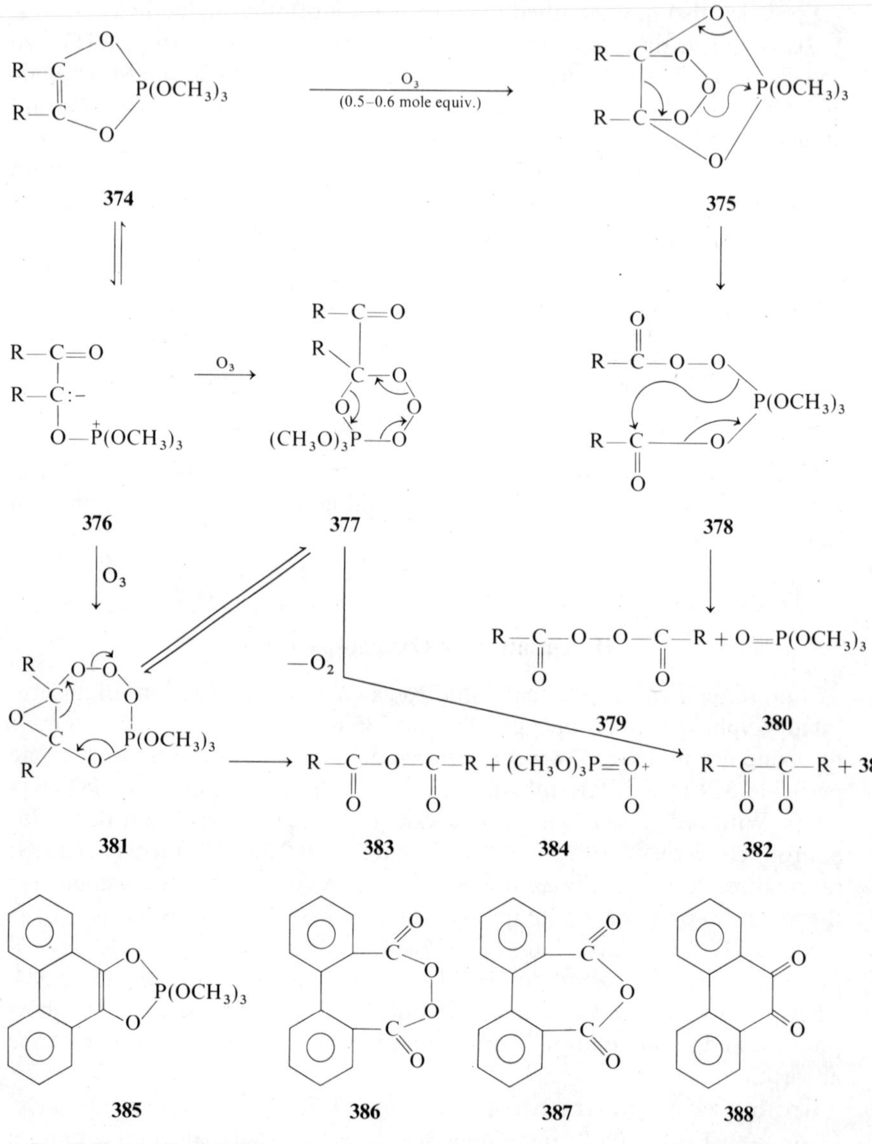

a: R = C_6H_5; **b**: R = CH_3

SCHEME 76. Ozonation of unsaturated oxyphosphoranes.

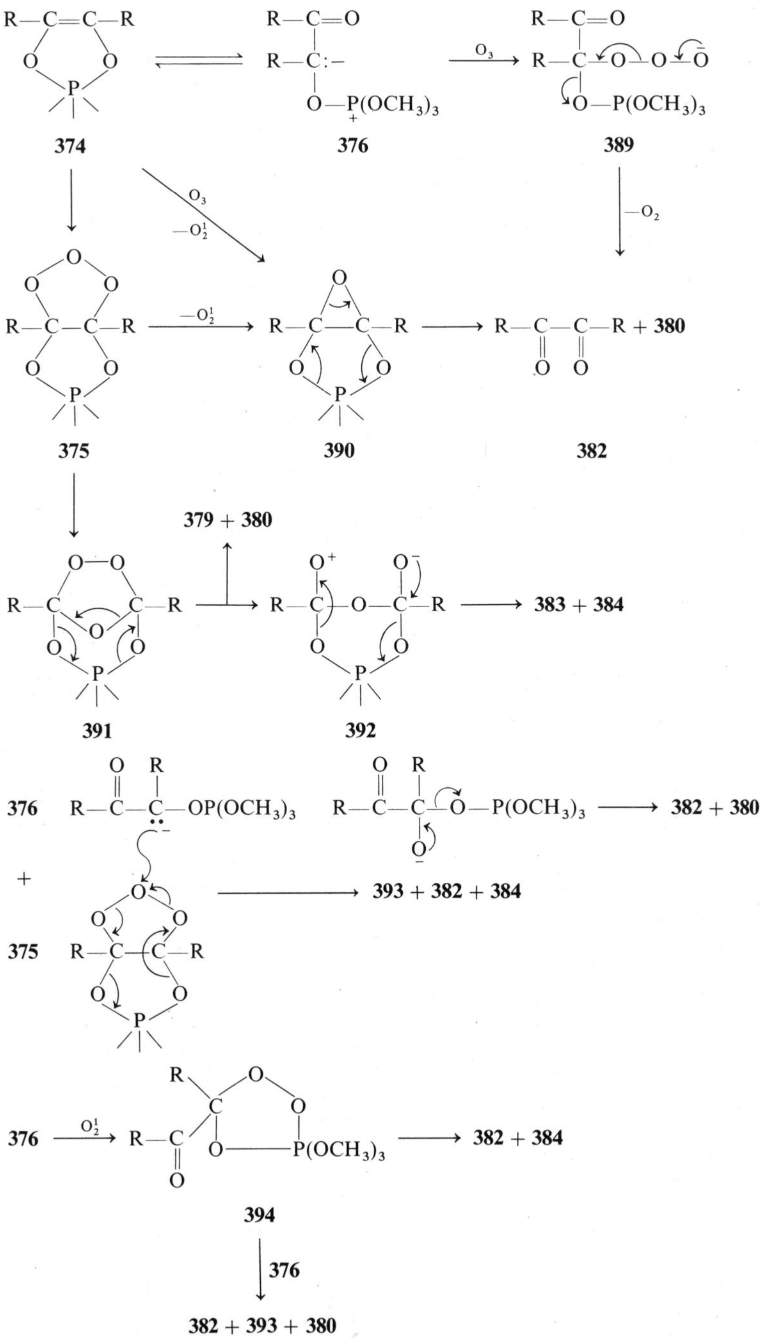

SCHEME 77: Suggested mechanisms of ozonation of unsaturated oxyphosphoranes.

Ozone attack on **374** should yield first **375** and then **391** (Scheme 77). Decomposition of **391**, as shown, would yield **379**, while decomposition via **392** would give anhydride **383**. Alternatively, the cleavage could be homolytic and proceed as described in Chapter VIII, Section V. Carles and Fliszár (*406*) suggested that diketone **382** arose via **390**, which they believed to be derived from ozonide **391**. This appears to be much less likely than origin of **390** from primary ozonide **375** or by direct ozonation of **374** as shown in Scheme 77 (see Chapter XI). Alternatively, the diketone could arise as described by **376** → **389** → **382**.

The requirement of only 0.5–0.6 mole of ozone per mole of compounds **374** at −75°C perhaps is due to oxidation of **376** by primary ozonide **375**, rather than by **377** as proposed by Ramirez (*583*). This would proceed via **393** to give 2 moles of diketone **382**. Another possibility is oxidation of **376** by singlet oxygen, which is evolved in several of the proposed reactions. This should proceed much faster than oxidation by ordinary oxygen, to give **394**, which could either decompose to diketone **382** or react with **376** to give **382** plus **393**. The route involving the interaction of **375** and **376** is probably the most important, however, since it can better explain the differences in ozone consumption and product ratios observed between oxyphosphoranes **374** and **385**. Interaction of the primary ozonide of **385** with the tautomer of **385** equivalent to **376** should be difficult for steric reasons, thus accounting for the essentially 1:1 reaction of **385** with ozone. The facts that at temperatures above −75°C even less ozone (0.2–0.36 mole equivalent) is required to consume **374a** and that the proportion of diketone **382** to diacyl peroxide **379** increases (*583*), would also appear to favor the additional oxidation route of **375** plus **376**. Consumption of **375** by this reaction would decrease the importance of the route to diacyl peroxide, **375** → **391** → **379**.

Ozonolysis of another type of phosphorus heterocycle, **395**, is also of passing interest. The reaction proceeded with the spontaneous evolution of carbon dioxide (*584*), possibly by the route shown in Eq. (23).

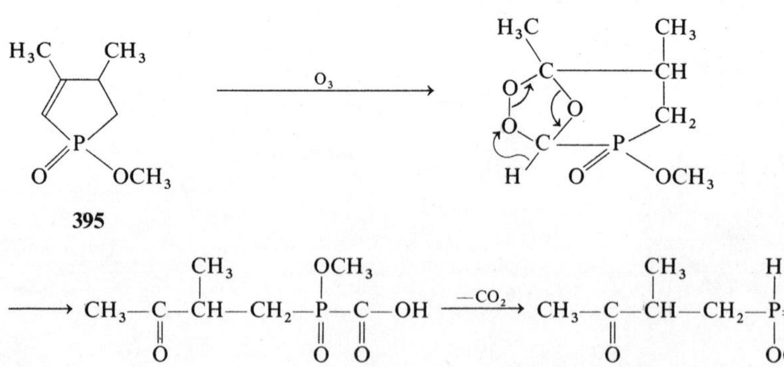

395

(23)

III. Allenes

Ozonolysis has been used frequently to prove the structure of allenes (*585–593*). When excess ozone was employed, both double bonds appeared to be cleaved (*585,587–591,593*) and, in some cases, carbon dioxide was produced, presumably from the central carbon of the allenic functional group (*585,588,593*). Smith *et al.* (*588*) reported an anomalous product, acetaldehyde, from ozonolyses of 1,2-pentadiene, 1,2-hexadiene, and 1,2-heptadiene, which they presumed to result from rearrangement of the allenes to 2-alkynes. Since, however, ozonation of alkynes does not produce aldehydes (to be discussed in Vol. II) the reported results either were erroneous or were due to the presence of small amounts of impurities, presumably other allenes.

Hartzler (*586*) reported that ozonolysis of allene **396** in ethanol yielded acetone and compound **397**; only 1 mole equivalent of ozone reacted. He rationalized the reaction as illustrated in Scheme 78.

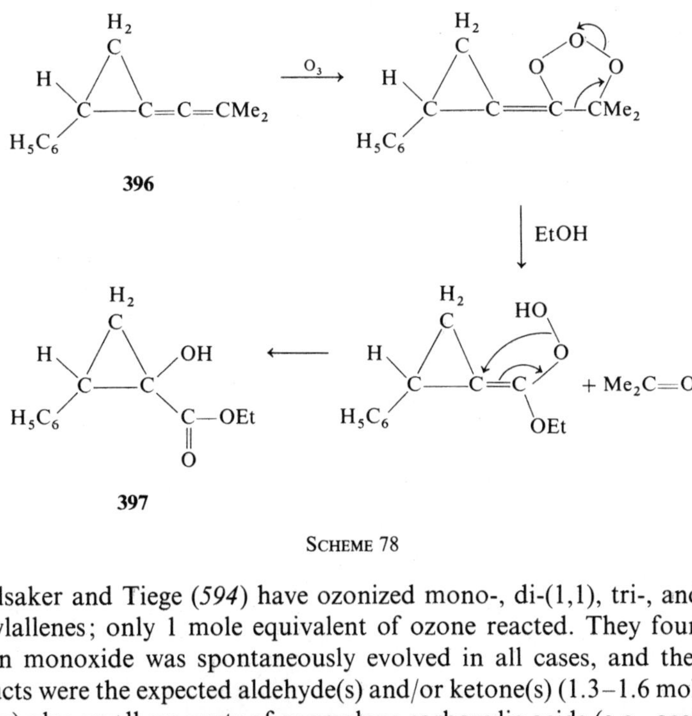

SCHEME 78

Kolsaker and Tiege (*594*) have ozonized mono-, di-(1,1), tri-, and tetramethylallenes; only 1 mole equivalent of ozone reacted. They found that carbon monoxide was spontaneously evolved in all cases, and the major products were the expected aldehyde(s) and/or ketone(s) (1.3–1.6 mole equivalents) plus small amounts of anomalous carboxylic acids (e.g., acetic acid from tetramethylallene). There was no difference in results observed depending on whether methylene chloride or methanol was the solvent. Illustrating with 2,4-dimethyl-2,3-pentadiene (i.e., tetramethylallene, **398**),

Kolsaker and Tiege (*594*) proposed the mechanism shown in Scheme 79. The cleavage of primary ozonide **399** could not have involved **401** and **402** since no methyl isobutyrate was found (from reaction of **401**) when methanol was the solvent. Neither is the attractive decomposition through **400** likely, since neither **405** nor **408** should spontaneously decompose to carbon monoxide and 2 moles of acetone. Thus, they suggested breakdown of **399** to acetone and **403** as the only feasible reaction course. Compound **403** must spontaneously decompose to carbon monoxide and acetone, probably via transition state **404**, since only traces of **407** (from **403** via **406**) were found when the ozonolysis was performed in methanol.

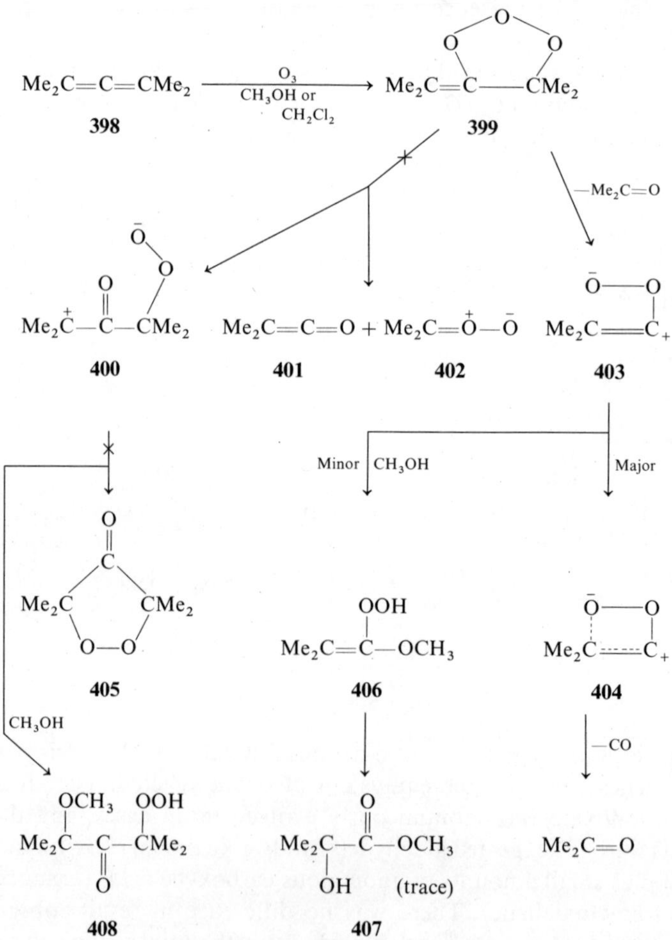

SCHEME 79. Substituted allene ozonation.

Crandall and Conover (*595*) have ozonized sterically hindered allenes **409**. Only 1 mole equivalent of ozone reacted with **409a**, whereas two were required for **409b**. From **409a** was obtained 62% di-*tert*-butyl ketone (**415**), 27% di-*tert*-butylcyclopropanone (**414**), and formaldehyde. The products from **409b** were the diepoxide (**416**), and traces of the monoepoxide (**413**). Crandall and Conover (*595*) rationalized these facts as shown in Scheme 80. First, ozone attacks **409** electrophilically to give **410**, which can either lose singlet oxygen, to give **411**, or cyclize to primary ozonide **412** and the other possible primary ozonide, which would yield formaldehyde. An intermediate

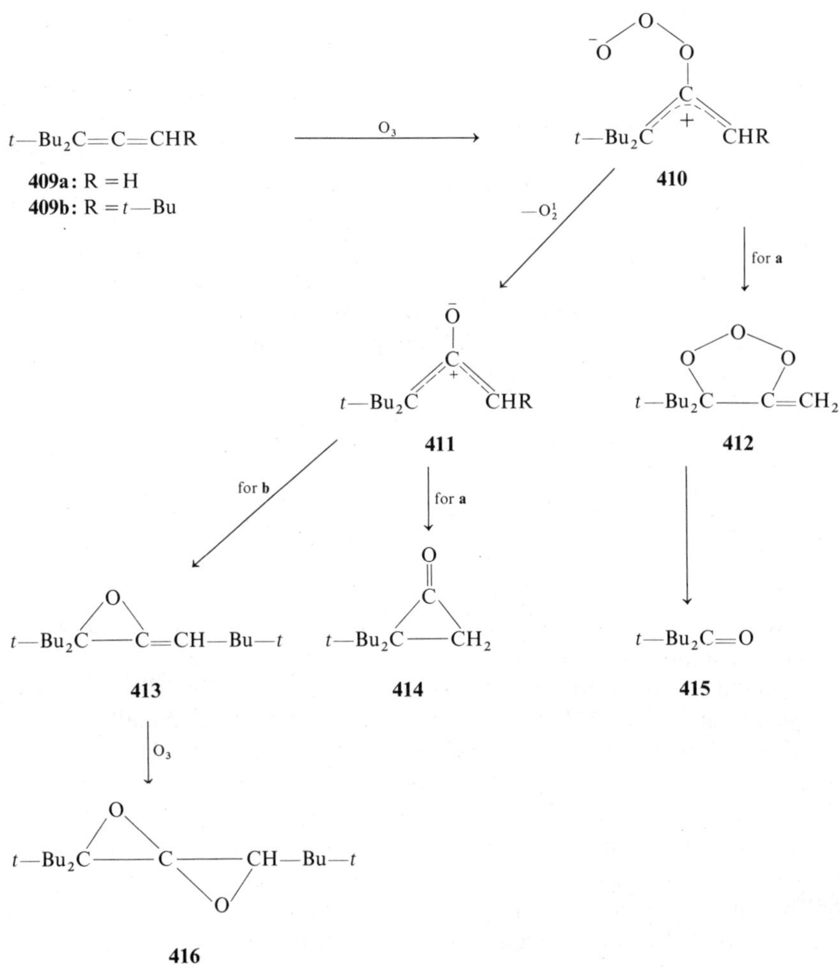

SCHEME 80. Hindered allene ozonations.

such as **410** is also likely for the reactions of Kolsaker and Tiege (Scheme 79). Intermediate **411** also has two possible cyclization choices, to epoxide **413** or to cyclopropanone **414**. Epoxide **413** then reacts further with ozone, probably as described for epoxide formation in Chapter XI, to give diepoxide **416**. What determines the preferred reactions involving **410** and **411** is not clear at present.

IV. Miscellaneous

A. CYCLOHEPTATRIENES AND CYCLOOCTATETRAENE

For a long time it was unclear whether certain substituted "cycloheptatrienes" had the triene structure, e.g., **417**, or a norcaradiene structure, e.g., **418**, or both via a valence tautomeric equilibrium (*596*). It is now evident that, except for certain benzo derivatives and other special types, the norcaradiene structure is unimportant (*596,597*). However, many cycloheptatriene systems react with various reagents, including ozone, as though they had norcaradiene structures (*596*). For example, **417** reacts with only 2 mole equivalents of ozone, as expected of **418** (*598*), and compound **419** gives a good yield of *cis*-caronic acid, **420**, upon ozonolysis (*599*). Other similar examples also are known (*599,600*).

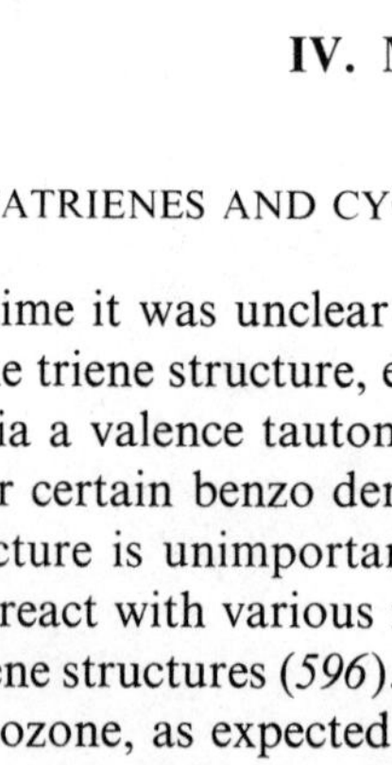

417

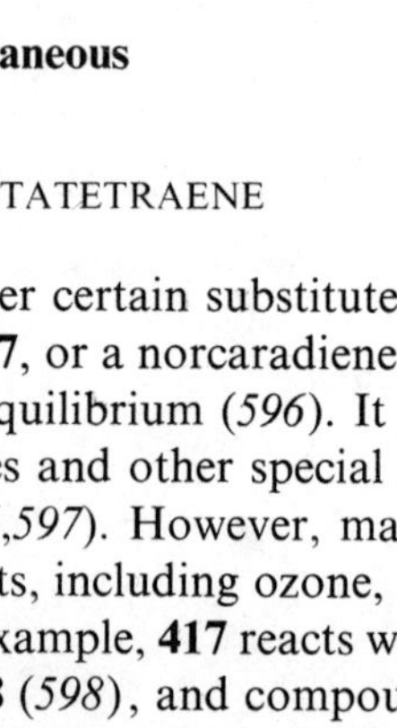

418

These reactions can be reconciled by an electrophilic ozone attack which induces bridging (*601*), as illustrated in Scheme 81.

A similar situation appears to exist to some extent during the ozonolysis of cyclooctatetraene (**421**, Scheme 81) in chloroform (*602*). Small amounts of 2,3-dicarboxybutanedioic acid (**425**) and succinic acid were obtained along with oxalic acid (**423**) (*602*). This can be explained by a process similar to that proposed by Friess and Boekelheide (*603*) for halogenation of cyclooctatetraene, involving a competition in the ozone attack to give either **422** or **424**. In constrast, Milas *et al.* (*604*) reported a nearly quantitative yield of glyoxal from ozonolysis of cyclooctatetraene in methanol followed by a reductive work-up, indicating that only the route involving **424** was operative under their conditions.

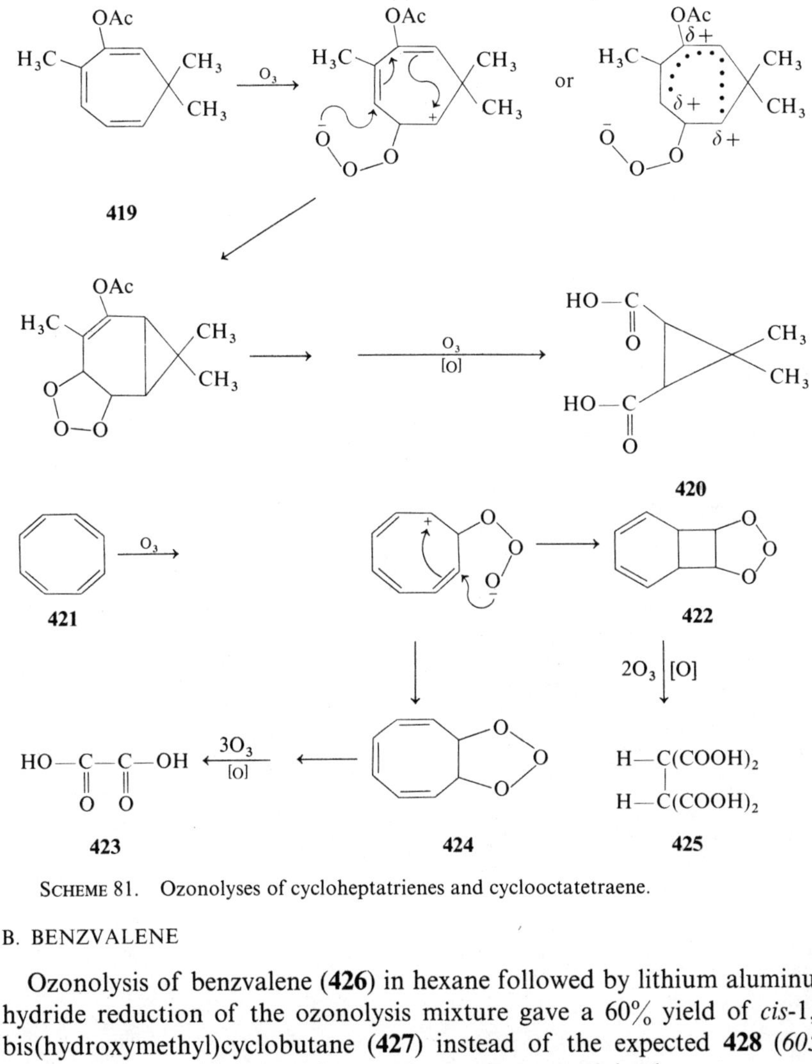

SCHEME 81. Ozonolyses of cycloheptatrienes and cyclooctatetraene.

B. BENZVALENE

Ozonolysis of benzvalene (**426**) in hexane followed by lithium aluminum hydride reduction of the ozonolysis mixture gave a 60% yield of *cis*-1,3-bis(hydroxymethyl)cyclobutane (**427**) instead of the expected **428** (*605*). At the moment, this interesting result remains unexplained.

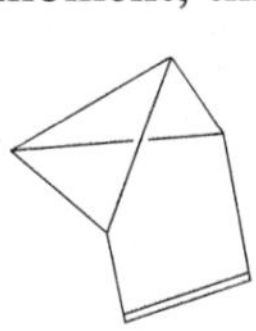

426

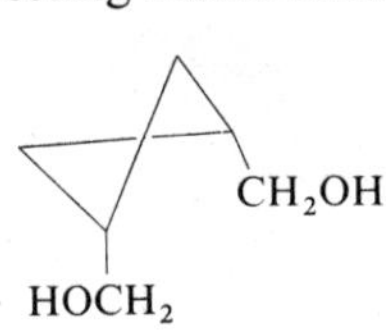

427

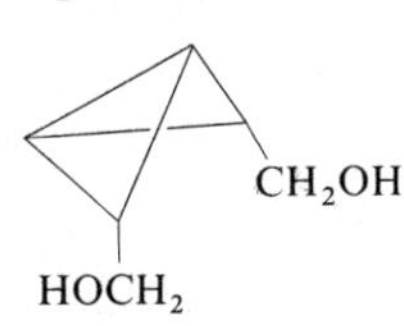

428

CHAPTER **XI**

Electrophilic Ozone Attack on Olefins. Epoxides and Other "Partial Cleavage" Products

I. Introduction

In the ozonolysis of olefins, which was the subject of Chapters III through X, all three atoms of the ozone molecule are utilized in product formation. There are many examples of olefins, however, that react, in varying degrees, with only one atom of the ozone molecule and release molecular oxygen. The products, which are isolated along with, or instead of, ozonolysis products, are epoxides, rearrangement products of an epoxide, products from reactions of other functional groups with an epoxide, or other types of products from electrophilic ozone attack followed by loss of molecular oxygen and additional reactions. Since only the π bond of the double bond is broken in the ozonation, these products are called "partial cleavage" products (*129*).

The earliest example of an olefin reacting with ozone in the manner just described appears to be tetrachloroethylene, which yielded trichloroacetyl chloride along with the expected phosgene ($O{=}CCl_2$) (*24,33,35*). The reaction was formulated as shown in Eq. (24).

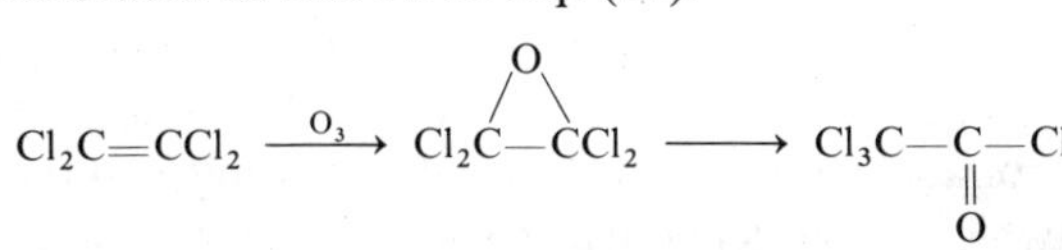
(24)

Next, and a very notorious example (*26,43*), was α-fenchene (**429**), which gave, in addition to a 50% yield of the expected ketone, a mixture of aldehyde **431** and the corresponding carboxylic acid (*606,607*). These, also, probably arose from rearrangement of an epoxide (**430**), as illustrated in Scheme 82.

The first report of the isolation of an epoxide ozonation product appears to be by Backer (*608*) in the case of the highly hindered olefin shown in Eq. (25). The epoxide was the major product.

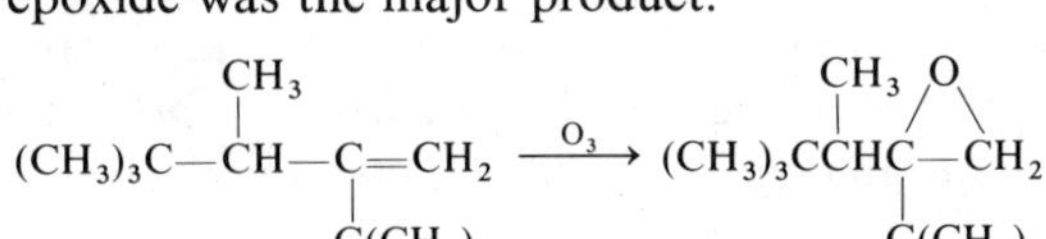
(25)

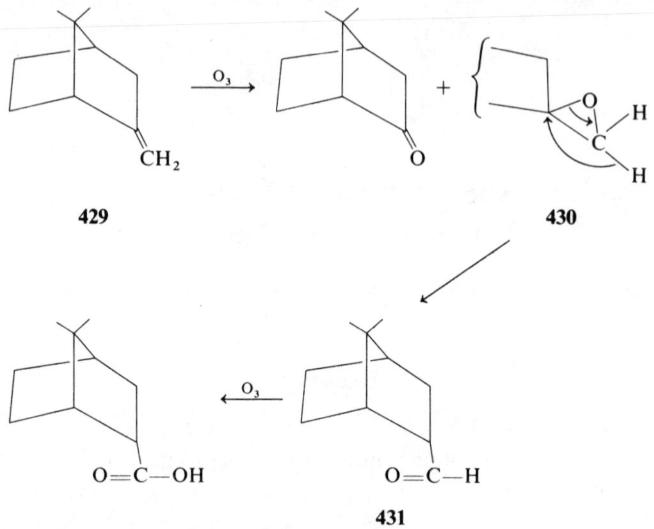

SCHEME 82. Ozonation of α-fenchene.

II. Examples

Table XXXI shows some epoxides obtained from ozonation of various olefinic compounds. Other examples also are in the literature (*129,192,475, 514,609,616,618–625*).

There are also many examples in the literature of ozonations leading to partial cleavage products derivable from epoxides by rearrangements similar to that illustrated by **430** in Scheme 82. Some examples are shown in Table XXXII. (*129,538,610,626–633*). Example 3 involves a ring enlargement, as shown in **432** of Scheme 83. The rearrangement of the epoxide in example 4 to give a vinyl alcohol is illustrated by **433**. The epoxide was isolated under nonacidic conditions; it easily rearranges to the vinyl alcohol and aldehyde, which are independently stable (*129*). Likewise, the products of examples 5 and 6 can be rationalized as shown by **434** and **435** in Scheme 83. Example 7 appears to be analogous to combined examples 4 and 5 in which a vinyl alcohol is first formed, after which further ozone attack and rearrangement occur; the authors, however, suggested a much more complicated mechanism (*632*). The diketone must have arisen via oxidation of the hydroxyketone (to be discussed in **Vol. II**).

The epoxide of longifolene (example 2, Table XXXI) was obtained only under nonacidic conditions. Ozonation in acetic acid gave the products shown in example 8 of Table XXXII. The first two products can be rationalized by reactions analogous to those shown in Schemes 82 and 83. The

TABLE XXXI

EPOXIDES FROM OZONATION OF OLEFINS

Example	Olefin	Epoxide yield (%)	References
1	$(t\text{—Bu})_2C(CH_3)\text{—}C(CH_3)\text{=}CH_2$	81	*140* (cf. *609*)
2	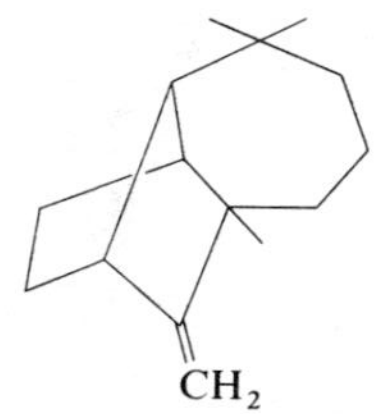	70	*610*
3	$(Neopent)_2C\text{=}CH\text{—}Bu\text{—}t$	91	*140*
4	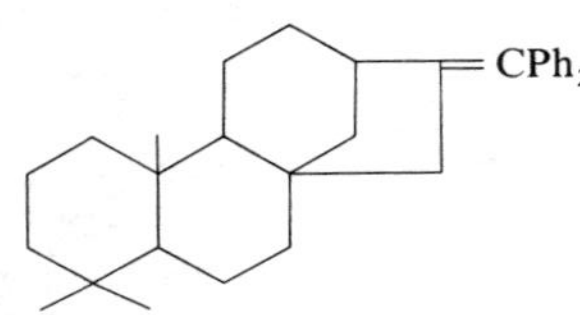	29[a]	*611*
5	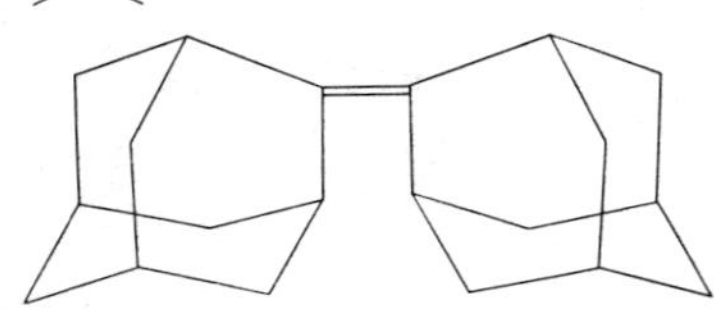	60	*192,339a*

TABLE XXXI (*Continued*)

Example	Olefin		Epoxide yield (%)	References
6	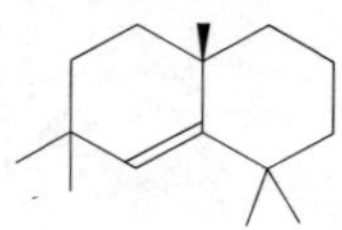	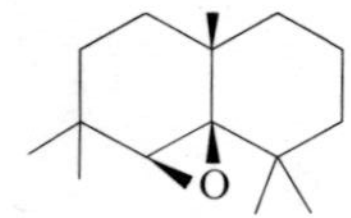 β-epoxide	92	*612*
7	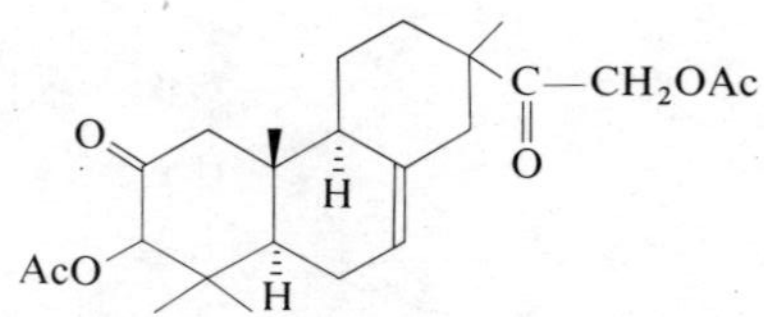	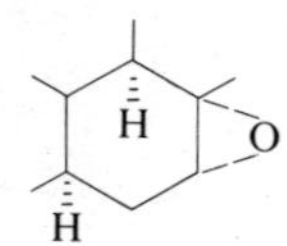 α-epoxide	65	*613,614*
8	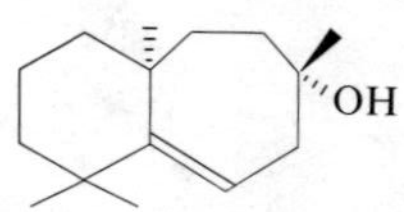 widdrol	α-epoxide	18	*615*
9	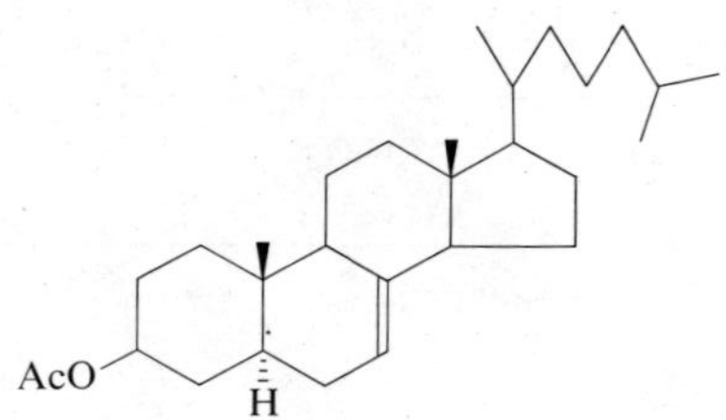	α-epoxide	20	*616*

10	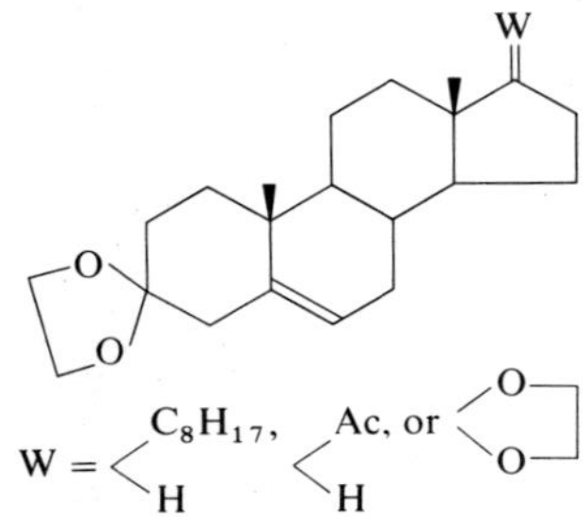	β-epoxides	40–56	*617*
11	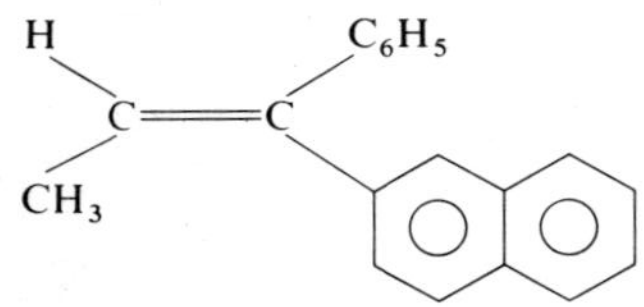 Z-olefin	Z-epoxide	31	*239*
12	E isomer of 11	E-epoxide	24	*239*

[a] Only product.

TABLE XXXII

REARRANGEMENT PRODUCTS OF EPOXIDES VIA OZONATION[a]

Example	Olefin	Product(s)	Yield (%)	References
1	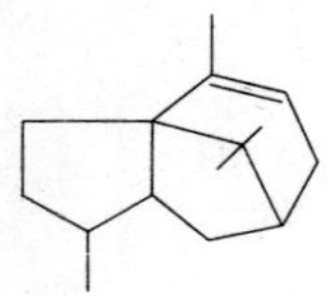	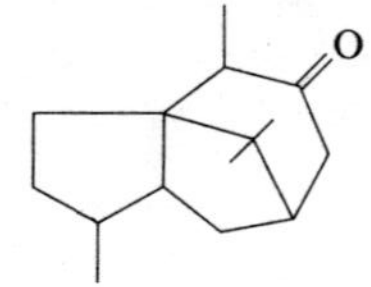	23	*538*
2	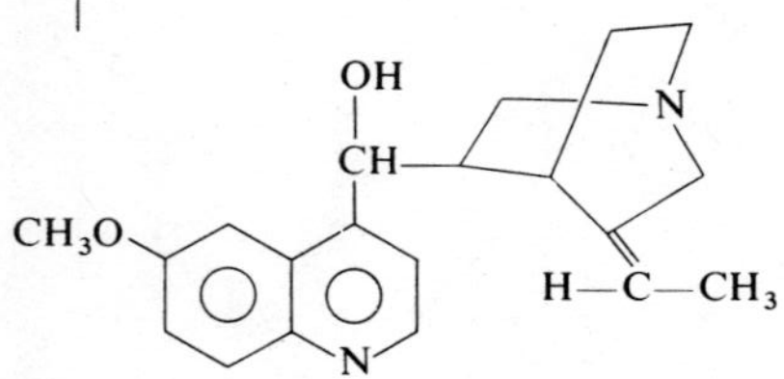	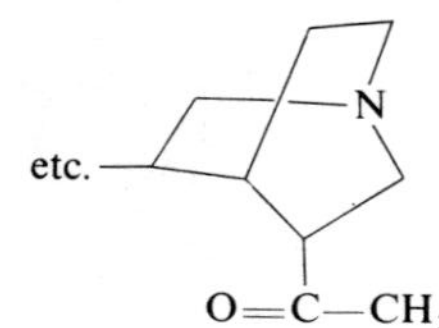	66	*626*
3	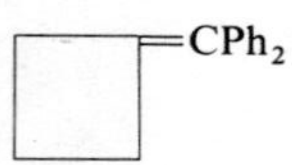	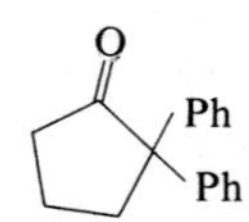	23	*627*
4	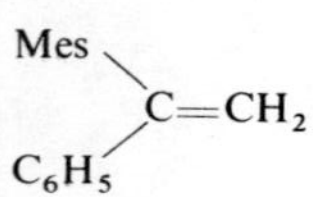	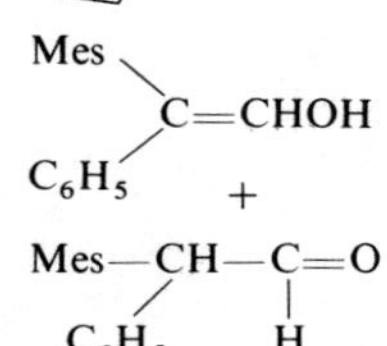	75–91[b]	*129,628*

5	$(Mes)_2C{=}CHOH$	$Mes{-}C({=}O){-}CH(OH)Mes$	High	*629,630*
6	$Mes{-}C(C_6H_5){=}CHOH$	$Mes{-}C({=}O){-}CHOHC_6H_5$	47	*631*
		$MesCH(C_6H_5){-}C({=}O){-}OH$	12	
7	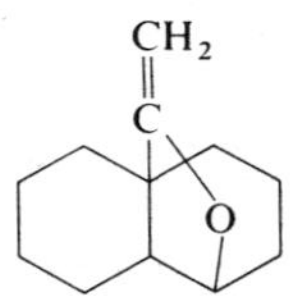	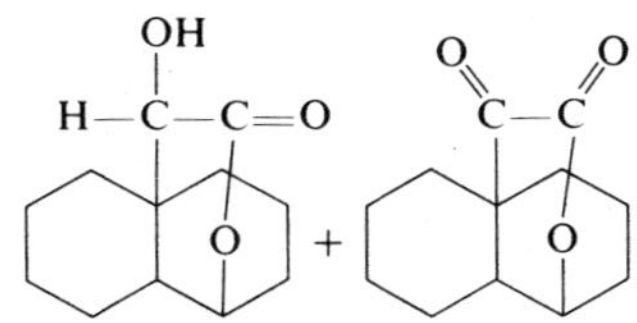	20	*632*
8	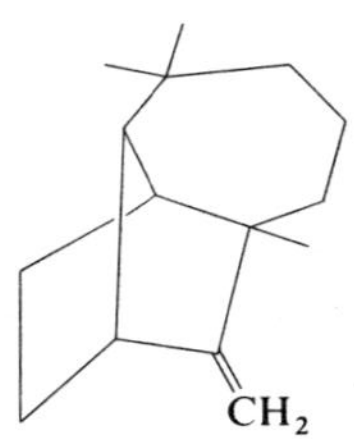	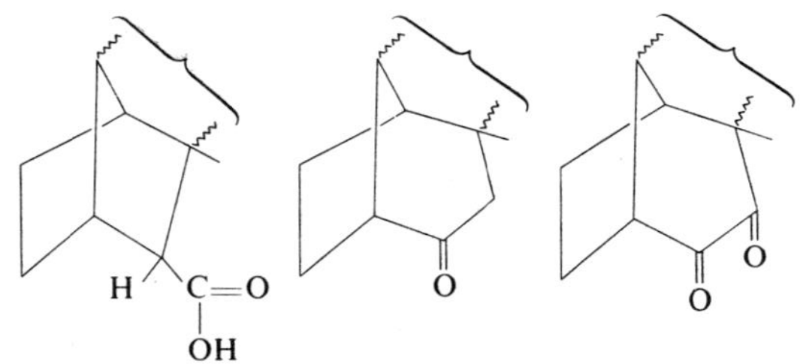		*610,633*

[a] Mes = mesityl.
[b] 75% isolated yield; higher yields by vpc; yields were higher in CH_3OH than in pentane or CH_2Cl_2.

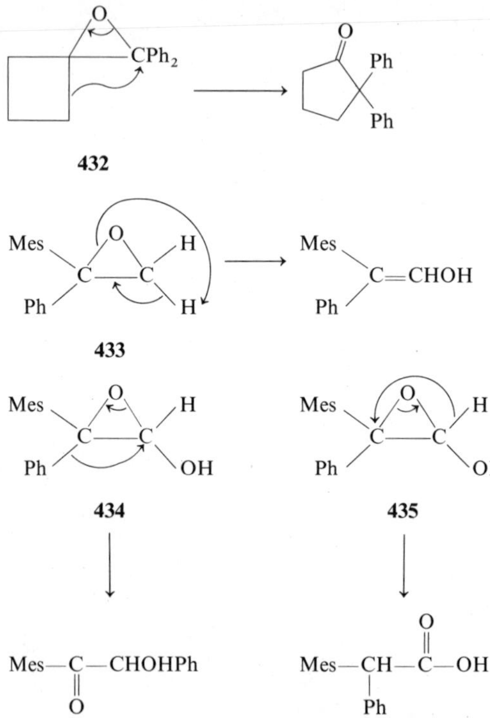

SCHEME 83. Epoxide rearrangement products.

Mes = mesityl

third product, which is more difficult to explain, perhaps arose via the enol of the aldehyde corresponding to the carboxylic acid product, similar to the route just described for example 7.

Other ozonation products of the type shown in Table XXXII also are known (*129,623,634–638*).

Partial cleavage ozonation products derivable from epoxides through solvolysis, intramolecular reactions similar to solvolysis, and/or other interactions are also obtained and are illustrated in Table XXXIII. (*129,613,614*, (*639–645*).The product of example 1 can be rationalized by an intramolecular reaction of the carboxyl group with the epoxide ring formed by ozonation. Similar interactions appear to have occurred in other instances (*646,647*). Example 2 involves simple solvolyses of the supposed epoxide intermediates. Other such examples also are known (e.g., *648*). The products of examples 3 and 4 can be rationalized through slightly more complicated solvolyses of epoxide intermediates, as shown in **436** and **437** of Scheme 84. Examples 5 and 6 involve a different kind of rearrangement of the epoxide

TABLE XXXIII

PARTIAL CLEAVAGE OZONATION PRODUCTS DERIVABLE FROM EPOXIDES VIA SOLVOLYSIS AND OTHER INTERACTIONS

Example	Olefin	Product	Yield (%)	References
1	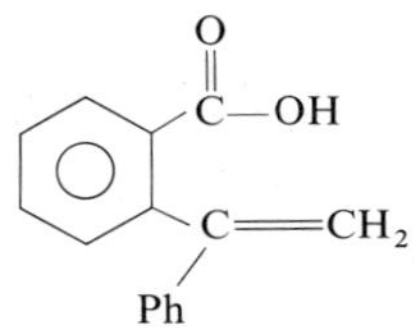	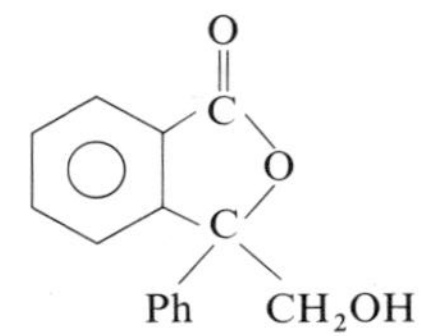	40	*129*
2	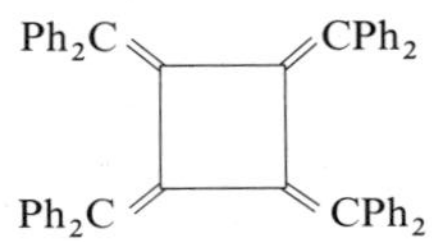 (MeOH solvent)	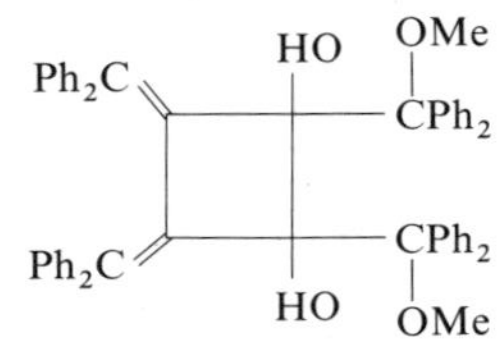	37	*639*
3	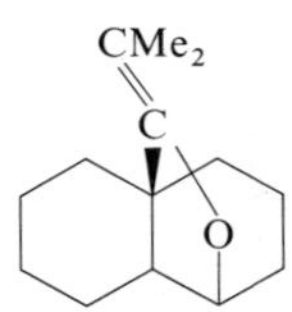	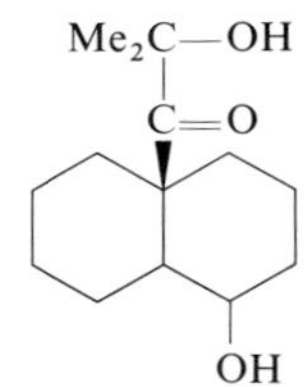	13	*640*
4	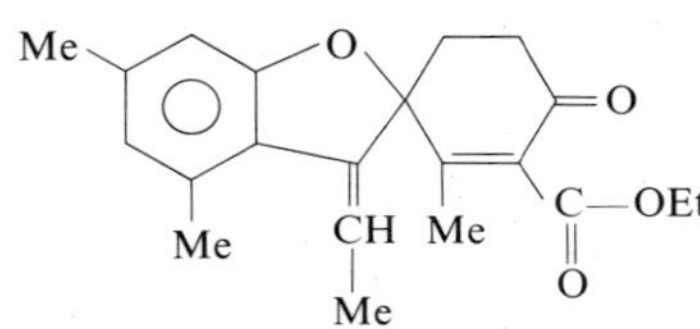 (MeOH solvent)	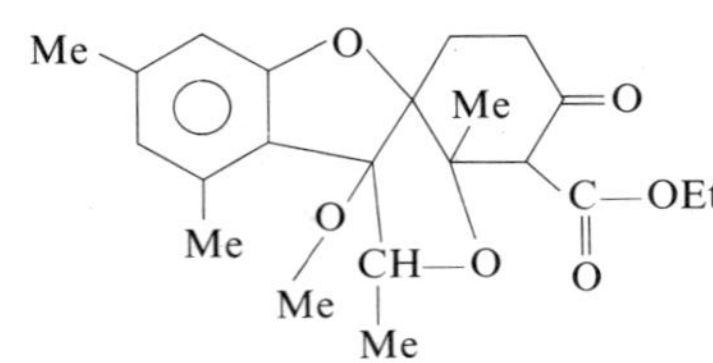	55	*641–643*

TABLE XXXIII (*Continued*)

Example	Olefin	Product	Yield (%)	References
5	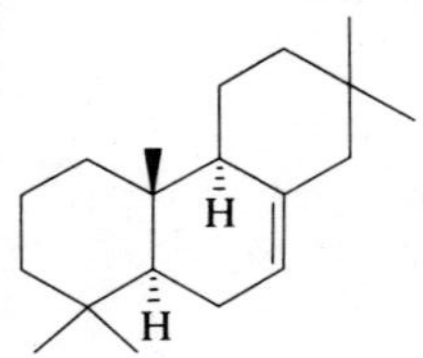	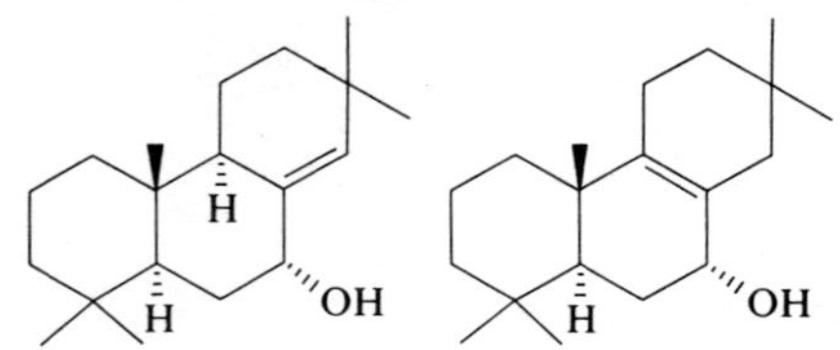		*613,614*
6	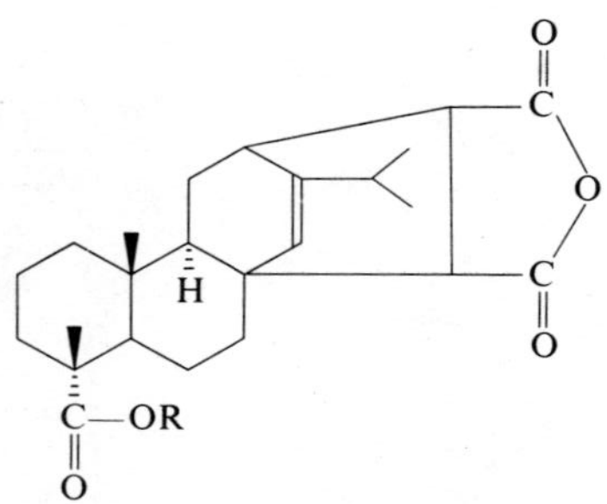	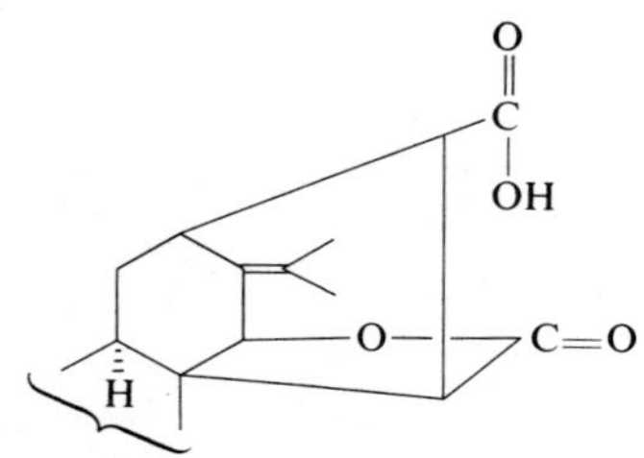		*644*
7	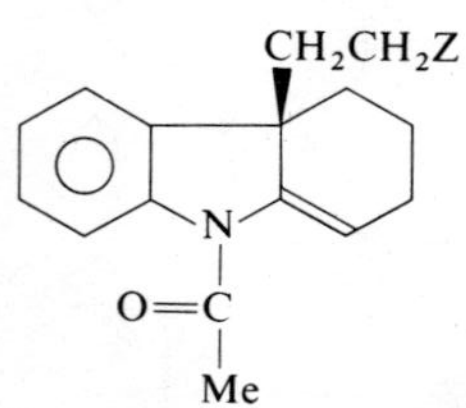 Z = H or Br	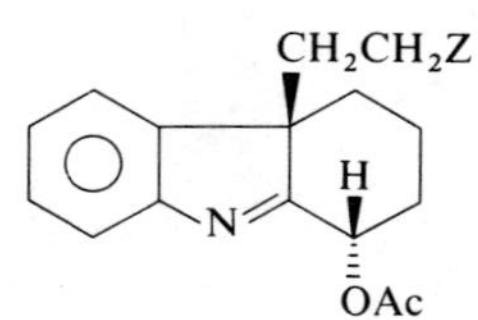	80–85	*645*

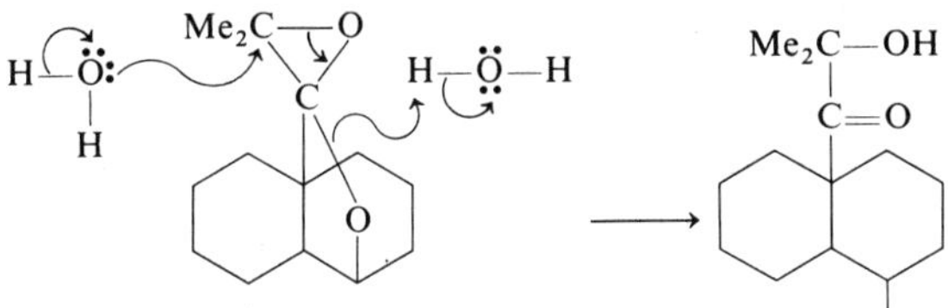

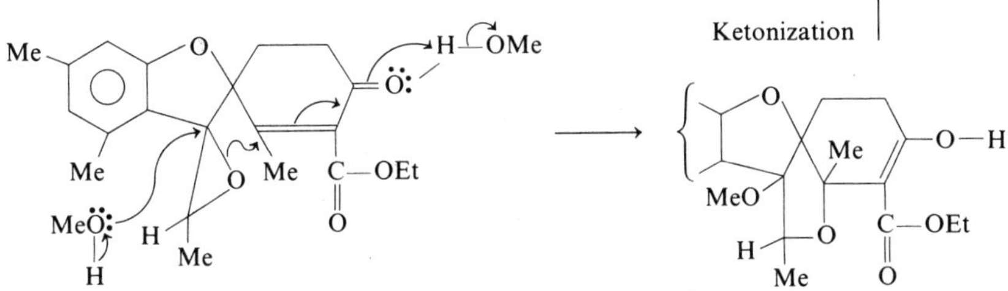

437

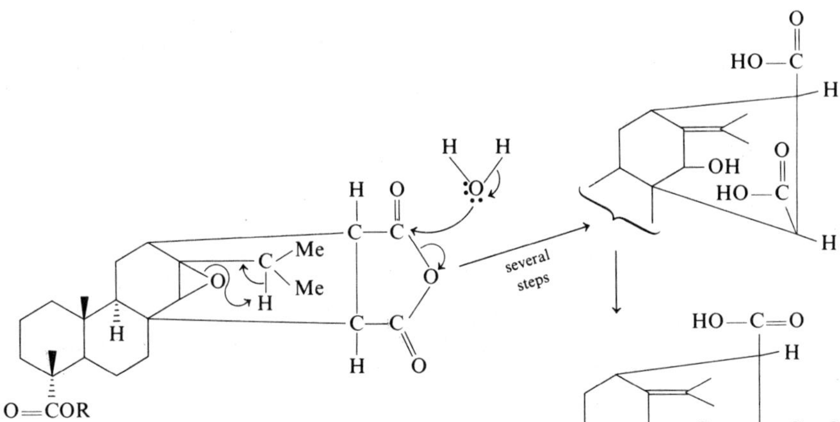

438

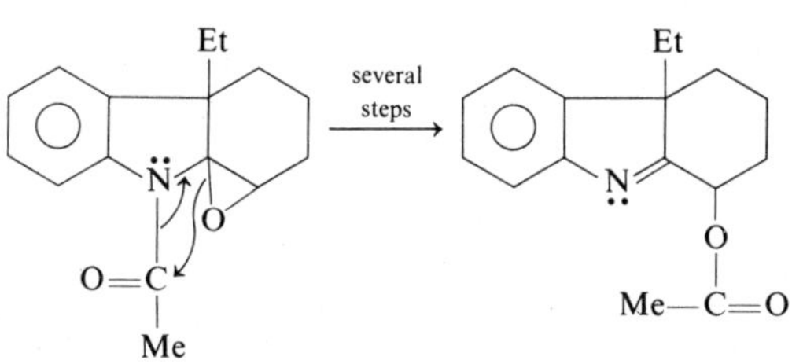

439

SCHEME 84. Solvolyses, etc., of epoxides from ozonation.

intermediate, as also illustrated in Scheme 84 (i.e., **438**, for example 6). Such a reaction also appears to have occurred in another instance (*647*). The rationale for example 7 is outlined in **439** of Scheme 84.

III. Theory of Competition between Ozonolysis and "Partial Cleavage"

A. HINDERED OLEFINS

A perusal of the examples of "partial cleavage" in Tables XXXI–XXXIV indicates that steric factors are important in regard to the promotion of this reaction in competition with ozonolysis. In 1967, a systematic study was made of this competition (*129*). The results are shown in Table XXXIV. Preliminary to the study, an in-depth investigation of the ozonation of 1-mesityl-1-phenylethylene (example 4, Table XXXII) was carried out. It was shown that the epoxide was the precursor to the rearrangement products and that for each mole of partial cleavage product, 1 mole of molecular oxygen was evolved (*129*). The latter was shown to be true in other cases also; thus, some of the partial cleavage determinations in Table XXXIV were based on the molecular oxygen yield (*129*). The molecular oxygen was later shown to be singlet oxygen, as expected (*140,141*).

Table XXXIV clearly shows that as the bulk of the groups around a double bond increases, the percentage of partial cleavage product(s) increases at the expense of ozonolysis. In order to account for these results, Bailey and Lane (*129*) proposed that a π complex (**440**) is the first species formed between ozone and an olefin (or other π system). 1,3-Dipolar cycloaddition to a 1,2,3-trioxolane (**441**) (primary ozonide on the way to ozonolysis products) and loss of molecular oxygen, via **442**, **443**, or **444** to give an epoxide (**445**) or other partial cleavage products (e.g., **446**), are the competitive fates of the π complex. It was suggested (*129,140*) that as the bulk of the substituents around the double bond increases, 1,3-dipolar cycloaddition decreases in ease and loss of molecular oxygen to give partial cleavage products increases in importance. In the case of 1,1-dimesitylethylene (Table XXXIV) partial cleavage is the only observable reaction. Other authors also had noted the bulk relationship (*62,110,111*).

The evidence for the preceding theory is convincing. As will be discussed in more detail in Section IV of this chapter, the π complex of 1-mesityl-1-phenylethylene (Table XXXIV) has been characterized (*139*–141). It decomposed to the epoxide without going through any intermediate state stable enough to be observed (*140,141*). This would seem to eliminate an actual peroxy epoxide intermediate such as **443** and to limit the intermediate, or transition state, to **442** or **444**.

TABLE XXXIV

PERCENTAGE OF PARTIAL CLEAVAGE FROM OZONATION OF ARYL- AND ALKYL-SUBSTITUTED ETHYLENES

Aliphatic olefin	% Partial cleavage[a]	References	Aromatic olefin	% Partial cleavage[a]	References
$t\text{—Bu—}\underset{\displaystyle CH_3}{\underset{\vert}{C}}\text{=}CH_2$	10[b]	*129*	$Ph_2C\text{=}CH_2$	15[b]	*129*
			$Ph\text{—}\underset{\displaystyle o\text{—Tol}}{\underset{\vert}{C}}\text{=}CH_2$	30[b]	*129*
$(t\text{—Bu})_2C\text{=}CH_2$	15[c]	*129*	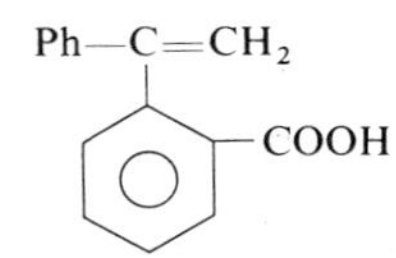	40[d]	Example 1, Table XXXIII
$(\text{Neopent})_2C\text{=}CH_2$	29[c]	*140*	$Ph\text{—}\underset{\displaystyle \text{Mes}}{\underset{\vert}{C}}\text{=}CH_2$[e]	82[f]	Example 4, Table XXXII
$(t\text{—Bu})_2\underset{\displaystyle CH_3}{\underset{\vert}{C}}\text{——}\underset{\displaystyle CH_3}{\underset{\vert}{C}}\text{=}CH_2$	81[c]	Example 1, Table XXXI	$\text{Mes}_2C\text{=}CH_2$[e]	95[f,g]	*140*
$(\text{Neopent})_2C\text{=}CH\text{—Bu—}t$	91[c]	Example 3, Table XXXI			

[a] Yield of partial cleavage products based on unrecovered olefin. Remaining products were from ozonolysis.
[b] Yield based on molecular oxygen evolved (*649*).
[c] Yield of epoxide after vapor-phase chromatography.
[d] Isolated product yield.
[e] Mes = mesityl
[f] Yield of rearrangement products after vapor-phase chromatography.
[g] No other products observed.

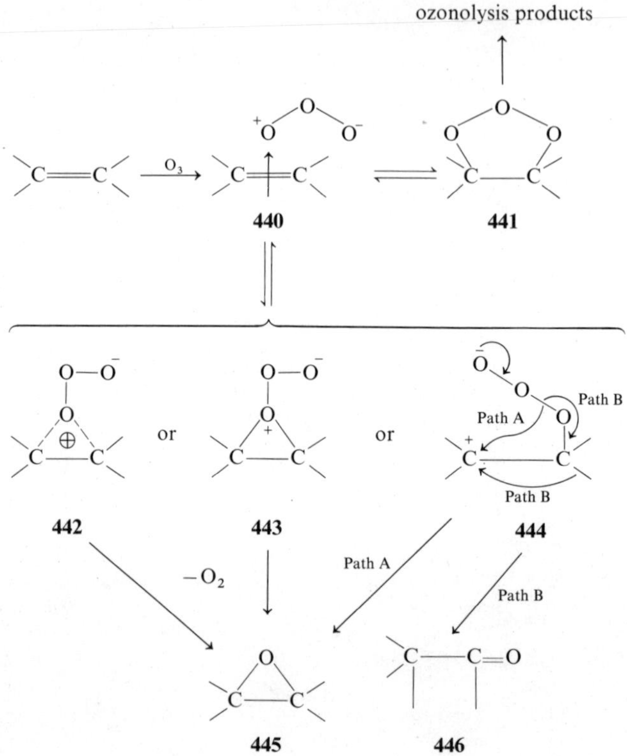

SCHEME 85. Competition between ozonolysis and partial cleavage.

With bulky olefins, at least, transition state **442** rather than intermediate **444** appears to be involved, since it has been shown in three instances that epoxide formation is stereospecific. *trans*-2,2-Dimethyl-3-hexene (1-*tert*-butyl-2-ethylethylene) afforded a small amount of the *trans*-epoxide along with ozonolysis products (*128*), and *cis*- and *trans*-1-(1-naphthyl)-1-phenyl-1-propene gave as major products their respective *cis*- or *trans*-epoxides uncontaminated with the other isomer (*239*) (examples 11 and 12 of Table XXXI).

Another interesting stereochemical observation is that epoxide formation during ozonation of steroid and other condensed ring olefinic systems appears to occur exclusively from one side or the other, as can be seen from examples 6–10 of Table XXXI. In the case of example 6, epoxidation by ozone, potassium permanganate, and sodium dichromate all occurred exclusively at the β face to yield the epoxide shown in Table XXXI, whereas peracid epoxidation showed a 70:30 preference for the other epoxide (*612*). Since peracids generally attack the least-hindered side, preferentially,

Hochstetler (*612*) assumed that ozone and the other oxidizing agents that gave the β-epoxide came in from the more hindered side. Further, in contrast to peracid epoxidation, this would appear to indicate that the transition states for β-epoxidation, including ozonation, resemble products more than reactants.

Similar results are reported for the compounds of example 10 (Table XXXI) and for another similarly constructed steroid (*616*). Ozonation produced exclusively the β-epoxides while peracid epoxidation afforded both α- and β-epoxides with the α-epoxide being predominant (*616,617*). Along with the β-epoxides from ozonation of these compounds, ozonolysis products were obtained (*617*). Likewise, ozonation of widdrol (example 8, Table XXXI) gave results different from those obtained by peracid epoxidation. The α-epoxide was obtained exclusively from ozonation, whereas a nearly 3:1 predominance of the β-epoxide resulted from peracid epoxidation (*615*). Ozonolysis was predominant over epoxidation to the extent of nearly 4:1 during ozonation of widdrol (*615*). Enzell (*615*) rationalized this on the basis that in the competition between 1,3-dipolar cycloaddition and epoxidation, attack of ozone on the less-hindered (β) side led to 1,3-dipolar cycloaddition, whereas attack on the more-hindered (α) side resulted in epoxidation. The same argument could be applied to the ozonations of example 10 (Table XXXI).

With examples 7 and 9 of Table XXXI, the α-epoxides were produced exclusively upon ozonation, but in contrast to the preceding results, the α-epoxide was also the predominant epoxide formed by peracid epoxidation (*613,614*). The same was shown to be true with example 5 of Table XXXIII, where the α-epoxide was established as the precursor to the products listed; peracid epoxidation gave a 3:1 predominance of the α-epoxide (*613,614*). This, perhaps, is better explained by the thought of Hochstetler (*612*) that, whereas peracid generally attacks the less-hindered side and the transition state resembles the reactants, the controlling factor in the case of epoxidation by ozone is that the transition state resembles the products and the more stable epoxide is produced. In the cases under discussion, the more stable epoxide might be the one formed on the less-hindered side. At this stage it is not possible to decide between the concepts of Hochstetler and Enzell.

Although the results of examples 11 and 12 of Table XXXI eliminate the possibility of a σ-complex intermediate (**444**) in these ozonations, most of the products of Table XXXII can be rationalized through such an intermediate just as well as via rearrangement of an epoxide. In certain instances this route appears to be more likely than the epoxide route. A case in point is example 3 of Table XXXII, with which there was obtained, in addition to the partial cleavage product shown, benzophenone (ozonolysis product,

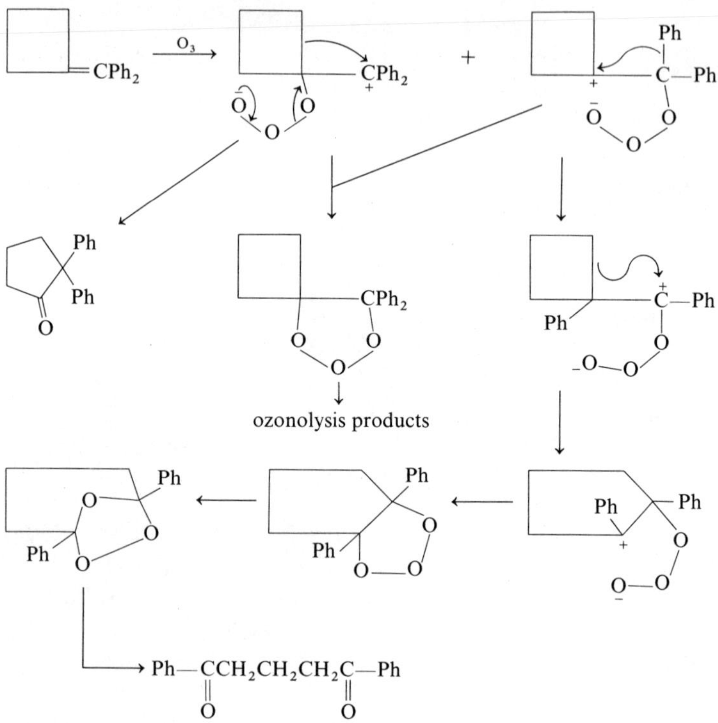

447

SCHEME 86

46%) and 1,3-dibenzoylpropane (**447**), 4% (*627*). These results indicate a carbonium ion intermediate, at least in the formation of the dibenzoylpropane (**447**). A plausible mechanism is outlined in Scheme 86 (cf. **432**, Scheme 83).

A σ complex rather than an epoxide route to the products of example 4 (Table XXXIII) also appears to be more direct, as previously suggested (*129*). Another possible instance is example 2 (Table XXXIII). From ozonation in chloroform, only ozonolysis products were isolated, whereas ozonation in a chloroform–methanol mixture gave the product shown as the major product, along with minor amounts of ozonolysis products (*639*). Perhaps in the less polar solvent, only 1,3-dipolar cycloaddition occurred, whereas in the more polar solvent σ complex formation and partial cleavage became predominant. Solvent effects also have been noted in other cases, perhaps indicating, if not a σ complex intermediate, at least some carbonium ion character to the transition state leading to partial-cleavage products. A case in point is example 4, Table XXXII, which afforded 75–81%

partial cleavage in pentane and methylene chloride, but 90–91% yields of such products in methanol (*129*). Even more striking is the finding that 1,1-dineopentylethylene (Table XXXIV) affords up to 31% epoxide in methanol solution, but less than 5% in pentane (*650*).

B. UNHINDERED OLEFINS

The solvent effect just discussed in the preceding subsection also was found in ozonations of *cis*- and *trans*-1,2-difluoroethylene (*580*). In methyl chloride solvent, epoxidation was predominant, whereas in isobutane, ozonides were the major products. In addition, there was no stereospecificity in the epoxide formation; both the *cis*- and the *trans*-olefins gave the *cis*-epoxide. This would appear to require a σ complex intermediate, i.e., **444**. These compounds are unique, however, in that, unlike the previous examples, steric hindrance may not be the factor that causes epoxidation to predominate. There are a number of other examples, also of compounds without bulky substituents at the double bond, that give partial cleavage products, sometimes predominantly. Some of these are listed in Table XXXV, along with the ones just discussed.

It is to be noted that in each of the examples in Table XXXV at least one electron-rich atom is directly attached to a double-bond carbon and that, at least with examples 1–5, the partial cleavage reaction is promoted by a

TABLE XXXV

PARTIAL CLEAVAGE DURING OZONATION OF UNHINDERED OLEFINS

Example	Olefin	Product	Yield (%)	References
1	$(MeO)_2C{=}C(OMe)_2$	$(MeO)_3C{-}C(=O){-}OMe$	33–65[a]	*651*
2	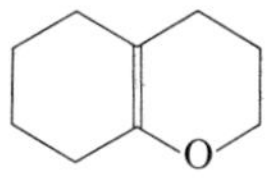 (solvent, ROH)	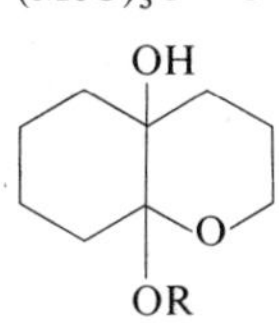 R = Me or Et	11–46	*652*
3	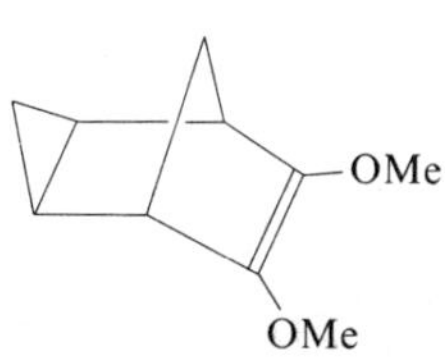	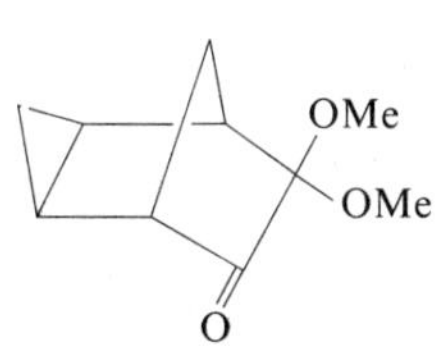	ca. 50[b]	*653*

TABLE XXXV (*Continued*)

Example	Olefin	Product	Yield (%)	References
4	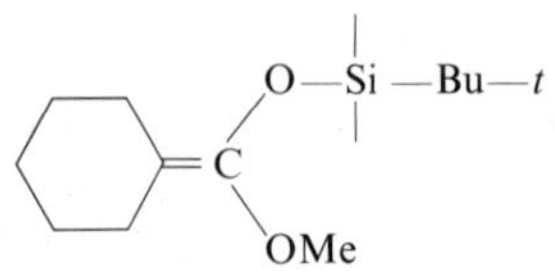	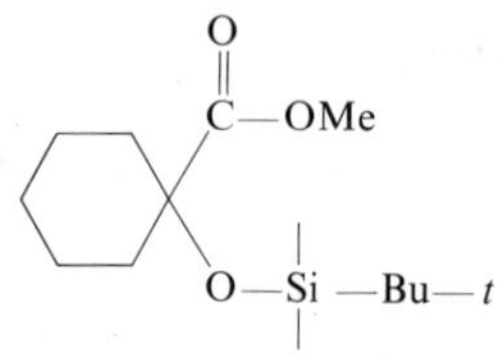	70[c]	*654*
5	FCH=CHF cis and trans	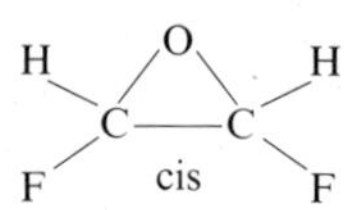	—[d]	*580*
6	$F_2C{=}CF_2$	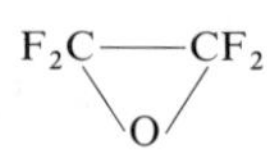	—[e]	*579*
7	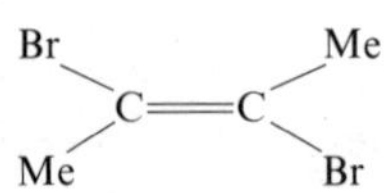	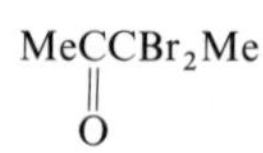	14–20[f]	*170,577*
8	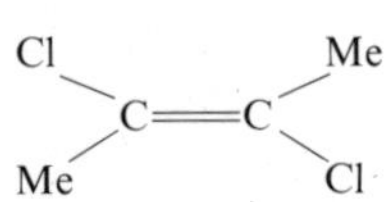	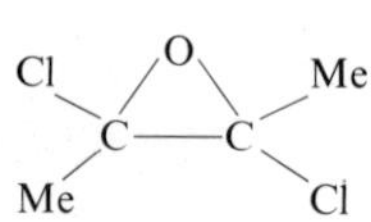	1–9	*578*
9	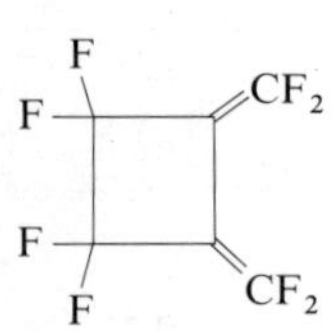	$(C_3F_4O)_x$[g]		*655*

[a] 33% in hexane, 39% in chloroform, and 59–65% in methanol; the other product was $(MeO)_2C{=}O$.

[b] Solvent: methanol. Exact yield not given.

[c] Solvent: methanol. Remaining product is cyclohexanone (from ozonolysis). Other similar examples are cited in this paper.

[d] Major product along with formyl fluoride in CH_3Cl; minor product in isobutane.

[e] Exact yield not given, but ratio of epoxide to $F_2C{=}O$ was approximately 1:2 and the ratio of O_3:olefin reacting was approximately 0.5.

[f] No important solvent effect observed. Ratio of O_3:olefin reacting was 0.5.

[g] Yield and structure unspecified. Probably a diepoxide, although the authors made other suggestions.

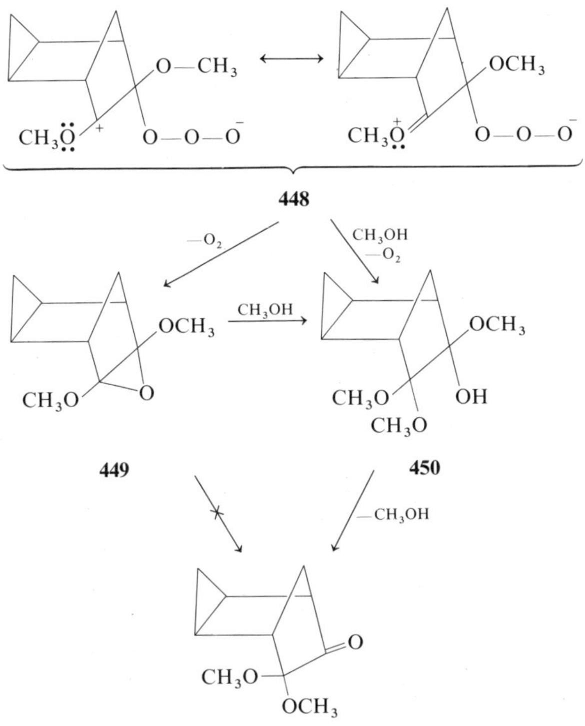

SCHEME 87

polar solvent. Thus, the driving force with these nonhindered olefins probably is the formation of a resonance stabilized σ complex, as illustrated by **448** in Scheme 87. Although the products of examples 1 and 3 could be rationalized by rearrangement either of a σ complex (e.g., **448**) or of an epoxide (e.g., **449**), it has been shown in the case of example 3 that no rearrangement is involved at all. Ozonation in the presence of methanol-d_4 resulted in the incorporation of $—OCD_3$ in place of one of the $—OCH_3$ groups (*653*). Thus, the reaction course involves either **449** → **450** → **451** or **448** → **450** → **451**. Example 4 of Table XXXV probably involves a somewhat different type of rearrangement at the σ complex stage, as illustrated in Eq. (26).

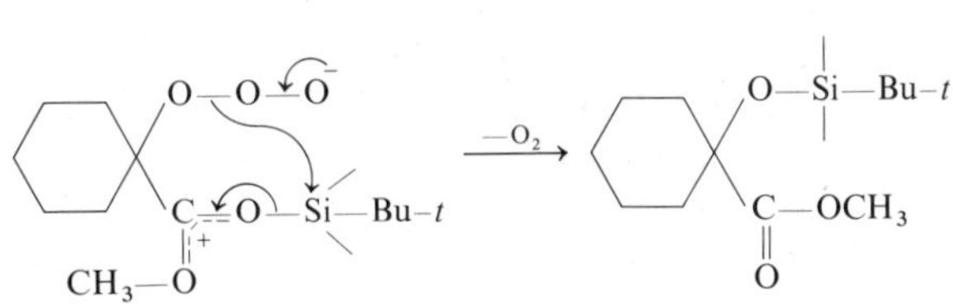

(26)

Although the evidence is strong that the reaction course for examples 1–5 of Table XXXV involves a σ complex (e.g., **448**) and/or an epoxide, this may or may not be the route with examples 6–8. With examples 6 and 7, about 0.5 mole, and with example 8, 0.8–0.9 mole, of ozone per mole of olefin reacted (*577–579*). For these reactions, epoxidation of the olefin by the Criegee carbonyl oxide zwitterion was suggested (*170,577–579*). Further evidence cited was a decrease in epoxide yield when ozonations were carried out in the presence of tetracyanoethylene, accompanied by a proportionate yield of the epoxide of tetracyanoethylene (*170,577*, Chapter VIII, Section II, C). Such an epoxidation mechanism is analogous to the mechanism proposed by Kwart *et al.* (*360,425*) for peracid epoxidations and involves 1,3-dipolar cycloaddition of the carbonyl oxide to the olefin. Kwart *et al.* (*360,425*), however, found that for most olefins an oxy carbonyl oxide zwitterion (e.g., **452**) is necessary. For this reason, and also because the haloolefins would be poor dipolarophiles (see Chapter IV, Section II), such a mechanism seems unlikely, and the previously discussed mechanism involving a species such as **444** appears probable for all examples of Table XXXV. For further discussion of ozonations of haloolefins see Chapter X, Section I.

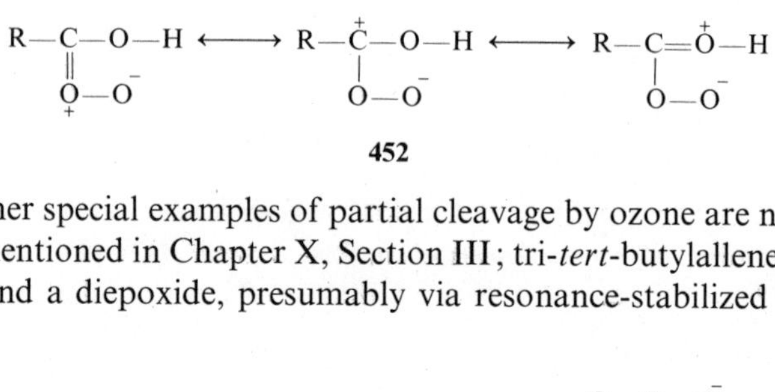

452

Four other special examples of partial cleavage by ozone are noteworthy. One was mentioned in Chapter X, Section III; tri-*tert*-butylallene (**453**) gave a mono- and a diepoxide, presumably via resonance-stabilized σ complex **454** (*595*).

453

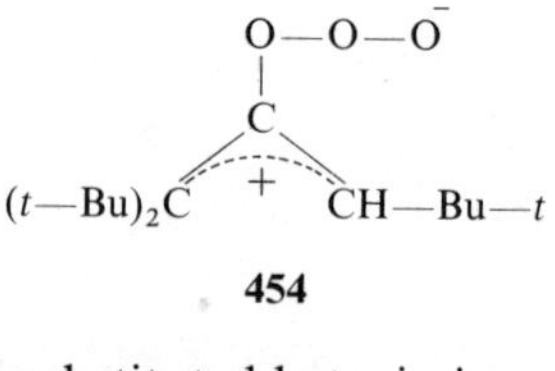

454

Two disubstituted ketenes (*656*) and two substituted ketenimines (*657*) also have been ozonized and shown to afford unstable partial cleavage products. The results and rationale are outlined in Schemes 88 and 89.

Ozonation of ketenes **455** gave unstable partial cleavage products which are thought to be equilibrium mixtures **457** (*656*). In the case of **455a** the material polymerized to **456a** immediately and could be trapped only by ozonation at $-78°C$ in the presence of methanol or hexafluoroacetone to

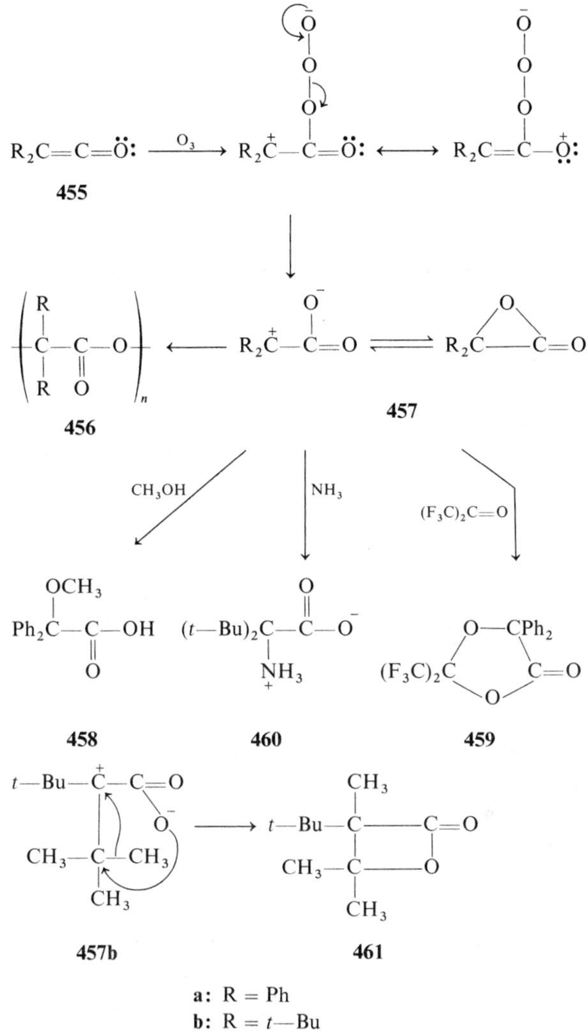

SCHEME 88. Ketene ozonations.

give **458** and **459**, respectively, as shown. Intermediate **457b** was more stable and did not polymerize (to **456b**) until the reaction mixture temperature rose to -20°C. Treatment with ammonia at -78°C gave **460b**, and, in the presence of hexafluoroacetone, rearrangement of **457b** to **461** occurred (*656*).

Ozonation of ketenimine **462** produced, presumably, unstable epoxide **463**, which decomposed to acetone and phenyl isocyanide in 90% and 34%

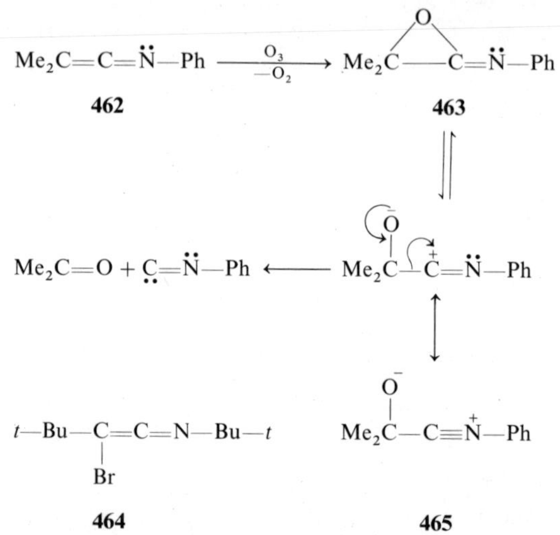

SCHEME 89. Ketenimine ozonation.

yields, respectively, plus some polymeric material (*657*). It seems likely that the true intermediate was resonance-stabilized 465. Ketenimine **464** reacted similarly (*657*).

Strong chemiluminescence was observed during ozonation of 10,10′-dimethyl-9,9′-biacridylidene (**466**) and for a period of time thereafter (*658*). Although the cause was not established, it was surmised that an epoxide is produced with release of singlet oxygen, which reacts with unchanged substrate to give the dioxetane **467**. The latter reaction is known to be chemiluminescent (*658*).

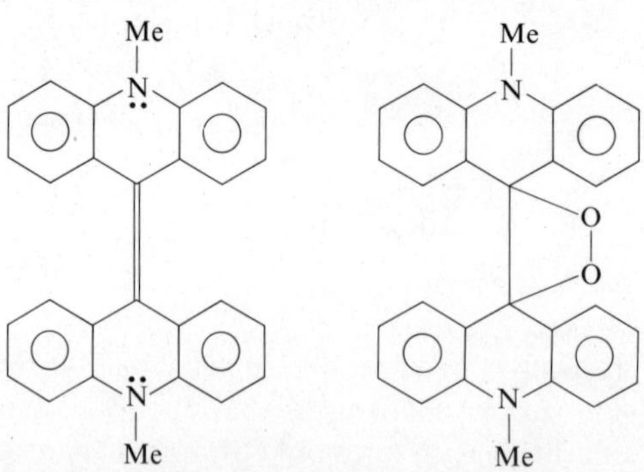

466 **467**

Finally, in this subsection on mechanism, it should be mentioned that a closed σ complex, such as **442**, or an actual peroxy epoxide (**443**) has been suggested for certain normal ozonolyses, particularly for *cis*-olefins (*127*, *128*) (see Chapter VI, Section III), and for certain phenyl-substituted cyclobutenes (*130*) (see Chapter VII, Section I, C). If so, it must be an actual peroxy epoxide (**443**), which does not appear to be an intermediate in the partial cleavage reaction.

IV. Low-Temperature Studies. π Complexes

The possibility of a π-complex precursor for both ozonolysis and partial cleavage by ozone was discussed in Section III, A of this chapter. Pi complexes have been proposed periodically since 1957 (*62,97,118,120,127–129,157,659*). The actual existence of such complexes was first established in 1971 when, at −150°C or below in dichlorodifluoromethane (Freon 12) or isopentane, interaction between ozone and 1-mesityl-1-phenylethylene resulted in a burgundy-red solution that gave essentially the same NMR spectrum as 1-mesityl-1-phenylethylene (*139*). Absorption in the visible region at 195°C was at 460 nm, whereas pure ozone absorption is centered around 620 nm (*139–141*). The complex began to decompose, directly to the epoxide, around −140°C; no intermediate species was observed by NMR (*140,141*). This discovery led to the search for other π complexes, and such were found with aromatic systems substituted with alkyl and other groups that activate for electrophilic attack (*141*). It was found that the absorption in the visible region increased in wavelength as the degree of alkyl substitution increased and the ionization potential of the aromatic system decreased, indicating that the more highly substituted is the system, the less is the energy required to promote an electron from the donor to the acceptor (*141*). Further discussion of the aromatic complexes will be given in Vol. II. Since in 1-mesityl-1-phenylethylene the styryl system is planar whereas the mesityl group cannot be in the same plane, and since the mesitylphenylethylene complex absorbs at higher wavelength than the mesitylene complex, it was suggested that complexation in mesitylphenylethylene was centered in the styryl system (*141*). Loss of molecular oxygen would then result directly in the epoxide, perhaps via a transition state such as **442** (Scheme 85). On the other hand, it is possible that the π complexes are only incidental, rather than being precursors to further reactions with ozone, such as ozonolysis and epoxidation, especially since complexation was shown to be reversible with the benzene homologs (*141*).

Following the discovery of these π complexes of aryl-substituted olefins and of aromatics, the observance of π complexes of purely aliphatic olefinic systems was reported (*102,176*). The supposed π complex of Heicklen and co-workers (*102*) appeared to decompose to starting materials rather than to rearrange to the primary ozonide and, therefore, to be only incidental.

Alcock and Mile (*176*) observed an actual red complex with 2,4,4-trimethyl-1-pentene, which appeared to rearrange to the primary ozonide; they suggest that Heicklen's species was only dissolved or trapped ozone, rather than a complex.

V. Highly Hindered Olefins. Radical Formation with Vinyl Alcohols

Table XXXIV shows that 1, 1-dineopentyl-2-tert-butylethylene gives 91% epoxide and that 1,1-dimesitylethylene yields at least 95% partial cleavage products (*140*). Attempts to ozonize tetramesitylethylene apparently resulted in attack solely at the mesityl groups; no products from double-bond attack were observed (*140*). Mosher reported that highly branched aliphatic olefins react very slowly with ozone (*660*); Keul (*149*) showed that whereas adamantane derivatives **468a–d** react normally with ozone, derivatives **468e** and **f** do not react at all; and Criegee and Askani (*661*) observed that only the less-hindered double bond of **469** reacts with ozone.

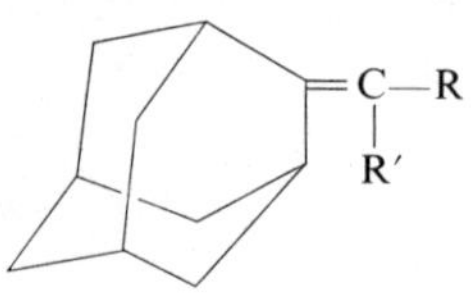

468

a: R = R′ = H
b: R = H, R′ = CH_3
c: R = H, R′ = *t*—Bu
d: R = H, R′ = C_6H_5
e: R = R′ = C_6H_5
f: R = R′ = *t*—Bu

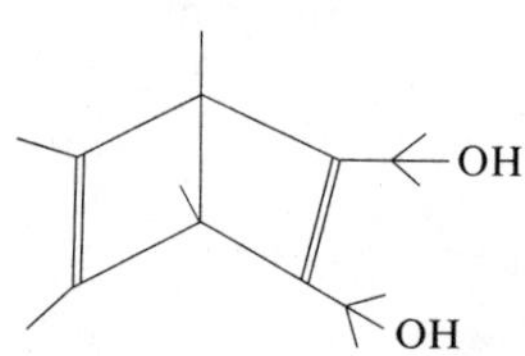

469

In order to gain further information concerning the reactivity of highly hindered olefins, the ozonation of trimesitylvinyl alcohol (**470**) was studied at −78°C. The result was the first observation and characterization of a radical produced by ozonation of an olefin (*630,659*). The radical, which was violet to black in solution, showed an ESR singlet signal (heptet in dilute CS_2 solution) with a "*g*" value of 2.004. The radical was destroyed by galvinoxyl or, at room temperature in a protic solvent, it reverted back to the vinyl alcohol. The structure assigned, **472**, is consistent with the ESR signal, the heptet arising via through-space interaction of the oxy radical with the ortho methyl groups of one of the mesityl rings (*630*). The suggested mechanism for the radical formation, **470** → **471** → **473** → **472** (Scheme 90), is based on the fact that ozonation of the vinyl alcohol at −150°C gave rise

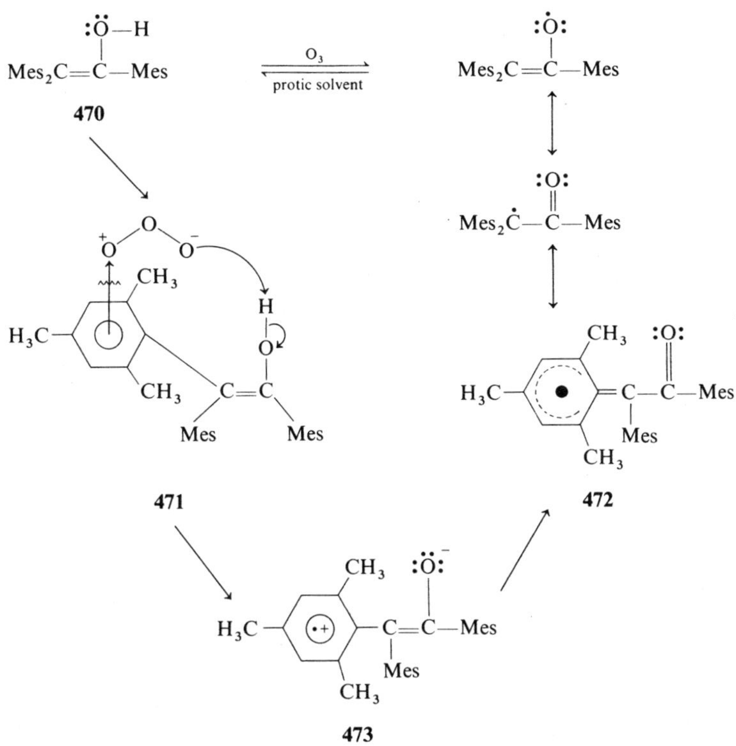

SCHEME 90. Radical formation via ozonation.

first to a π complex that decomposed to the radical (*630*). No olefinic double-bond ozonolysis or partial cleavage products were observed from the ozonation of trimesitylvinyl alcohol (*630*).

Ozonation of the less-hindered vinyl alcohol, 2,2-dimesitylvinyl alcohol (**474**) at -78°C gave only the product shown in Table XXXII example 5, although a faint pink color was evident throughout the ozonation (*630*). Ozonation at -150°C, however, produced a deeper pink to purple color, which disappeared when the reaction mixture was allowed to come to room temperature; the products were mesitoin (**477**) (see Table XXXII) in 45% yield, and the dimer (**476**) of radical **475** in 53% yield (*630*). The color of the radical reappeared when the reaction mixture was warmed to 45°C and disappeared again on cooling to room temperature. The rationalization of these reactions is shown in Scheme 91 (*630*).

Judging from these results, radical formation by ozonation is possible with hindered vinyl alcohols and, possibly, with other species having a protic group similarly located. It is closely associated with π-complex formation and with the decreased reactivity of the double bond of hindered olefins.

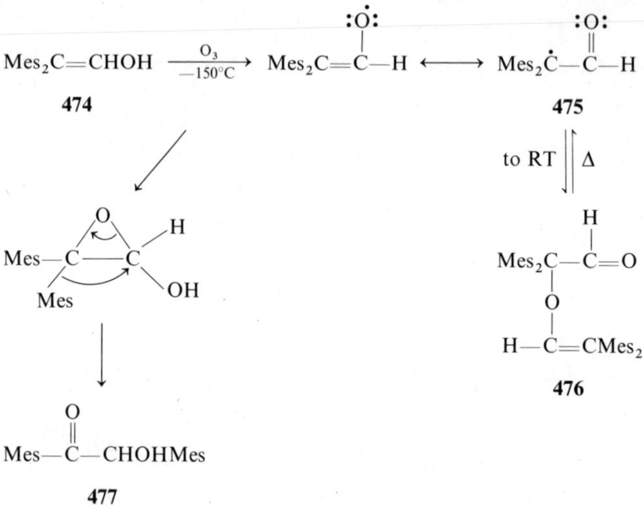

Mes=mesityl

SCHEME 91. Radical formation with 1,1-dimesitylethylene. Reprinted in part, with permission, from Bailey *et al.* (*630*). Copyright by the American Chemical Society.

CHAPTER **XII**

Gas-Phase Ozonation of Olefins

The first ozonolysis, that of ethylene, was carried out by Schönbein in the gas phase (*21*), as also were later, more detailed studies (*34,541,662–665*) with ethylene and other simple olefins. The present interest in vapor-phase ozonations began with the realization that ozone is a product of photochemical decomposition of nitrogen dioxide in air and that it reacts in the atmosphere with various hydrocarbons and other pollutants to produce smog, eye irritants, and plant toxicants (*666–672*). Leighton (*673*) has published an excellent review of studies in this field, covering the literature up to about 1960. Since then, other reviews have at least touched upon ozone in this regard (*673a*).

I. Kinetics

Beginning with Cadle and Schadt (*155*), the kinetics of ozonation of various olefinic compounds in the vapor phase have been studied by many investigators under a variety of conditions, particularly in regard to reactant pressures, oxygen pressure, and total pressure (*152–155,670,673,674–689*). Most workers reported 1:1 olefin:ozone stoichiometry, but others found that 1.5–3.0 moles of olefin per mole of ozone reacted (*153–155,675,677*). The stoichiometry appears to vary with reactant pressures (*154,155,677,682, 690*), oxygen pressure (*153*), and other reaction conditions (*673*). Several investigators (*154,155,677*) reported that olefin to ozone consumption increased with increasing olefin to ozone concentration ratios. Saltzman (*690*) found that elevated concentrations of either reactant caused a greater consumption of that particular reactant. According to Wei and Cvetanović (*153*), greater than 1:1 olefin:ozone stoichiometry occurs only in the presence of molecular oxygen. However, Becker *et al.* (*679*) reported variable ozone:olefin ratios (0.8–1.4) at low oxygen pressures, and both DeMore (*676*) and Stedman *et al.* (*678*) found that with simple olefins the stoichiometry was 1:1, regardless of whether oxygen was present.

In most cases second-order kinetics were observed, but several investigators reported variable and complex kinetics (*155,675,677,683,684*). Most of the latter results were from ozonations of 1,3-butadiene, allene, and chlorinated olefins. However, Toby and Toby (*684*) believe that the kinetics

TABLE XXXVI

SUMMARY OF KINETIC STUDIES ON GAS-PHASE OZONOLYSES OF CERTAIN SIMPLE OLEFINS[a]

Initial concentration (mTorr)		Diluent pressure (Torr)	Stoichiometry, $\Delta[HC]/\Delta[O_3]$	Rate constant "k" (ambient temp.), $10^{18} \times cm^3\ molecule^{-1}\ sec^{-1}$	References
$[HC]_0$	$[O_3]_0$				
		Ethylene			
$\sim(2\text{–}3) \times 10^3$	$\sim(3\text{–}5) \times 10^2$	~150–650 (air)	1.9–3.2	3.0 ± 1.3	*155*
~15	~15	~760 (air)	0.9	1.3	*670*
~6	~2	~760 (air)	1.6	2.6 ± 0.4	*154*
$\sim(2\text{–}4) \times 10^4$	$\sim(0.3\text{–}2) \times 10^4$	~ 50–500 (O_2)	—	4.0	*152*
$\sim 2.4 \times 10^4$	$\sim 4 \times 10^3$	~100 (O_2)	1.4–2.0	2.6	*153*
$\sim 2.4 \times 10^4$	$\sim 4 \times 10^3$	~100 (N_2)	1	2	*153*
[Other "k" values[b]: 1.2 (*676*); 1.6 (*678*); 2.8 (*679*); 1.7 (*680,691*); 1.9 (*682*)]					
		Propene			
~8–16	~0.8	~760 (air)	—	6.2	*155*
~30	~8	~760 (air)	1.06	8.0	*670*
$\sim(2\text{–}4) \times 10^4$	$\sim(0.3\text{–}2) \times 10^4$	~ 50–500 (O_2)	—	11.2	*152*
$\sim 2.4 \times 10^4$	$\sim 4 \times 10^3$	~100 (O_2)	1.4–2.0	10.6	*153*
$\sim 2.4 \times 10^4$	$\sim 4 \times 10^3$	~100 (N_2)	1	9.2	*153*
[Other "k" values[b]: 7.5 (*676*); 12.5 (*678*); 14.5 (*679*); 10.5 (*680*); 13 (*682,688*)]					
		trans-2-Butene			
—	—	~760 (air)	—	166	*673,689*
~0.2	~0.3	~760 (air)	1.5	430	*154*
$\sim(2\text{–}4) \times 10^4$	$\sim(0.3\text{–}2) \times 10^4$	~ 50–500 (O_2)	—	38	*152*
$\sim 2.4 \times 10^4$	$\sim 4 \times 10^3$	~100 (O_2)	1.4–2.0	35	*153*
$\sim 2.4 \times 10^4$	$\sim 4 \times 10^3$	~100 (N_2)	1	21	*153*
[Other "k" values[b]: 275 (*678*); 260 (*682,688*)]					

1-Hexene					
~0.8–8	~0.8–8	~760 (air)	1.6–2.5	9.0 ± 0.2	*155*
—	—	~760 (air)	—	10.0	*670*
~0.8–30	~0.2–60	~760 (air)	—	9.2	*674*
—	—	~760 (air)	1	11.0	*673,689*
~0.65	~0.5	~760 (air)	1.8	11.0 ± 0.2	*154*
[Other "*k*" values[b]: 11 (*678*); 11.1 (*682*); 16 (*688*)]					

"*k*" Values for Other Olefins[b]

1-Butene: 9.0 (*153*); 10.0 (*154*); 12.3 (*682*)
1-Pentene: 5.3 (*155*); 7.5 (*670*); 9.0 (*153*); 10.7 (*682*)
Isobutene: 6.2 (*670*); 8.4 (*153*); 13.6 (*682*); 15 (*673,689*); 16 ± 7 (*692*); 18 (*679*); 23 (*154*)
cis-2-Butene: 28 (*153*); 50 (*670*); 125 ± 25 (*692*); 140 (*688*); 161 (*682*); 340 (*154*)
cis-3-Methyl-2-pentene: 456 ± 8 (*682*)
trans-3-Methyl-2-pentene: 563 ± 17 (*682*)
2,3-Dimethyl-2-butene: 39 (*153*); 750 (*154*); 1510 ± 80 (*682*)
Cyclopentene: 813 ± 79 (*682*)
Cyclohexene: 30 (*52*); 59 (*155*); 169 ± 15 (*682*)
1,3-Butadiene: 8.2 (*670*); 8.4 (*682*); 9.1 (*152*)
1,1-Difluoroethylene: 0.0008 (*679*)
1,1-Dichloroethylene: 0.0037 (*677*)

[a] Reprinted, in part, with permission, from Stedman *et al.* (*678*). Copyright by the American Chemical Society.
[b] Rate constant $\times 10^{-13} \times cm^3$ molecule^{-1} sec^{-1}, usually in presence of oxygen.

of all gas-phase ozonolyses of olefins are complex. Some of the results with simple olefins are summarized in Table XXXVI.

In several researches, activation energies were determined (*154,676,679, 685*); the reported values were: ethylene (4.2–4.9), propene (3.9), isobutene (2.8), *trans*-2-butene (0.2), *cis*- and *trans*-2-butene mixture (2.3), and 1,3-butadiene (5.3). Huie and Herron (*685*) found a good linear correlation between activation energies and ionization potentials for the olefins studied.

The data of Table XXXVI, as well as the activation energies, show that, in general, the rate of ozone attack increases as the number of alkyl substituents at a double bond increases and that chloro substituents have the opposite effect. In this respect, therefore, vapor-phase and liquid-phase ozonolyses of olefins are similar (see Chapter IV, Section II). This effect, which is indicative of the electrophilic character of ozone attack, perhaps can be seen more clearly from Table XXXVII, which summarizes the results

TABLE XXXVII

RELATIVE RATE CONSTANTS FOR VAPOR-PHASE OZONOLYSES[a]

Olefin	Relative rate constants	
	k_{O_2}[b]	k_{N_2}[c]
Ethylene	1.00	1.00
Propene	4.06	4.52
1-Butene	3.44	4.05
1-Pentene	3.44	4.05
1-Hexene	3.91	5.00
3-Methyl-1-butene	2.97	3.57
3-Methyl-1-pentene	2.81	3.57
4-Methyl-1-pentene	3.59	4.05
Isobutene	3.13	4.76
2-Methyl-1-butene	4.06	5.95
2-Methyl-1-pentene	4.06	5.95
cis-2-Butene	10.6	9.52
trans-2-Butene	13.4	10.5
cis-2-Pentene	10.6	11.4
trans-2-Pentene	12.8	12.9
4-Methyl-2-pentene	10.9	11.0
2-Methyl-2-butene	10.9	15.2
2-Methyl-2-pentene	10.9	15.2
3-Methyl-2-pentene	10.6	15.2
2,3-Dimethyl-2-butene	15.0	26.2

[a] Values from Wei and Cvetanović (*153*) converted to values relative to ethylene.

[b] In presence of O_2.

[c] In absence of O_2.

of Wei and Cvetanović (*153*). Also, analogous to liquid-phase ozonolysis, a steric effect can be seen in Table XXXVII; for example, 1,2-dialkylethylenes react faster than do 1,1-dialkylethylenes. All of this indicates that vapor-phase, as well as liquid-phase, ozonolysis involves 1,3-dipolar cycloaddition in the initial ozone attack.

The data of Table XXXVII show that, in most cases, the reaction rates are higher in the absence of molecular oxygen than in its presence. This also has been observed by other workers (*677,679–682,684,685,692*) and has been interpreted to indicate the presence of radical intermediates that are scavenged by oxygen, if present, or react with ozone or olefin in the absence of oxygen. Herron and Huie (*680*), therefore, suggested that true rate constants are obtained only in the presence of molecular oxygen. Toby and Toby (*684*) proposed that steps are involved in vapor-phase ozonolyses in which ozone reacts with oxy radicals to give molecular oxygen.

II. Products

In varying degrees many investigators have studied gas-phase ozonolyses of olefins from the viewpoint of product identification and determination (*153,154,313,541,664,665,668–670,675,677,679,683,688,690,692–702*). A review of the results obtained through about 1959 has been given by Leighton (*673*). In addition to the expected aldehydes, ketones, and carboxylic acids from normal ozonolysis, there also was obtained carbon monoxide and carbon dioxide, principally from 1-alkenes, and various "anomalous" products. The yields of expected aldehydes and ketones generally were 50% or less. Table XXXVIII lists all products isolated from vapor-phase ozonolyses of ethylene, isobutene, and *cis*-2-butene (*696*). In Table XXXIX, are shown the various "anomalous" products reported from vapor-phase ozonolyses of eighteen other simple olefins. The yields were low, but significant. Analyses were mostly by gas–liquid chromatography and mass spectrometry (*153, 313*). There appears to be a peroxidic intermediate that leads to major portions of the products of vapor-phase ozonolyses which, however, is not an ozonide (*313,673,703*). Ozonides have been reported in only two instances, in yields of 8–19% from vapor-phase ozonolyses of 1-pentene, 1-hexene, and 3-heptene (*670*) and in undetermined yield from propene (*676*). Identifications and determinations were by infrared spectra (*670,676*).

In addition to the products just discussed, chemiluminescent products have been observed from vapor-phase ozonations of simple olefins at total pressures below 50 Torr (*679,691,692,695,696,704*). The emitting species have been identified as formaldehyde in its first excited singlet state, α-dicarbonyl compounds (glyoxal and methylglyoxal, but not biacetyl), and electronically and vibrationally excited OH radicals (*679,691,692,703,704*). Table XL lists the major examples.

TABLE XXXVIII

STABLE PRODUCTS OBSERVED IN THE REACTIONS OF OZONE WITH ETHYLENE, ISOBUTENE, AND *cis*-2-BUTENE[a,b]

Olefin	HCHO	CH_3CHO	CH_3COCH_3	CH_2CO	i-C_3H_7CHO	MF[c]	$(CHO)_2$	C_3H_6	C_3H_8	HCOOH	CH_3OH	CO_2	CO	References
C_2H_4	—	0.019	—	—	—	—	—	—	—	0.25	—	—	—	*152,313*
	0.53	—	—	—	—	—	—	—	—	<0.03	—	0.13	0.88	*669*
	Major	—	—	—	—	—	—	—	—	—	—	—	—	*696*
i-C_4H_8	—	—	0.53	—	0.005	0.005	—	0.005	0.009	0.66	0.013	0.37	—	*152,313*
	—	0.008	0.52	—	—	—	—	—	—	—	—	0.35	—	*153*
	Major	—	Major	—	—	—	—	—	—	—	—	—	—	*696*
cis-C_4H_8-2	—	1.02	—	—	—	—	—	0.023	—	0.20	0.152	0.42	—	*152,153*
	—	0.98	—	—	—	—	—	—	—	—	—	0.35	—	*153*
	Major	Major	—	Major	—	—	Minor	—	—	Minor	Minor	—	—	*696*

[a] Reprinted, with permission, from Atkinson *et al.* (*696*). Copyright by the American Chemical Society.

[b] Product yields are given as the number of moles of product per mole of ozone consumed.

[c] MF = methyl formate.

TABLE XXXIX

ANOMALOUS PRODUCTS FROM GAS-PHASE OZONOLYSES OF SIMPLE OLEFINS

Example	Olefin	Products	References
1	C—C=C	CH_3OH, H—C—OCH_3, $H_2C=C=O$, CH_4 ‖ O	*313,669,670*
2	C—C—C=C	CH_3CHO, $H_2C=CH_2$, CH_3CH_3, CH_3OH, H—C—OCH_3 ‖ O	*153,154,313*
3	C—C—C—C=C	CH_3CHO, CH_3CH_2CHO, $CH_3CH=CH_2$, $CH_3CH_2CH_3$	*153,313*
4	C—C=C—C—C	$H_2C=CH_2$, CH_3CH_3, HCOOH, CH_3OH, $H_2C=C=O$, CO_2	*153,313,669*
5	C=C—C—C | C	CH_3OH, CH_3CHO	*153,313*
6	C=C—C—C | C	$CH_3CH=CH_2$, $CH_3CH_2CH_3$, CH_3OH, $Me_2C=O$, CH_3CHO	*153,313*
7	C—C=C—C | C	CO_2, C_4H_8, HCOOH, CH_3OH	*153,313*
8	C=C—C—C—C—C	C_4H_8, C_4H_{10}, $Me_2C=O$, $CH_3CH_2CH_2CHO$ CH_3 CH_2CHO, CH_3OH, a hydroperoxide	*313,688,690*
9	C—C=C—C—C—C	CO_2, CH_3CH_2CHO	*153*
10	C=C—C—C—C | C	CH_3CHO, CH_3CH_2CHO, CH_3C—CH_2CH_3 ‖ O	*153*
11	C=C—C—C—C | C	CH_3CHO, CH_3CH—CHO, $Me_2C=O$ | CH_3	*153,688*
12	C=C—C—C—C | C	CH_3CHO, CH_3CH_2CHO, $Me_2C=O$	*153*
13	C—C=C—C—C | C	CO_2 CH_3CH_2CHO	*153*
14	C—C=C—C—C | C	CO_2, CH_3CHO	*153*

TABLE XXXIX (*Continued*)

Example	Olefin	Products	References
15	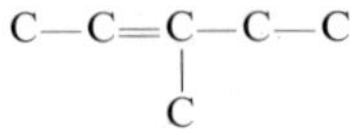	CO_2	*153*
16	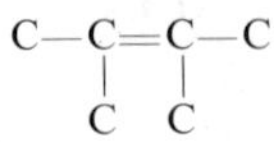	CO_2, HCOOH, CH_3OH, CH_3CHO, CH_3COOH, $H_2C{=}O$, CH_3COOCH_3	*153,313, 694*
17		HCOOH, CH_3OH, 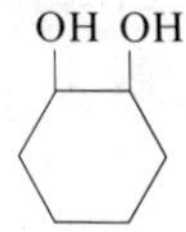 $H_2C{=}CH_2$, CH_3CHO	*313, 693*
18	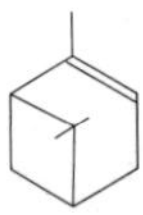	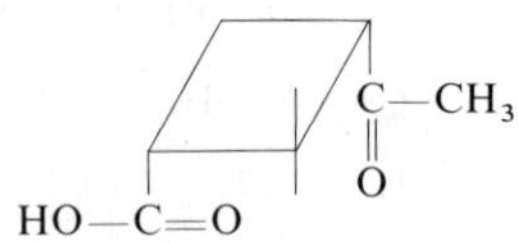	*697*
19	C=C—C=C	$H_2C{=}CH_2$, $CH_3CH{=}CH_2$, CH_3CHO	*313*

TABLE XL

CHEMILUMINESCENCE SPECIES IDENTIFIED IN OZONE–OLEFIN REACTIONS[a]

	Chemiluminescing species				
Olefin	·OH[b]	$H_2C{=}O$	$H\overset{O}{\overset{\|}{C}}-\overset{O}{\overset{\|}{C}}-H$	$CH_3-\overset{O}{\overset{\|}{C}}CHO$	References
Ethylene	+	+			*692*
Propene	+	+	+		*679,692,704*
1-Butene	+	+	+		*692,704*
cis-2-Butene	+	+	+		*692,704*
trans-2-Butene	+	+	+		*692,704*
Isobutene	+	+		+	*692,704*
2-Methyl-2-butene	+	+		+	*692,704*
2,3-Dimethyl-2-butene	+	+		+	*692,704*

[a] Reprinted, with permission, from Finlayson *et al.* (*692*). Copyright by the American Chemical Society.

[b] Both vibrationally and electronically excited OH were observed.

III. Mechanisms

A. THE INITIAL OZONE ATTACK. THE CRIEGEE MECHANISM

Previous to 1973, it was assumed by most workers in the field that gas-phase ozonolysis proceeded by the same mechanism as does liquid-phase ozonolysis (i.e., the Criegee mechanism as outlined in Chapter III, Section I, C and discussed in Chapters VI and VII), with the exception that additional fates for the Criegee carbonyl oxide zwitterion occurred and that recombination of the zwitterion with the aldehyde or ketone moiety to give an ozonide was a minor fate, if it occurred at all (*153,154,313,669,670,673,688,694*). As mentioned earlier, ozonides were reported in only two instances and the yields were low (*670,676*).

The effects of double-bond substituents on the rates of vapor-phase ozonolyses are much the same as for liquid-phase ozonolyses (*152,153*) (cf. Section I of this chapter with Chapter IV, Section II, A) and strongly suggest that the initial interaction is a 1,3-dipolar cycloaddition to give a 1,2,3-trioxolane primary ozonide. Calculations by DeMore (*676*) bear out the validity of this hypothesis. O'Neal and Blumstein (*703*) calculate the enthalpy of the cycloaddition with 1-butene to be approximately 47 kcal/mole. Some investigators believe that 1,3-dipolar cycloaddition is preceded by π complexation of the double bond by ozone (*152,155,313,679*), as also suggested for liquid-phase ozonolysis (Chapter 4, Section II, A).

Some of the reactions proposed for the Criegee carbonyl oxide in gas-phase ozonolysis are outlined in Eqs. (27)–(35) (*153,154,313,669,670,673, 688,691,703*). As can be seen, Eqs. (27)–(33) were designed to explain some of the anomalous products of Tables XXXVIII and XXXIX. Reactions (34) and (35) were proposed to account for the reported greater than unity olefin to ozone ratios (*153–155,675,677,678*). Also, the radical produced in Eq. (35) could proceed to an epoxide and account for the 1,2-cyclohexanediol produced in the ozonation of cyclohexene (*693*) (see example 17, Table XXXIX).

$$R{-}\underset{\displaystyle H}{\overset{+}{C}}{-}O{-}\bar{O} \longrightarrow RH + CO_2 \qquad (27)$$

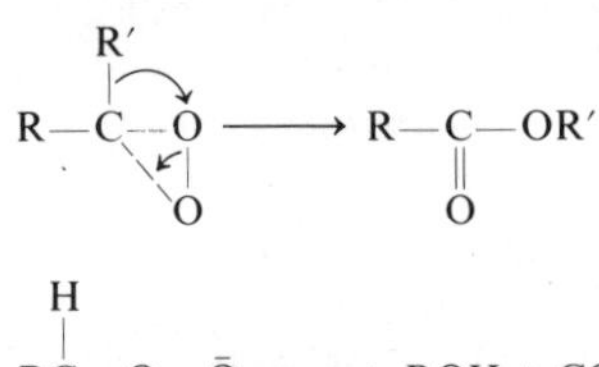

(28)

$$R\underset{+}{\overset{\displaystyle H}{C}}{-}O{-}\bar{O} \longrightarrow ROH + CO \text{ (via } R{-}O{-}\underset{\displaystyle \overset{\|}{O}}{C}{-}H\text{?)} \qquad (29)$$

$$RCH_2{-}\underset{\displaystyle H}{\overset{+}{C}}{-}O{-}\bar{O} \longrightarrow R{-}\underset{\displaystyle H}{C}{=}C{=}O + H_2O \qquad (30)$$

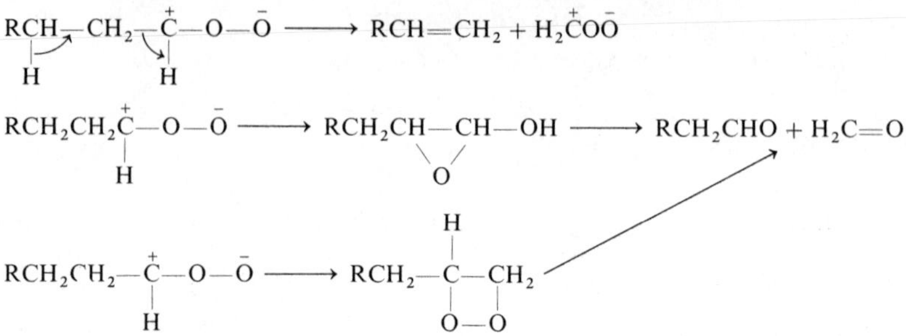

(31)

(32)

(33)

$$R_2\overset{+}{C}\text{—O—}\overset{-}{O} + O_2 \longrightarrow R_2C{=}O + O_3 \tag{34}$$

$$R'_2\overset{+}{C}\text{—O—}\overset{-}{O} + R_2C{=}CR_2 \longrightarrow R'_2C{=}O + R_2C(\dot{O})\text{—}\dot{C}R_2 \longrightarrow \tag{35}$$

In passing, it is noteworthy that the carbonyl oxide has been theorized to be the unstable ozonolysis intermediate that causes plant damage (*671*).

B. THE O'NEAL–BLUMSTEIN MECHANISM

Some of the reactions suggested in Eqs. (27)–(35) appear unlikely, especially Eqs. (27), (30) and (32). Because of this and other failures of the Criegee zwitterion to explain vapor-phase ozonation results, O'Neal and Blumstein (*703*) published in 1973 a new mechanism for gas-phase ozone–olefin reactions. 1,3-Dipolar cycloaddition is kept as the first step, but the breakdown of the primary ozonide is described as two-step, involving a diradical intermediate, as first suggested by Benson (*183*). The first stages of the mechanism are outlined in Scheme 92. One fate of the biradical formed from the primary ozonide corresponds to the Criegee split to an aldehyde or ketone and a carbonyl oxide. The carbonyl oxide is shown as a diradical, following the original suggestion of Benson (*183*). Although O'Neal and Blumstein (*703*) did not commit themselves in this respect, other workers feel that the diradical representation is more logical for vapor-phase ozonolyses (*78,691, 692,696,705*). Regardless, O'Neal and Blumstein (*703*) consider the Criegee split to be a minor fate in vapor-phase ozonations, at least for olefins of four carbons or more. The suggested major fates involve so-called α, β, and γ hydrogen extractions as outlined in Scheme 92. Further details of the α and β extractions are shown in Scheme 93.

The reasons for minimizing the importance of the Criegee route in the O'Neal–Blumstein mechanism are severalfold: (a) calculations showed it to be energetically less favorable than hydrogen extraction (*78,692,703*); (b) there appears to be a peroxidic precursor to many of the products that is not an ozonide (*313,541,664,665,673,703*); (c) anomalous aldehydes and ketones, from cleavage of a bond next to the double bond, are obtained in consider-

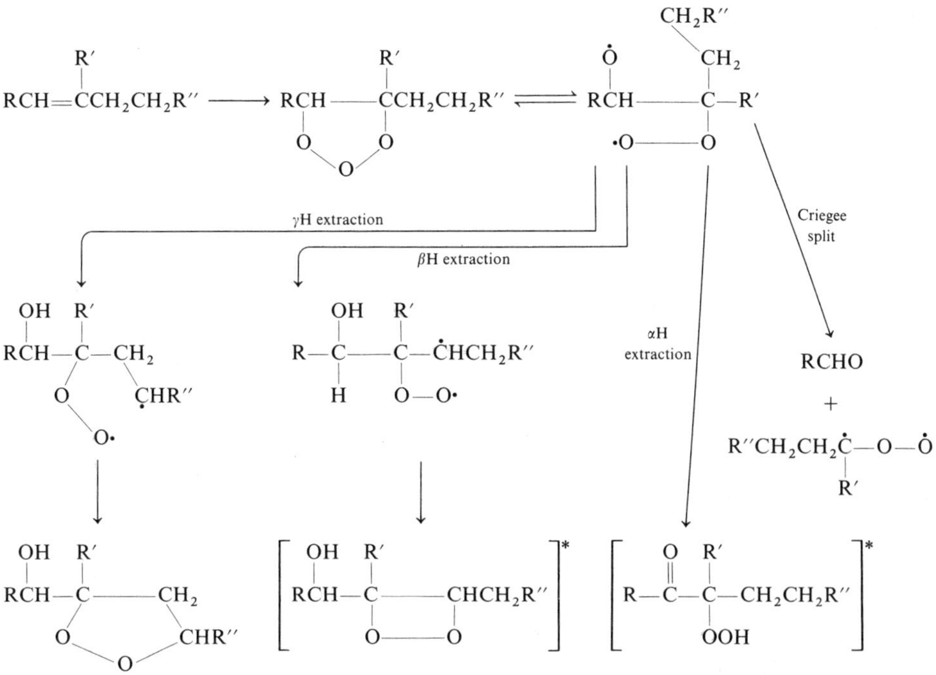

478 479 480

SCHEME 92. O'Neal–Blumstein mechanism for vapor-phase ozonation of olefins.

*Denotes excited-state intermediates.

able yield (*313,673,703*) (carbonyl oxide rearrangement should give an ester, alcohol, and/or carboxylic acid, but not an aldehyde or ketone); (d) α-dicarbonyl compounds are products which are difficult to explain by the Criegee route; (e) chemiluminescence occurs at low pressures, as described earlier; (f) the overall yield of products is generally low.

The O'Neal–Blumstein mechanism accounts for most of the products and facts. The α-hydrogen extraction route leads to a keto hydroperoxide (**481**) (*703*). Decomposition of it, or its vibrationally excited precursor (**480**), could be the source of the normal ozonolysis products, **482** and **483**, as well as of the α-dicarbonyl product, **484**. Decomposition to radical **485** was proposed as a source of other products (*703*). Although not mentioned specifically by O'Neal and Blumstein (*703*), the decomposition of **485** to **486** and **487** could explain most of the hydrocarbon products shown in Table XXXIX. For example, propane and propene, from ozonation of 1-pentene, could result from a radical such as **487**. Production of radicals such as **485** and **487** might also account for the "oxygen effect" mentioned in Section I of this chapter.

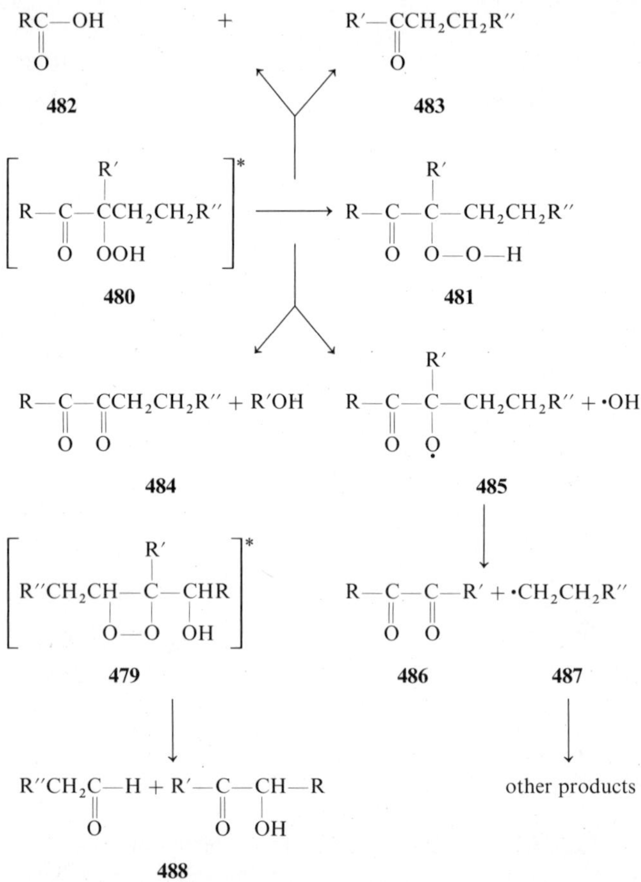

SCHEME 93. α- and β-Hydrogen extraction fates.

*Denotes excited-state intermediates.

Oxygen can scavenge these radicals which otherwise could react with ozone or the alkene.

Decomposition of the β-hydrogen extraction product (**479**) (Schemes 92 and 93), could lead to the anomalous aldehydes and/or ketones resulting from adjacent carbon–carbon bond cleavage (e.g., **488**). γ-Hydrogen extraction should result in **478** (Scheme 92). Although no such product has been identified, O'Neal and Blumstein (*703*) believe that this reaction accounts for the low overall yield of products isolated from vapor-phase ozonation of olefins such as 1-hexene.

Pitts and co-workers (*692,696*) have discussed both the ordinary and the chemiluminescent products obtained from vapor-phase ozonations of ethylene, isobutene, and *cis*-2-butene (see Tables XXXVIII and XL) on the

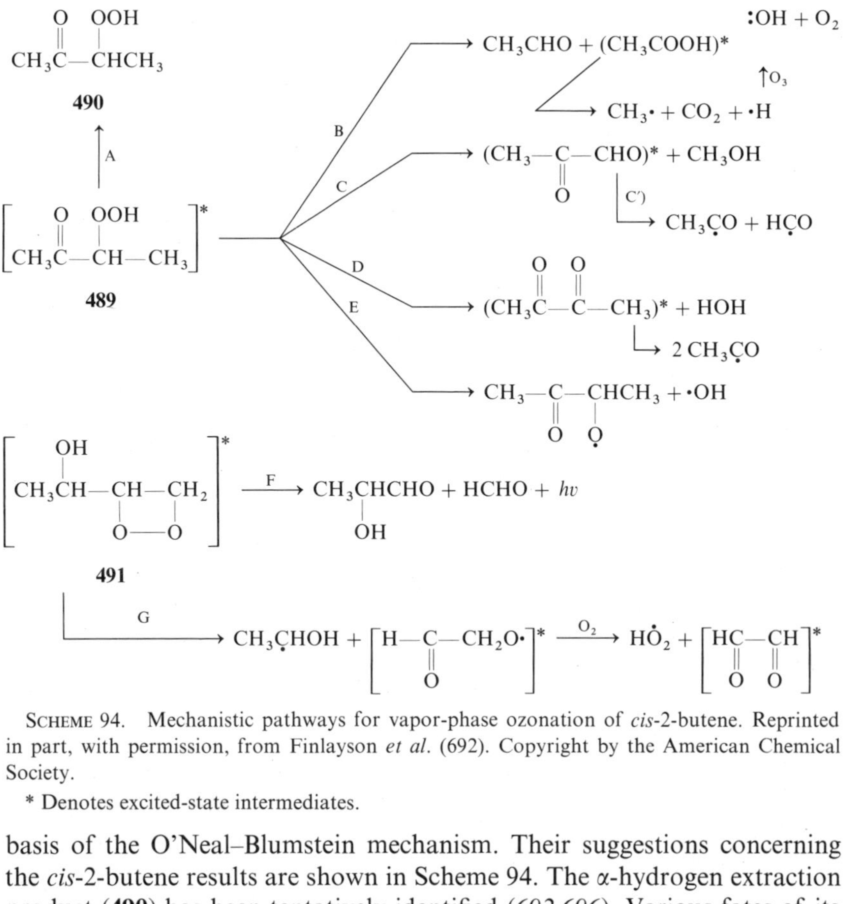

SCHEME 94. Mechanistic pathways for vapor-phase ozonation of *cis*-2-butene. Reprinted in part, with permission, from Finlayson *et al.* (692). Copyright by the American Chemical Society.

* Denotes excited-state intermediates.

basis of the O'Neal–Blumstein mechanism. Their suggestions concerning the *cis*-2-butene results are shown in Scheme 94. The α-hydrogen extraction product (**490**) has been tentatively identified (*692,696*). Various fates of its precursor (**489**) are shown by reactions A–E (Scheme 94). Acetic acid, methylglyoxal, and biacetyl, predicted by reactions B–D (Scheme 94), were not observed as products (Table XXXVIII). Pitts *et al.* (*692*) suggest that the reaction exothermicity, from decomposition of the excited hydroperoxide, **489**, is enough to decompose these also, as shown. The vibrationally excited OH chemiluminescence observed (Table XL) probably arises from the hydrogen atom in reaction B (Scheme 94) reacting with ozone, as shown, since this emission was identical with the Meinel band previously observed from reaction of H with O_3 (*692*). The electronically excited OH probably arises indirectly from further reactions of the vibrationally excited OH (*692*). The suggested (*692*) sources of the formaldehyde and glyoxal chemiluminescence (Table XL) are the excited β-hydrogen extraction product (**491**) and reactions F) and G (Scheme 94). Another possibility, combination of two formyl radicals from reaction C′) is not likely in an oxygen atmosphere. These

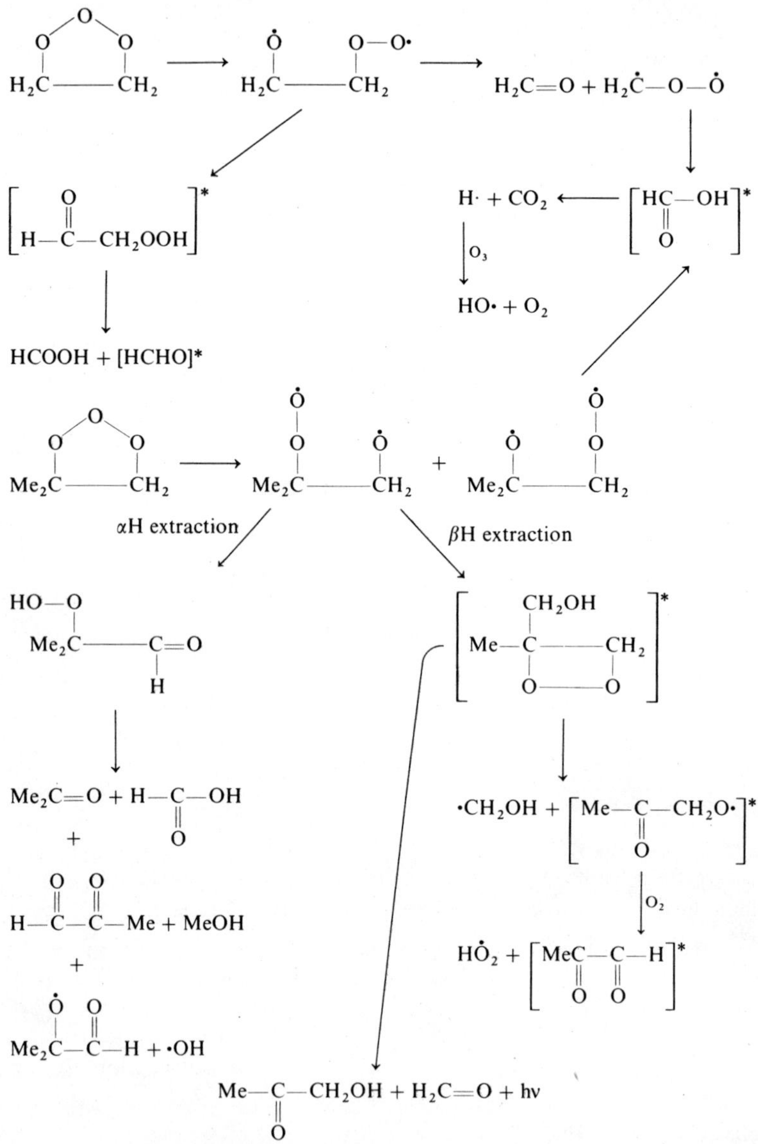

SCHEME 95. Mechanistic pathways for vapor-phase ozonations of ethylene and isobutene. *Denotes excited-state intermediates.

reactions, F) and G), also explain formaldehyde and glyoxal as reaction products (Table XXXVIII). Ketene (Table XXXVIII) was assumed to arise from direct decomposition of the primary ozonide (*696*). The source of formic acid was not discussed (*692,696*).

Scheme 95 outlines the proposed (*692*) chemiluminescence sources for ethylene and isobutene (Table XL). Experiment has shown that oxygen is required for the methylglyoxal excitation (*692*), as shown. The scheme also accounts for the major products from vapor-phase ozonations of ethylene and isobutene (Table XXXVIII). It does not rationalize, however, the production of acetaldehyde from ethylene or acetaldehyde, isobutyraldehyde, methyl formate, propene, and propane from isobutene.

It is conceivable that acetaldehyde arises from ethylene ozonation via an addition of the diradical form of the carbonyl oxide to ethylene, followed by decomposition and rearrangement to formaldehyde and acetaldehyde. Similarly, isobutyraldehyde could arise from isobutene.

Schemes 94 and 95 include several different radicals produced during the ozonations, again providing an explanation for the "oxygen effect" cited earlier. In general, it appears that the O'Neal–Blumstein mechanism (*692, 696,703*), which embodies parts of the Criegee mechanism in a minor role, provides, at least, a good qualitative description of the complex vapor-phase ozone-olefin reaction routes.

IV. Special Types

A. 1,3-BUTADIENE AND ALLENE

Some studies have been made concerning vapor-phase ozonations of 1,3-butadiene and allene (*313,683,684,699,706*), although the only real product identification carried out was with 1,3-butadiene (*313*). The work dealt primarily with the complex kinetics of the reactions, especially in regard to ozone (*683,684*), and the chemiluminescence of the allene reaction (*699,706*).

Molecular oxygen was an important product in these ozonations and, because of this and the complex kinetics, Toby and Toby (*683,684*) suggested mechanistic steps that involve ozone reacting with oxy radicals to produce molecular oxygen. They attempted to apply the O'Neal–Blumstein (*703*) mechanism to the butadiene reaction and suggested that the key intermediate was hydroperoxide **493** from α-H extraction involving **492**. They stated that **493** should decompose to acrolein and formic acid, major products of 1,3-butadiene ozonolysis (*313*). Actually, however, decomposition of **493** would be expected to produce acrylic acid and formaldehyde, via **494** or **495** and the hydroxyl radical. Most likely, the Criegee mechanism was operating in this case.

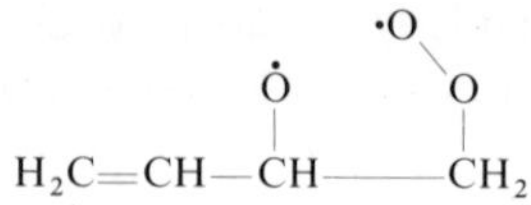

492

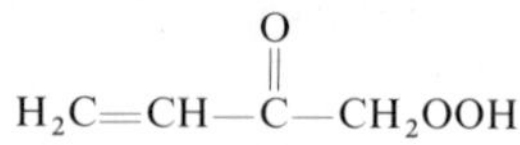

493

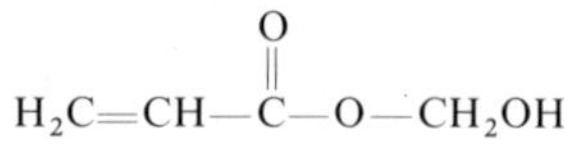

494

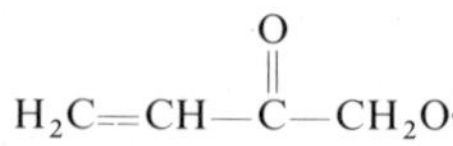

495

The chemiluminescence observed in the allene ozonation was due to the production of electronically excited formaldehyde (*699,706*), and electronically and vibrationally excited OH (*699*).

B. HALOETHYLENES

Considerable study has been given to the ozonation of haloethylenes (*579, 675,677,700–702,707–710*), beginning in 1966 (*675,700*). The work of Heicklen and co-workers has dealt with both the kinetics and the products of ozonation of 1,1-dichloroethylene (*677*), *cis*- and *trans*-1,2-dichloroethylene (*709*), and tetrachloroethylene (*702*). They reviewed their work and that of others in a separate report (*710*).

The products from the gas-phase ozonation of tetrachloroethylene at 25°C were oxygen, phosgene ($Cl_2C{=}O$), and trichloroacetyl chloride, as major products, along with traces of tetrachloroethylene epoxide, carbon monoxide, carbon dioxide, and chlorine (*702*). The stoichiometry was not determined, but the average value of the ratio of phosgene to trichloroacetyl chloride was 1.3. The trichloroacetyl chloride perhaps originated via rearrangement of the epoxide, as described earlier for liquid-phase ozonations of chloroethylenes (Chapter X, Section I and Chapter XI, Section I and III). The ozonation was strongly inhibited by oxygen, which was a product, thus making reliable kinetic studies impossible (*702*). However, the initial kinetics approximated second order with respect to olefin. When oxygen was present at the beginning of the reaction, the initial reaction rate was reduced by a factor of 10 or more. Heicklen *et al.* (*702*) proposed a diradical chain reaction mechanism to explain these results [Eqs. (36)–(42), with Eqs. (41) and (42) being chain-termination steps].

$$Cl_2C{=}CCl_2 + O_3 \longrightarrow Cl_2C{=}O + Cl_2\dot{C}{-}O{-}\dot{O} \tag{36}$$

$$Cl_2\dot{C}{-}O{-}\dot{O} + O_3 \longrightarrow Cl_2C{=}O + 2\,O_2 \tag{37}$$

$$Cl_2\dot{C}{-}O{-}\dot{O} + Cl_2C{=}CCl_2 \longrightarrow Cl_2\dot{C}{-}O{-}O{-}CCl_2{-}CCl_2{\cdot} \tag{38}$$

496

$$\mathbf{496} + O_3 \longrightarrow Cl_2\dot{C}{-}O{-}O\wr C(Cl_2){-}C(Cl_2){-}O\wr O{-}O\cdot \longrightarrow$$

497

$$Cl_2\dot{C}{-}O{-}\dot{O} + Cl_2\overset{O}{\widehat{C{-}C}}Cl_2 + Cl_3C{-}\overset{Cl}{\overset{|}{C}}{=}O + O_2 \qquad (39)$$

498 **499**

$$\mathbf{497} + Cl_2C{=}CCl_2 \longrightarrow Cl_2\dot{C}OO\wr C(Cl_2)C(Cl_2)O\wr OOC(Cl_2)\dot{C}Cl_2$$

$$\longrightarrow Cl_2\dot{C}O\dot{O} + \mathbf{498} + \mathbf{499} + 2\ Cl_2C{=}O \qquad (40)$$

$$\mathbf{496} + O_2 \longrightarrow Cl_2\dot{C}OOC(Cl_2)C(Cl_2)O{-}\dot{O} \qquad (41)$$

500

$$\mathbf{500} + Cl_2C{=}CCl_2 \longrightarrow Cl_2\dot{C}O\wr OC(Cl_2)\wr C(Cl_2)O\wr OC(Cl_2)\dot{C}Cl_2$$

$$\longrightarrow 3\ Cl_2C{=}O + \mathbf{498} + \mathbf{499} \qquad (42)$$

The products of the vapor–phase ozonation of 1,1-dichloroethylene at 25°C were phosgene, chloroacetyl chloride, formic acid, carbon monoxide, carbon dioxide, molecular oxygen, and hydrogen chloride (*677*). The olefin to ozone consumption ratio was approximately equal to 1 when the initial pressures were equal, but approached 2, with excess olefin, and 5, with excess olefin and an oxygen atmosphere. The yield of phosgene per mole of ozone was 0.25–0.4 in the absence of oxygen and 1 when the olefin was in excess and oxygen was the diluent. The kinetics were complex in the absence of oxygen, but appeared to be second order in an oxygen atmosphere. The presence of oxygen reduced the reaction rate, but did not stop the reaction. The proposed mechanism was similar to that for tetrachloroethylene, as shown by Eqs. 43–51 (*677*). Equation (44) supposedly accounts for the absence of formaldehyde as a product.

$$Cl_2C{=}CH_2 + O_3 \longrightarrow H_2C{=}O + Cl_2\dot{C}{-}O{-}\dot{O} \qquad (43)$$

$$H_2C{=}O + O_3 \longrightarrow O_2 + HCOOH\ (\text{or } CO + H_2O) \qquad (44)$$

$$Cl_2\dot{C}O\dot{O} + O_3 \longrightarrow Cl_2C{=}O + 2\ O_2 \qquad (45)$$

$$Cl_2\dot{C}O\dot{O} + Cl_2C{=}CH_2 \longrightarrow Cl_2\dot{C}OOCH_2\dot{C}Cl_2 \qquad (46)$$

501

$$\mathbf{501} \longrightarrow Cl_2C{=}O + ClCH_2{-}\underset{\underset{O}{\|}}{C}{-}Cl \qquad (47)$$

$$\mathbf{501} + O_3 \longrightarrow Cl_2\dot{C}OOCH_2C(Cl_2)OOO\cdot \longrightarrow Cl_2\dot{C}O\dot{O} + Cl_2C{=}O + HCOOH \qquad (48)$$

502

$$\mathbf{502} + 2\,Cl_2C{=}CH_2 \longrightarrow Cl_2\dot{C}{-}O{-}\dot{O} + 3\,ClCH_2{-}\overset{\overset{\displaystyle O}{\|}}{C}{-}Cl \qquad (49)$$

$$\mathbf{501} + O_2 \longrightarrow \underset{\mathbf{503}}{Cl_2\dot{C}OOCH_2C(Cl_2)OO\cdot} \qquad (50)$$

$$\mathbf{503} + Cl_2C{=}CH_2 \longrightarrow Cl_2\dot{C}{-}O{-}\dot{O} + 2\,ClCH_2{-}\overset{\overset{\displaystyle O}{\|}}{C}{-}Cl \qquad (51)$$

The vapor-phase ozonation of *cis*- and *trans*-1,2-dichloroethylene also was complex (*709*). The only initial products were formyl chloride and oxygen, with stoichiometry according to Eq. (52). The kinetics were complicated,

$$2\,ClCH{=}CHCl + 2\,O_3 \longrightarrow 4\,H{-}\overset{\overset{\displaystyle Cl}{|}}{C}{=}O + O_2 \qquad (52)$$

and varied with the relative proportions of the reactants. Again, there was molecular oxygen inhibition, and, because of this, a very complicated chain mechanism was proposed (*709*).

Since the rates of vapor-phase ozonations with simple, nonhalogenated olefins also were slower in the presence of oxygen than in its absence (Section I of this chapter), Heicklen *et al.* (*710*) have suggested that the mechanisms of those reactions might also be more complicated than as discussed in Section III of this chapter.

In regard to Eqs (36) and (43), Heicklen and co-workers (*710*) observed no primary ozonide by infrared techniques (*102*) and, therefore, suggested that these reactions proceed without an addition intermediate, possibly via a π complex. No primary ozonide has been observed with *cis*-1,2-di-*tert*-butylethylene either, but the possibility still remains in such cases that the primary ozonide is extremely unstable and short-lived, or that it is essentially only a transition state.

Vapor-phase ozonolysis of tetrafluoroethylene at room temperature gave carbonyl fluoride ($F_2C{=}O$) and oxygen as the only products (*675*). Under the same conditions hexafluoropropene afforded a mixture of carbonyl fluoride and trifluoroacetyl fluoride in a ratio of 4:1, respectively, and octafluoro-2-butene gave the same products in a ratio of 2.8:1, respectively; oxygen also was produced (*675*). The reaction velocities were in the order $C_2F_4 > C_3F_6 > C_4F_8$-2 to the extent of approximately 74:12:1. Since, in contrast to the results with tetrachloroethylene, no epoxide or trihaloacetyl halide was obtained from the tetrafluoroethylene ozonation, Heicklen (*675*) assumed that no carbonyl oxide diradical was involved in the reaction. His proposed mechanism is outlined in Eqs. 53–55 (*675*). It is noteworthy that CF_2 emission has recently been reported from ozonation of tetrafluoroethylene under chemiluminescent conditions (*701*).

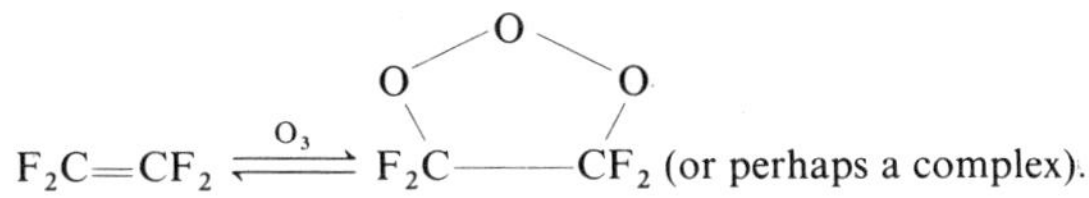
(53)

504

$$\mathbf{504} + F_2C{=}CF_2 \longrightarrow 3\,F_2C{=}O + F_2C \quad (54)$$

$$F_2C + O_3 \longrightarrow F_2C{=}O + O_2 \quad (55)$$

A similar mechanism was proposed for hexafluoropropene and octafluorobutene-2, with the addition of Eq. (56) to explain the conversion of trifluoroacetyl fluoride to carbonyl fluoride (*675*). To this reviewer, a more reasonable explanation for the production of excess carbonyl fluoride from hexafluoropropene and octafluorobutene-2 might be that shown in Eqs. (57) and (58).

$$F_3C{-}CF{=}O^* + O_3 \longrightarrow 2\,F_2C{=}O + O_2 \qquad \text{*denotes excitation} \quad (56)$$

$$F_3C{-}CF{=}CF{-}CF_3 \xrightarrow{O_3} \text{adduct} \longrightarrow F_3C{-}CF{=}O + F_3C{-}\dot{C}F{-}O{-}\dot{O} \quad (57)$$

505

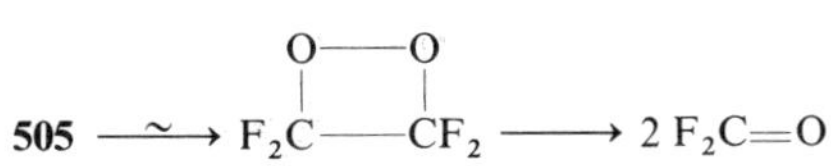
(58)

Ozone-initiated oxidation of tetrafluoroethylene also has been studied (*700,707*). The products were carbonyl fluoride, tetrafluoroethylene oxide, trifluoroacetyl fluoride, and polymeric material.

V. Solid–Gas-Phase Ozonolyses

The reactions of gaseous ozone with crystalline *trans*-stilbene (**506a**) and *trans*-3,4-di-(*p*-hydroxyphenyl)-3-hexene (**506b**) have been studied (*711,712*). The reaction with *trans*-stilbene was quite fast, the crystals disappearing within 3 minutes and producing a liquid composed largely of benzaldehyde plus minor quantities of benzoic acid (*711*). In contrast, the more hindered **506b** required 4 hours for only 12% conversion to **507b**. This is in contrast to rapid reaction of **506b** with ozone in methanol solution to give a 70% yield of aldehyde **507b** (*711*). The difference is said to result from the inaccessibility of the double bond of **506b** to ozone attack in the solid crystal, due not only to steric effects, but also, largely, to extensive hydrogen bonding, which prevents penetration of the gaseous ozone into the crystal (*712*). The ozone attack is initiated at penetrative paths or dislocations. As product accumulates with each solid reactant, stresses occur and generate more dislocations along slip planes which, in turn, function as new centers for ozone attack. Thus, the ozonolysis accelerates as the reaction progresses. This occurs more rapidly in *trans*-stilbene (**506a**) than in its substituted analog (**506b**) (*712*).

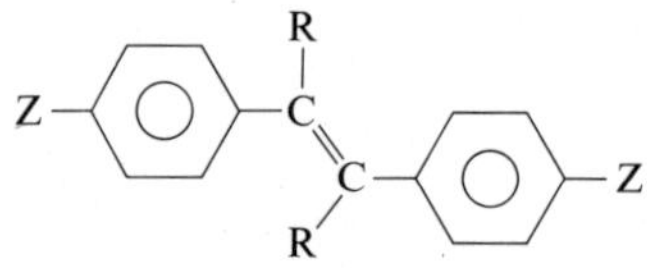

506

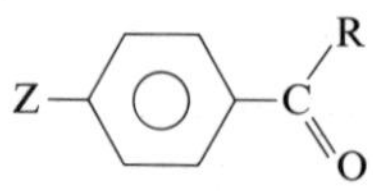

507

a: R = Z = H
b: R = Et, Z = OH

EPILOGUE

Overview and New Developments

As implied in the Preface, an attempt has been made in this monograph to present in one place a more or less "state of the union" discussion of our present-day knowledge concerning ozone and its reactions with olefins. Other recent reviews have been either narrower or less comprehensive or detailed in their coverage (*14,27,70,110–113,115,116,131,132,310,713–715*).

Much has been learned in the nearly 20 years since the author's earlier review (*62*). An attempt has been made to show the development of this knowledge. It appears to this reviewer that we are finally on the right road, after several detours, in regard to the broad, general mechanism for liquid phase ozonolysis of olefins. There still are details, however, which are not yet fully understood. These, in the mind of this reviewer, deal largely with the conformations of the primary and final ozonides, the nature of the transition states and intermediates leading from the primary to the final ozonide, and the equilibrium involving *syn*- and *anti*-carbonyl oxide zwitterions (or diradicals?). In what way do minor changes in reaction conditions affect these steps and, consequently, the stereochemistry of ozonide formation? More specifically, what roles do solvent polarity and nucleophilicity, etc., complexing agents, reaction temperature, rate of warming the reaction mixture, concentration changes, presence of excess aldehyde, etc., play in determining cis:trans ozonide ratios? What roles do these factors play in competitions between ozonide formation and the production of other peroxidic ozonolysis products? What are the various factors that determine the result of competition between ozonolysis and epoxide formation (or other partial cleavage reactions)? What factors are responsible for so-called anomalous reactions between ozone and olefins? Can a carbonyl oxide moiety ever be observed and characterized as such? Are the cycloaddition and cycloreversion processes occurring during liquid phase ozonolyses always concerted, or are there competing nonconcerted, perhaps radical-type, processes? These and other questions await complete answers. In addition, the author recalls that oftentimes in the past, just as we thought we had complete answers, new information suddenly appeared which sent us scurrying back to the laboratory for further study.

The twelve chapters of this book were completed in November 1976. Between that date and the type-setting date, some new and important information concerning liquid-phase ozonolysis has been disclosed and should be briefly reviewed.

Hiberty (*716*) has tested various primary ozonide possibilities in the ethylene-ozone reaction by means of *ab initio* S.C.F.M.O. theory. His results negate all possibilities except the 1,2,3-trioxolane, for which he finds the Kuczkowski O-envelope conformation (Chapter 6, Section IV) to be slightly more stable than the C–C half-chair. Because of this he prefers the Kuczkowski approach to that of Bauld–Bailey (Chapter 6, Section IV). However, the Bauld–Bailey scheme has been expanded (*717,717a*) in such a way as to give it a fllexibility which, in the opinion of this author, makes it superior to the Kuczkowski scheme.

The most important concept in the expanded Bauld–Bailey mechanism (*717,717a*) is an extension of the least motion principle already invoked (Chapter 6, Section IV). A Newman C–O projection of the C–C half-chair conformation of the primary ozonide, **508** (Scheme 96), shows why an equatorial substituent is more easily incorporated into a carbonyl oxide (which becomes anti) than is an axial substituent (which would become syn) (Bauld–Bailey Rule 2). An equatorial bond already has nearly the required 180°C relationship with the oxygen–oxygen bond of the incipient carbonyl oxide, whereas an axial bond is about 80° away from the required parallel relationship with the O–O bond, **508**. For a syn carbonyl oxide to be produced during the cleavage and rehybridization processes, an axial substituent must move inward toward the ring oxygens, and this should involve more repulsive forces than for a hydrogen to make the same movement in the formation of an anti carbonyl oxide. Thus, anti carbonyl oxide formation is kinetically more favorable than syn carbonyl oxide formation.

This concept can be extended to explain why *trans*-1,2-diisopropylethylene gives cis:trans ozonide ratios greater than unity with a slow warm-up (Chapter 7, Figure 7 and Table 16), whereas the ratio is less than unity with a fast warm-up (Fig. 7). An additional principle involved states that the primary ozonide conformation decomposing is not necessarily the predominant nor the thermodynamically more stable one, but may be the one in an equilibrium mixture of conformers which kinetically decomposes most readily to syn and/or anti carbonyl oxides. In the case of the *trans*-diisopropylethylene primary ozonide, the diaxial conformer is the more thermodynamically stable and predominant C–C half chair conformer and, during a fast warm-up, it decomposes rapidly to give, predominantly, syn carbonyl oxide. With a slow warm-up, however, the diequatorial conformer decomposes first, and there is now time for the equilibrium to shift (Eq. 59), providing more diequatorial conformer, more anti carbonyl oxide, and a

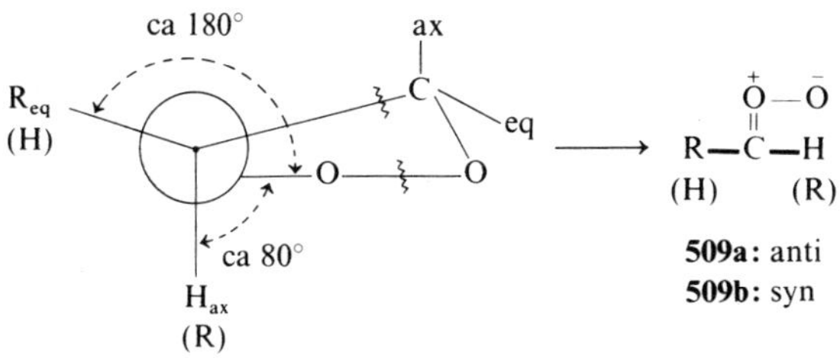

509a: anti
509b: syn

508

Anti-carbonyl oxide ⇌ diequatorial conformer ⇌ diaxial conformer ⇌ *syn*-carbonyl oxide (59)

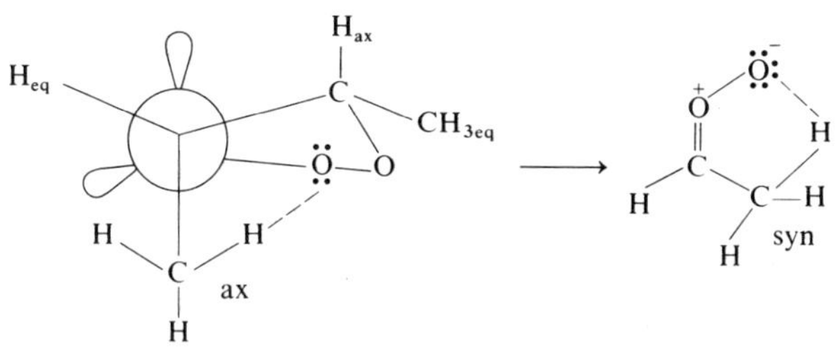

510 **511**

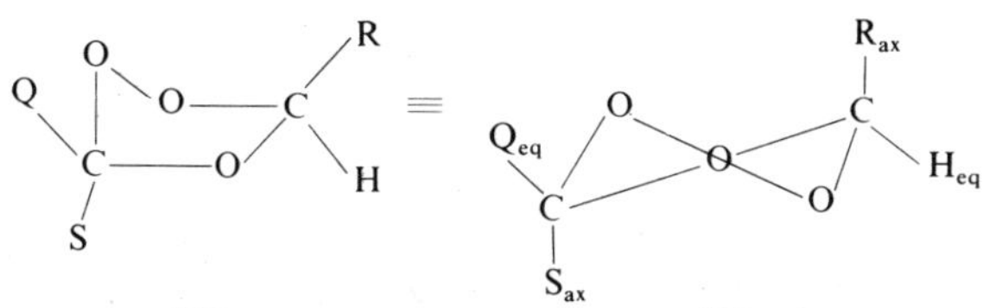

512 **513**

SCHEME 96. Expanded Bauld–Bailey Concepts.

predominance of cis final ozonide. This principle has also been extended to other examples (717a).

The expanded Bauld–Bailey concepts (*717*,*717a*), however, allow for one exception to the principle that equatorial substituents are preferentially incorporated into carbonyl oxides. This is the anomalous case of *cis*-2-butene (Chapter 6, Section IV, C). Fliszár's (*338*) calculations and studies had indicated that an axial rather than an equatorial methyl substituent is preferentially incorporated into a carbonyl oxide. A logical explanation is that a syn methylcarbonyl oxide is perferentially stabilized over its anti stereoisomer through hydrogen bonding, **511**, which should be more important with a methyl than any other alkyl group. This effect, conceivably, could begin to be felt in the transition state (**510**) for the primary ozonide decomposition and, thus, account for this heretofore anomalous result.

The Bauld–Bailey expanded mechanism (*717,717a*) emphasizes the necessary supra-supra orbital alignment during cycloadditions and reversions (Chapter 6, Scheme 16) just as much as does the Kuczkowski scheme. The transition state for primary ozonide decomposition does not have to be a perfect envelope, but, more likely, would resemble **508**, since in the cleavage process the weaker O–O bond should lead the way. Likewise, in the recombination of carbonyl oxide and carbonyl moieties, supra-supra orbital alignment is necessary and the transition state must be similar to that of Kuczkowski's (Scheme 17, Chapter 6). However, if one follows this through to the Kuczkowski envelope (**512**) and compares a model of this with a model of the Bauld–Bailey O–O half-chair conformation (**513**) for the final ozonide, it can be seen that there is very little difference between the two. Therefore, Bailey and Ferrell (*717a*) prefer the Bauld–Bailey rationale to that of Kuczkowski, especially since it is uncertain exactly where in the coordinate the transition state should lie.

Vapor-phase ozonation of olefins appears to be much more complicated than its liquid-phase counterpart. Although recent developments seem to be pointing the way, an understanding of the complicated mechanistic details is still in its infancy.

The "state of the union" today, therefore, can best be described by the closing statement of Criegee in his final publication (*132*), in which he first quoted Otto as describing (in 1844!) the ozone field as a rich field for further research, and then stated "I consider this statement to have lost none of its validity" (in 1975!).

Perhaps in the second monograph of this treatise, some additional answers to questions concerning the mechanisms of ozonation of olefins can be given. There, the story of ozone and the details of the variety of ways in which it reacts with many different types of organic compounds will be continued.

Bibliography

1. Homer, "Iliad," **8**, 135; **14**, 415; "Odyssey," **12**, 417; **14**, 307; "The Complete Works of Homer." Modern Library Edition, Random House, New York, 1950; F. Mohr, *Ann. Phys.* (*Leipzig*) [2] **91**, 625 (1854).
2. T. C. Manley and S. J. Niegowski, *Kirk-Othmer Encycl. Chem. Technol., 2nd Ed.* **14**, 410 (1967).
3. M. J. Molina and F. S. Rowland, *Nature* (*London*) **249**, 810 (1974); R. J. Cicerone, R. S. Stolarski, and S. Walters, *Science* **185**, 1165 (1974); S. C. Wofsy, M. B. McElroy, and N. D. Sze, *ibid.* **187**, 535 (1975); J. Basuk, *Bull. Am. Meteorol. Soc.* **56**, 589 (1975); *Chem. Eng. News* **54**, No. 40, 4 and 16 (1976).
4. M. B. McElroy, S. C. Wofsy, J. E. Penner, and J. C. McConnell, *J. Atmos. Sci.* **31**, 287 (1974); F. N. Alyea, D. M. Cunnold, and R. G. Prinn, *Science* **188**, 117 (1975); H. S. Johnston, *Acc. Chem. Res.* **8**, 289 (1975).
5. R. C. Whitten, W. J. Borucki, and R. P. Turco, *Nature* (*London*) **257**, 38 (1975).
6. R. S. Stolarski and R. J. Cicerone, *Can. J. Chem.* **52**, 1610 (1974).
7. Anonymous, *Chem. Eng.* (*N.Y.*) **59**, No. 9, 246 (1952); *Am. Inkmaker* **45**, No. 10, 36 (1967).
8. R. J. Bender, *Power* **114**, No. 8, 58 (1970); E. W. J. Diaper, *Chem. Technol.* **368**, 498 (1972); H. W. Prengle, Jr., C. E. Mauk, R. W. Legan, and C. G. Hewes, III, *Hydrocarbon Process.* **54**, No. 10, 82 (1975).
9. F. L. Evans, III, ed., "Ozone in Water and Wastewater Treatment." Ann Arbor Sci. Publ., Ann Arbor, Michigan, 1972.
10. C. Nebel, R. D. Gottschling, R. L. Hutchison, T. J. McBride, D. M. Taylor, J. L. Pavoni, M. E. Tittlebaum, H. E. Spencer, and M. Fleischman, *J. Water Pollut. Control Fed.* **44**, 212 (1972).
11. J. J. McCarthy and C. H. Smith, *J. Am. Water Works Assoc.* **66**, 718 (1974).
12. A. Vosmaer, "Ozone, Its Manufacture, Properties, and Uses," p. 118. Van Nostrand-Reinhold, Princeton, New Jersey, 1916.
13. F. Leh, *J. Chem. Educ.* **50**, 404 (1973).
14. R. W. Murray, *Tech. Methods Org. Organomet. Chem.* 1 (1969).
15. B. Weinstock (Symposium Chairman), *Adv. Chem. Ser.* **113**, 1–186 (1972).
16. M. S. Peters, *Chem. Eng. Prog., Symp. Ser.* **67**, No. 115, 1 (1971).
17. C. F. Schönbein, *C. R. Hebd. Seances Acad. Sci.* **10**, 706 (1840).
18. J. W. Mellor, "A Comprehensive Treatise on Inorganic and Theoretical Chemistry," Vol. 1, pp. 877–878. Longmans, Green, New York, 1922.
19. "Gmelin's Handbuch der Anorganischen Chemie, Sauerstoff, System Number 3," 8th ed., Vol. 4, pp. 984–987. Verlag Chemie, Weinheim, 1960.
20. C. F. Schönbein, *J. Prakt. Chem.* [1], **34**, 492 (1845); **52**, 135 (1851); **105**, 198 (1868); others in between.

21. C. F. Schönbein, *J. Prakt. Chem.* [1] **105**, 232 (1868).
22. C. F. Schönbein, *J. Prakt. Chem.* [1], **66**, 282 (1855).
23. C. D. Harries, *Justus Liebigs Ann. Chem.* **343**, 311 (1905).
24. E. Fonrobert, "Das Ozon." Enke, Stuttgart, 1916.
25. C. Rath, *Z. Angew. Chem.* **35**, 717 (1922).
26. L. Long, Jr., *Chem. Rev.* **27**, 437 (1940).
27. R. W. Murray, *Acc. Chem. Res.* **1**, 313 (1968).
27a. R. E. Oesper, *J. Chem. Educ.* **6**, 432 and 677 (1929).
28. E. von Gorup-Besanez, *Justus Liebigs Ann. Chem.* **110**, 86 (1859); **125**, 207 (1863).
29. A. Houzeau, *C. R. Hebd. Seances Acad. Sci.* **75**, 142 (1872); A. Houzeau and A. Renard, *ibid.* **76**, 572 (1873); *Justus Liebigs Ann. Chem.* **170**, 123 (1873); *Bull. Soc. Chim. Fr.* [2] **19**, 408 (1873).
30. M. Berthelot, *C. R. Hebd. Seances Acad. Sci.* **92**, 895 (1881).
31. M. Mailfert, *C. R. Hebd. Seances Acad. Sci.* **94**, 1186 (1882).
32. L. Legler, *Ber. Dtsch. Chem. Ges.* **18**, 3351 (1885).
33. A. Besson, *C. R. Hebd. Seances Acad. Sci.* **118**, 1347 (1894); **121**, 125 (1895).
34. M. Otto, *Ann. Chim. Phys.* [7] **13**, 106 (1898).
35. F. Swarts, *Bull. Acad. R. Belg.* [3] **36**, 532 (1898).
36. A. Trillat, *C. R. Hebd. Seances Acad. Sci.* **133**, 822 (1901).
37. R. A. Worstell, *J. Chem. Educ.* **9**, 291 (1932).
38. C. Engler, *Leopoldina* **15**, 124 and 141 (1879); **16**, 91 and 104 (1880).
39. A. Ladenberg, *Z. Angew. Chem.* **21**, 1057 (1908).
40. A. Vosmaer, "Ozone, Its Manufacture, Properties, and Uses." Van Nostrand-Reinhold, Princeton, New Jersey, 1916.
41. E. K. Rideal, "Ozone." Constable Press, London, 1920.
42. M. Moeller, "Das Ozon." Verlag Vieweg, Braunsweig, 1921.
43. E. Fonrobert, *in* "Die Methoden der organischen Chemie" (J. Houben, ed.), 3rd ed., Vol. III, pp. 406–457. Thieme, Leipzig, 1930.
44. M. L. Josien and G. Sourisseau, *in* "Nouveau Traité de Chemie Minérale" (P. Pascal, ed.), Vol. XIII, pp. 243–248. Masson, Paris, 1960.
45. C. D. Harries, *Ber. Dtsch. Chem. Ges.* **36**, 1933 (1903).
46. C. D. Harries, "Untersuchungen Über das Ozon und Seine Einwirkung auf Organische Verbindungen." Springer-Verlag, Berlin and New York, 1916.
46a. C. D. Harries, "Untersuchungen Über die Natürlichen and Künstlichen Kautschukarten." Springer-Verlag Berlin and New York, 1919.
47. C. Harries, *Z. Anorg. Allg. Chem.* **99**, 195 (1917).
48. W. Nagel, *Z. Angew. Chem.* **37**, 105 (1924).
48a. R. Wilstätter, *Ber. Dtsch. Chem. Ges. A* **59**, 123 (1926).
49. E. Molinari and E. Soncini, *Ber. Dtsch. Chem. Ges.* **39**, 2735 (1906); E. Molinari and C. Barosi, *ibid.* **41**, 2794 (1908); other papers in between.
50. R. G. Kadesch, *Prog. Chem. Fats Other Lipids* **6**, 291 1963; cf. T. Wehl, *Ber. Dtsch. Chem. Ges.* **39**, 3347 (1906).
51. H. Staudinger, *Ber. Dtsch. Chem. Ges.* **58**, 1088 (1925).
52. A. Rieche and R. Meister, *Ber. Dtsch. Chem. Ges.* **65**, 1274 (1932).
53. A. Rieche, R. Meister, and H. Sauthoff, *Justus Liebigs Ann. Chem.* **553**, 187 (1942).
54. A. Rieche, "Alkyl Peroxyde und Ozonide." Steinkopff, Darmstadt, 1931.
55. R. Pummerer, *Z. Angew. Chem.* **47**, 366 (1934); R. Pummerer and H. Richtzenhain, *Justus Liebigs Ann. Chem.* **529**, 33 (1937).
56. E. Briner, R. Patry, and E. De Luserna, *Helv. Chim. Acta* **7**, 62 (1924).
57. E. Briner, G. P. Rossetti, and S. Fliszár, *Helv. Chim. Acta* **48**, 1076 (1965).
58. R. Koetschau, *Z. Angew. Chem.* **35**, 509 (1922).

59. R. Koetschau, *Z. Angew. Chem.* **37**, 110 (1924).
59a. G. Maier, *Chem. Ber.* **110**, 27 (1977).
60. R. Criegee and G. Wenner, *Justus Liebigs Ann. Chem.* **564**, 9 (1949).
61. R. Criegee, *Justus Liebigs Ann. Chem.* **583**, 1 (1953).
62. P. S. Bailey, *Chem. Rev.* **58**, 925 (1958).
63. G. M. Schwab, *Z. Phys. Chem., Stoechiom. Verwandschaftsl.* **110**, 599 (1924).
64. D. M. Simpson, *Trans. Faraday Soc.* **41**, 209 (1945).
65. R. H. Hughes, *J. Chem. Phys.* **24**, 131 (1956).
66. R. Trambarulo, S. N. Ghosh, C. A. Burrus, Jr., and W. Gordy, *J. Chem. Phys.* **21**, 851 (1953).
67. M. Ardon, "Oxygen," Chapter 3. Benjamin, New York, 1965.
68. C. E. Thorp, "Bibliography of Ozone Technology," Vol. 2, Part One. Armour Res. Found., Chicago, Illinois, 1955.
69. A. G. Streng, *J. Chem. Eng. Data* **6**, 431 (1961).
70. R. Criegee, *Chem.-Z.* **99**, 138 (1975).
71. W. J. Orville-Thomas, *J. Mol. Spectros.* **3**, 588 (1959).
72. I. Fischer-Hjalmars, *Ark. Fys.* **11**, 529 (1957).
73. J. S. Wright, *Can. J. Chem.* **51**, 139 (1973).
74. S. Shih, R. J. Buenker, and S. D. Peyerimhoff, *Chem. Phys. Lett.* **28**, 463 (1974).
75. D. Grimbert and A. Devaquet, *Mol. Phys.* **27**, 831 (1974).
76. W. A. Goddard, III, T. H. Dunning, Jr., W. J. Hunt, and P. J. Hay, *Acc. Chem. Res.* **6**, 368 (1973).
77. W. R. Wadt and W. A. Goddard, III, *J. Am. Chem. Soc.* **96**, 1689 (1974); P. J. Hay, T. H. Dunning, Jr., and W. A. Goddard, III, *J. Chem. Phys.* **62**, 3912 (1975).
78. W. R. Wadt and W. A. Goddard, III, *J. Am. Chem. Soc.* **97**, 3004 (1975).
79. L. Salem and C. Rowland, *Angew. Chem., Int. Ed. Engl.* **11**, 92 (1972); *Angew. Chem.* **84**, 86 (1972).
80. R. Huisgen, *Angew. Chem., Int. Ed. Engl.* **2**, 565 (1963); *Angew. Chem.* **75** 604 (1963).
81. J. Meinwald, *Chem. Ber.* **88**, 1889 (1955).
82. C. Harries and P. Reichard, *Ber. Dtsch. Chem. Ges.* **37**, 612 (1904).
83. C. Harries, *Ber. Dtsch. Chem. Ges.* **37**, 839 (1904).
84. C. Harries and K. Haeffner, *Ber. Dtsch. Chem. Ges.* **41**, 3098 (1908).
85. C. Harries, *Justus Liebigs Ann. Chem.* **374**, 288 (1910).
86. C. Harries and F. Evers, *Justus Liebigs Ann. Chem.* **390**, 238 (1912).
87. C. Harries and H. Türk, *Ber. Dtsch. Chem. Ges.* **38**, 1630 (1905).
88. C. Harries, *Ber. Dtsch. Chem. Ges.* **42**, 446 (1909).
89. C. Harries and B. J. Palmén, *Ber. Dtsch. Chem. Ges.* **43**, 1432 (1910).
90. H. Staudinger, *Ber. Dtsch. Chem. Ges.* **58**, 1075 (1925).
91. R. Criegee, *Justus Liebigs Ann. Chem.* **560**, 127 (1948).
92. J. E. Leffler, *Chem. Rev.* **45**, 399 (1949).
93. N. A. Milas, *in* "The Chemistry of Petroleum Hydrocarbons" (B. T. Brooks *et al.*, eds.), Vol. II, Chapter 37. Van Nostrand-Reinhold, Princeton, New Jersey, 1955.
94. N. A. Milas, P. Davis, and J. T. Nolan, Jr., *J. Am. Chem. Soc.* **77**, 2536 (1955).
95. N. A. Milas and J. T. Nolan, Jr., *Adv. Chem. Ser.* **21**, 136 (1959).
96. P. R. Story, J. A. Alford, J. R. Burgess, and W. C. Ray, *J. Am. Chem. Soc.* **93**, 3042 (1971).
97. P. R. Story, J. A. Alford, W. C. Ray, and J. R. Burgess, *J. Am. Chem. Soc.* **93**, 3044 (1971).
98. P. R. Story, E. A. Whited, and J. A. Alford, *J. Am. Chem. Soc.* **94**, 2143 (1972).
99. P. S. Bailey, T. P. Carter, Jr., C. M. Fischer, and J. A. Thompson, *Can. J. Chem.* **51**, 1278 (1973).
100. K. R. Kopecky, P. A. Lockwood, J. E. Filby, and R. W. Reid, *Can. J. Chem.* **51**, 468 (1973).

101. D. R. Kerur and D. G. M. Diaper, *Can. J. Chem.* **51**, 3110 (1973).
102. L. A. Hull, I. C. Hisatsune, and J. Heicklen, *J. Am. Chem. Soc.* **94**, 4856 (1972).
103. R. Criegee, *Abst., 120th Nat. Meet. Am. Chem. Soc., New York* p. 22M (1951).
104. R. Criegee and G. Lohaus, *Chem. Ber.* **86**, 1 (1953).
105. R. Criegee, G. Blust, and H. Zincke, *Chem. Ber.* **87**, 766 (1954).
106. P. S. Bailey, *Chem. Ber.* **87**, 993 (1954).
107. G. Lohaus, *Chem. Ber.* **87**, 1708 (1954).
108. P. S. Bailey, *Chem. Ber.* **88**, 795 (1955).
109. R. Criegee, A. Kerckow, and H. Zinke, *Chem. Ber.* **88**, 1878 (1955).
110. R. Criegee, *Rec. Chem. Prog.* **18**, 111 (1957).
111. R. Criegee, *Adv. Chem. Ser.* **21**, 133 (1959).
112. R. Criegee, *in* "Peroxide Reaction Mechanisms" (J. O. Edwards, ed.), pp. 29–39. Wiley (Interscience), New York, 1962.
113. R. Criegee, *Chimia* **22**, 392 (1968).
114. A. T. Menyailo and M. V. Pospelov, *Usp. Khim.* **36**, 662 (1967); *Russ. Chem. Rev. (Engl. Transl.)* **36**, 284 (1967)..
115. J. S. Below, *in* "Oxidation" (R. L. Augustine, ed.), Vol. 1, pp. 259–335. Dekker, New York, 1969.
116. E. H. Pryde and J. C. Cowan, *Top. Lipid Chem.* **2**, 1–98 (1971).
117. P. S. Bailey and S. S. Bath, *J. Am. Chem. Soc.* **79**, 3120 (1957).
118. P. S. Bailey, *Chem. Ind. (London)* p. 1148 (1957).
119. P. S. Bailey, S. S. Bath, and J. B. Ashton, *Adv. Chem. Ser.* **21**, 143 (1959).
120. P. S. Bailey, S. B. Mainthia, and C. J. Abshire, *J. Am. Chem. Soc.* **82**, 6136 (1960).
121. R. E. Lutz and C. K. Dien, *J. Org. Chem.* **23**, 1861 (1958).
122. R. H. Perry, Jr., *J. Org. Chem.* **24**, 829 (1959).
123. E. Bernatek, P. Kolsaker, and T. Ledaal, *Tetrahedron Lett.* p. 1317 (1963).
124. E. Bernatek, P. Kolsaker, and T. Ledaal, *Skr. Nor. Vidensk. Akad. Oslo, 1* [N.S.] **11**, 1 (1963).
125. T. Ledaal, *Tetrahedron Lett.* p. 3861 (1966).
126. G. M. Badger, *Recl. Trav. Chim. Pays Bas* **71**, 468 (1952).
127. P. R. Story, R. W. Murray, and R. D. Youssefyeh, *J. Am. Chem. Soc.* **88**, 3144 (1966).
128. R. W. Murray, R. D. Youssefyeh, and P. R. Story, *J. Am. Chem. Soc.* **89**, 2429 (1967).
129. P. S. Bailey and A. G. Lane, *J. Am. Chem. Soc.* **89**, 4473 (1967).
130. H. G. Reinhardt, G. A. Doorakian, and H. H. Freedman, *J. Am. Chem. Soc.* **90**, 5934 (1968).
131. R. Criegee, *Chem. Unserer Zeit* **7**, 75 (1973).
132. R. Criegee, *Angew. Chem., Int. Ed. Engl.* **14**, 745 (1975); *Angew. Chem.* **87**, 765 (1975).
133. R. Criegee, *Angew. Chem.* **71**, 525 (1959).
134. R. Criegee and G. Schröder, *Chem. Ber.* **93**, 689 (1960).
135. F. L. Greenwood and S. Cohen, *J. Org. Chem.* **28**, 1159 (1963).
136. F. L. Greenwood, *J. Org. Chem.* **29**, 1321 (1964).
137. F. L. Greenwood, *J. Org. Chem.* **30**, 3108 (1965).
138. P. S. Bailey, J. A. Thompson, and B. A. Shoulders, *J. Am. Chem. Soc.* **88**, 4098 (1966).
139. P. S. Bailey, J. W. Ward, and R. E. Hornish, *J. Am. Chem. Soc.* **93**, 3552 (1971).
140. P. S. Bailey, J. W. Ward, R. E. Hornish, and F. E. Potts, III, *Adv. Chem. Ser.* **112**, 1 (1972).
141. P. S. Bailey, J. W. Ward, T. P. Carter, Jr., E. Nieh, C. M. Fisher, and A. Y. Khashab, *J. Am. Chem. Soc.* **96**, 6136 (1974).
142. L. J. Durham and F. L. Greenwood, *Chem. Commun.* p. 843 (1967).
143. L. J. Durham and F. L. Greenwood, *J. Org. Chem.* **33** 1629 (1968).
144. W. G. Alcock and B. Mile, *J. Chem. Soc., Chem. Commun.* p. 575 (1973).

145. F. L. Greenwood and L. J. Durham, *J. Org. Chem.* **34**, 3363 (1969).
146. F. L. Greenwood and B. J. Haske, *Tetrahedron Lett.* p. 631 (1965).
147. L. J. Durham and F. L. Greenwood, *Chem. Commun.* p. 24 (1968).
148. N. L. Bauld, J. A. Thompson, C. E. Hudson, and P. S. Bailey, *J. Am. Chem. Soc.* **90**, 1822 (1968).
149. H. Keul, *Chem. Ber.* **108**, 1198 (1975).
150. R. Huisgen, *Angew. Chem., Int. Ed. Engl.* **2**, 633 (1963); *Angew. Chem.* **75**, 742 (1963).
151. T. W. Nakagawa, L. J. Andrews, and R. M. Keefer, *J. Am. Chem. Soc.* **82**, 269 (1960).
152. T. Vrbaski and R. J. Cvetanović, *Can. J. Chem.* **38**, 1053 (1960).
153. Y. K. Wei and R. J. Cvetanovic, *Can. J. Chem.* **41**, 913 (1963).
154. J. J. Bufalini and A. P. Altshuller, *Can. J. Chem.* **43**, 2243 (1965).
155. R. D. Cadle and C. Schadt, *J. Am. Chem. Soc.* **24**, 6002 (1952).
156. D. G. Williamson and R. J. Cvetanović, *J. Am. Chem. Soc.* **90**, 3668 (1968).
157. D. G. Williamson and R. J. Cvetanović, *J. Am. Chem. Soc.* **90**, 4248 (1968).
158. W. Pritzkow and G. Schöppe, *J. Prakt. Chem.* **311**, 689 (1969).
159. A. J. Whitworth, R. Ayoub, Y. Rousseau, and S. Fliszár, *J. Am. Chem. Soc.* **91**, 7128 (1969).
160. J. Carles and S. Fliszár, *Adv. Chem. Ser.* **112**, 35 (1972).
161. G. Klutsch and S. Fliszár, *Can. J. Chem.* **50**, 2841 (1972).
162. H. Henry, M. Zador, and S. Fliszár, *Can. J. Chem.* **51**, 3398 (1973).
163. S. D. Razumovskii and G. E. Zaikov, *Zh. Org. Khim.* **8**, 464 (1972); *J. Org. Chem. USSR* (*Engl. Transl.*) **8**, 468 (1972).
164. E. R. Altwicker and J. Basila, *Tetrahedron* **29**, 1969 (1973).
165. U. Schmidt and P. Grafen, *Justus Liebigs Ann. Chem.* **656**, 97 (1962).
166. C. R. Noller, J. F. Carson, H. Martin, and K. S. Hawkins, *J. Am. Chem. Soc.* **58**, 24 (1936).
167. A. Roedig and E. Klappert, *Justus Liebigs Ann. Chem.* **605**, 126 (1957).
168. A. Roedig and K. Kiepert, *Justus Liebigs Ann. Chem.* **593**, 55 (1955).
169. A. E. Akopyan and V. S. Aslamazyan, *Izv. Akad. Nauk Arm. SSR, Khim. Nauki* **13**, 155 (1960); *Chem. Abstr.* **55**, 6359 (1961); *Zh. Obshch. Khim.* **32**, 2443 (1962); *J. Gen. Chem.* USSR (*Engl. Transl.*) **32**, 2411 (1962); A. E. Akopyan and G. A. Gabrielyan, *Izv. Akad. Nauk Arm. SSR, Khim. Nauki* **13**, 165 (1960); *Chem. Abstr.* **55**, 6359 (1961).
169a. A. N. Akopyan, A. M. Saakyan, and E. A. Dzhuvadyam, *Zh. Obshch. Khim.* **33**, 2965 (1963); *J. Gen. Chem. USSR* (*Engl. Transl.*) **33**, 2894 (1963).
170. K. Griesbaum and J. Brüggemann, *Adv. Chem. Ser.* **112**, 50 (1972).
171. W. R. Nes and D. L. Ford, *J. Am. Chem. Soc.* **83**, 4811 (1961).
172. H. Q. Smith and E. S. Wallis, *J. Org. Chem.* **19**, 1628 (1954).
173. A. I. Yakubchik and N. G. Kasatkina, *Zh. Obshch. Khim.*, **26**, 699 (1956); *J. Gen. Chem. USSR* (*Engl. Transl.*) **26**, 803 (1956).
174. S. D. Razumovskii, L. V. Berozova, and G. E. Zaikov, *Izv. Akad. Nauk SSSR, Ser. Khim.* p. 223 (1974); *Bull. Acad. Sci. USSR, Div. Chem. Sci.* p. 218 (1974).
175. G. Wagner and A. Greiner, *Z. Phys. Chem.* (*Leipzig*) **215**, 92 (1960).
176. W. G. Alcock and B. Mile, *J. Chem. Soc., Chem. Commun.* p. 5 (1976).
177. J. G. Traynham, G. R. Franzen, G. A. Knesel, and D. J. Northington, Jr., *J. Org. Chem.* **32**, 3285 (1967).
178. M. Tits and A. Bruylants, *Bull. Soc. Chim. Belg.* **57**, 50 (1948).
179. S. K. Rakovskii, S. D. Razumovskii, and G. E. Zaikov, *Izv. Akad. Nauk SSSR, Ser. Khim.* **8**, 1766 (1974); *Bull. Acad. Sci. USSR, Div. Chem. Sci.* **8**, 1686 (1974).
180. C. R. Engel, S. Rakhit, and W. W. Huculak, *Can. J. Chem.* **40**, 921 (1962).
181. C. R. Engel and W. W. Huculak, *Can. J. Chem.* **37**, 2031 (1959).
182. S. D. Razumovskii, *Izv. Akad. Nauk SSSR, Ser. Khim.* p. 335 (1970); *Bull. Acad. Sci. USSR, Div. Chem. Sci.* p. 284 (1974).

183. S. W. Benson, *Adv. Chem. Ser.* **77**, 74 (1968).
184. R. E. Erickson and R. L. Clark, *Tetrahedron Lett.* p. 3997 (1969).
185. R. A. Firestone, *J. Org. Chem.* **33**, 2285 (1968); *J. Chem. Soc. A* p. 1570 (1970); *J. Org. Chem.* **37**, 2181 (1972); **41**, 2212 (1976).
186. R. Huisgen, *J. Org. Chem.* **33**, 2291 (1968); **41**, 403 (1976).
187. G. Schröder, *Chem. Ber.* **95**, 733 (1962).
188. L. D. Loan, R. W. Murray, and P. R. Story, *J. Am. Chem. Soc.* **87**, 737 (1965).
189. R. Criegee, S. S. Bath, and B. von Bornhaupt, *Chem. Ber.* **93**, 2891 (1960).
190. F. M. Dean, D. R. Randell, and G. Winfield, *J. Chem. Soc.* p. 1071 (1959).
191. R. F. C. Brown, *Aust. J. Chem.* **17**, 154 (1964).
192. H. Keul, *Chem. Ber.* **108**, 1207 (1975).
193. P. Kolsaker, *Acta Chem. Scand.* **19**, 223 (1965).
194. R. Criegee and G. Lohaus, *Justus Liebigs Ann. Chem.* **583**, 6 (1953).
195. R. Criegee and R. Huber, *Chem. Ber.* **103**, 1855 and 1862 (1970).
196. G. Laber, *Justus Liebigs Ann. Chem.* **588**, 79 (1954).
197. R. Askani, *Chem. Ber.* **98**, 2322 (1965).
198. R. Criegee, R. Askani, and H. Grüner, *Chem. Ber.* **100**, 3916 (1967).
199. R. Criegee, W. Eberius, and H. A. Brune, *Chem. Ber.* **101**, 94 (1968).
200. R. Criegee and H. G. Reinhardt, *Chem. Ber.* **101**, 102 (1968).
201. R. Criegee, G. Bolz, and R. Askani, *Chem. Ber.* **102**, 275 (1969).
202. R. Criegee and R. Rucktäschel, *Chem. Ber.* **103**, 50 (1970).
203. R. Criegee, C. Schweickhardt, and H. Knoche, *Chem. Ber.* **103**, 960 (1970).
204. R. Criegee and K. Noll, *Justus Liebigs Ann. Chem.* **627**, 1 (1959).
205. R. Criegee and H. Furrer, *Chem. Ber.* **97**, 2949 (1964).
206. P. Courtot, *Ann. Chim. (Paris)* [13] **8**, 197 (1963).
207. R. Criegee, G. Schröder, G. Maier, and H. G. Fischer, *Chem. Ber.* **93**, 1553 (1960).
208. R. Criegee, *Angew. Chem., Int. Ed. Engl.* **1**, 519 (1962); *Angew. Chem.* **74**, 703 (1962).
209. G. R. Evanega, W. Bergmann, and J. English, Jr., *J. Org. Chem.* **27**, 13 (1962).
210. R. Criegee and R. Askani, *Angew. Chem., Int. Ed. Engl.* **7**, 537 (1968); *Angew. Chem.* **80**, 531 (1968).
211. R. Criegee and H. Korber, *Justus Liebigs Ann. Chem.* **756**, 95 (1972).
212. H. Muxfeldt and G. Hardtmann, *Justus Liebigs Ann. Chem.* **699**, 113 (1963).
213. K. Miescher and H. Kägi, *Helv. Chim. Acta* **32**, 761 (1949).
214. B. Riniker, D. Arigoni, and O. Jeger, *Helv. Chim. Acta* **37**, 546 (1954).
215. Y. Kobayashi, I. Kumadaki, A. Ohsawa, Y. Hanzawa, M. Honda, and Y. Iitaka, *Tetrahedron Lett.* p. 3001 (1975).
216. W. J. Wechter and G. Slomp, Jr., *J. Org. Chem.* **27**, 2549 (1962).
217. R. W. Murray, P. R. Story, and L. D. Loan, *J. Am. Chem. Soc.* **87**, 3025 (1965).
218. R. Criegee and H. Korber, *Chem. Ber.* **104**, 1812 (1971).
219. P. R. Story and J. R. Burgess, *J. Am. Chem. Soc.* **89**, 5726 (1967).
220. P. R. Story and J. R. Burgess, *J. Am. Chem. Soc.* **90**, 1094 (1968).
221. E. Briner and E. Dallwigk, *C. R. Hebd. Seances Acad. Sci.* **243**, 630 (1956).
222. E. Briner and E. Dallwigk, *Helv. Chim. Acta* **39**, 1446 (1956).
223. E. Dallwigk and E. Briner, *Helv. Chim. Acta* **39**, 1826 (1956).
224. E. Briner and S. Fliszár, *Helv. Chim. Acta* **43**, 1113 (1960).
225. S. M. Goodwin, N. M. Johnson, and B. Witkop, *J. Am. Chem. Soc.* **75**, 4273 (1953).
226. D. Garvin and C. Schubert, *J. Phys. Chem.* **60**, 807 (1956).
227. O. S. Privett and E. C. Nickell, *J. Lipid Res.* **4**, 208 (1963).
228. O. S. Privett and E. C. Nickell, *J. Am. Oil Chem. Soc.* **41**, 72 (1964).
229. G. Riezebos, J. C. Grimmelikhuysen, and D. A. van Dorp, *Recl. Trav. Chim. Pays Bas* **82**, 1234 (1963).
230. A. T. Menyailo, M. V. Pospelov, and A. I. Shekuteva, *Neftekhimiya* **4**, 894 (1964).

231. O. Lorenz and C. R. Parks, *J. Org. Chem.* 30, 1976 (1965).
232. J. Carles and S. Fliszár, *Can. J. Chem.* **48**, 1309 (1970).
233. E. Briner, C. Christol, H. Christol, and S. Fliszár, *Helv. Chim. Acta* **44**, 1088 (1961).
234. S. D. Razumovskii, L. V. Berezova, N. D. Gil'chenok, O.M. Oranskaya, Y. N. Yur'ev, and V. K. Tsyskovskii, *Zh. Org. Khim.* **2**, 1942 (1966); *J. Gen. Chem. USSR (Engl. Transl.)* **2**, 1910 (1966).
235. L. J. Bellamy, "The Infrared Spectra of Complex Molecules," 2nd ed., Chapter 7. Wiley and Sons, New York, 1958.
236. J. E. Richmond and K. I. Altman, *J. Am. Chem. Soc.* **74**, 4368 (1952).
237. R. W. Murray and G. J. Williams, *J. Org. Chem.* **34**, 1891 (1969).
238. S. Fliszár and J. Carles, *Can. J. Chem.* **47**, 3921 (1969).
239. R. W. Murray and A. Suzui, *J. Am. Chem. Soc.* **95**, 3343 (1973).
240. H. R. Williams and H. S. Mosher, *J. Am. Chem. Soc.* **76**, 2987 (1954).
241. C. S. Marvel and V. Nichols, *J. Org. Chem.* **6**, 296 (1941).
242. R. Criegee, W. Schnorrenberg, and J. Becke, *Justus Liebigs Ann. Chem.* **565**, 7 (1949).
243. R. Criegee, *Fortschr. Chem. Forsch.* **1**, 508 (1950).
244. E. G. E. Hawkins, "Organic Peroxides." Van Nostrand-Reinhold, Princeton, New Jersey, 1961.
245. P. Groth, *Acta Chem. Scand.* **18**, 1301 and 1801 (1964).
246. P. Groth, *Acta Chem. Scand.* **19**, 1497 (1965).
247. R. W. Murray, P. R. Story, and M. L. Kaplan, *J. Am. Chem. Soc.* **88**, 526 (1966).
248. S. Fliszár, D. Gravel, and E. Cavalieri, *Can. J. Chem.* **44**, 1013 (1966).
249. H. M. White, H. O. Colomb, Jr., and P. S. Bailey, *J. Org. Chem.* **30**, 481 (1965).
250. N. A. Milas and I. Belič, *Vestn. Slov. Kem. Drus.* **10**, 37 (1963).
251. R. W. Murray, J. W. P. Lin, and D. A. Grumke, *Adv. Chem. Ser.* **112**, 9 (1972).
252. R. Pummerer, G. Ebermayer, and K. Gerlach, *Ber. Dtsch. Chem. Ges.* **64**, 804 and 809 (1931).
253. C. S. Marvel and V. E. Nichols, *J. Am. Chem. Soc.* **60**, 1455 (1938).
254. S. Fliszár, D. Gravel, and E. Cavalieri, *Can. J. Chem.* **44**, 67 (1966).
255. C. Harris, *Justus Liebigs Ann Chem.* **410**, 1 (1915).
256. A. Rieche, *Z. Angew. Chem.* **44**, 590 (1931).
257. A. Rieche, *Angew. Chem.* **45**, 441 (1932).
258. M. Berenstein, H. Paillard, and E. Briner, *Helv. Chim. Acta* **29**, 271 (1946).
259. P. S. Bailey and S. B. Mainthia, *J. Org. Chem.* **23**, 1089 (1958).
260. R. C. Slagel, *J. Org. Chem.* **31**, 593 (1966).
261. S. D. Razumovskii, G. A. Niazashvili, I. A. Tutorskii, and Y. N. Yur'ev, *Vysokomol. Soedin., Ser. A* **13**, 190 (1971); *Polym. Sci. USSR (Engl. Transl.)* **13**, 217 (1971).
262. J. E. Franz, W. S. Knowles, and C. Osuch, *J. Org. Chem.* **30**, 4328 (1965).
263. F. L. Greenwood and H. Rubinstein, *J. Org. Chem.* **32**, 3369 (1967).
264. N. Sonoda, Y. Fujiwara, and S. Tsutsumi, *Technol. Rep. Osaka Univ.* **17**, 215 (1967).
265. S. Murai, N. Sonoda, and S. Tsutsumi, *Bull. Chem. Soc. Jpn.* **37**, 1187 (1964).
266. L. Velluz, G. Muller, J. Mathieu, and A. Poittevin, *Tetrahedron* **9**, 145 (1960).
267. T. J. de Boer and J. C. van Velzen, *Recl. Trav. Chim. Pays Bas* **79**, 430 (1960).
268. R. Criegee and G. Lohaus, *Justus Liebigs Ann. Chem.* **583**, 12 (1953).
269. Q. E. Thompson, *J. Org. Chem.* **27**, 4498 (1962).
270. H. Plieninger and K. Suhr, *Chem. Ber.* **89**. 270 (1956).
271. W. P. Keaveney, M. G. Berger, and J. J. Pappas, *J. Org. Chem.* **32**, 1537 (1967).
272. E. S. Zelikman, Y. N. Yur'ev, L. V. Berezova, and V. K. Tsyskovskii, *Zh. Org. Khim.* **7**, 633 (1971); *J. Org. Chem. USSR (Engl. Transl.)* **7**, 641 (1971).
273. M. V. Pospelov, A. T. Menyailo, T. A. Bortyan, Y. A. Ustynyuk, and V. S. Petrosyan, *Zh. Org. Khim.* **9**, 311 (1973); *J. Org. Chem. USSR (Engl. Transl.)* **9**, 312 (1973).
274. R. Criegee, H. Pilz, and H. Flygare, *Ber. Dtsch. Chem. Ges.* **72**, 1799 (1939).

275. G. O. Schenck, H. D. Becker, K. H. Schulte-Elte, and C. H. Krauch, *Chem. Ber.* **96**, 509 (1963).
276. D. B. Sharp and T. M. Patrick, Jr., *J. Org. Chem.* **26**, 1389 (1961).
277. N. A. Milas, R. L. Peeler, Jr., and O. L. Mageli, *J. Am. Chem. Soc.* **76**, 2322 (1954).
278. A. Rieche and C. Bischoff, *Chem. Ber.* **94**, 2722 (1961).
279. A. Rieche and E. Schmitz, *Chem. Ber.* **90**, 1082 (1957).
280. R. Criegee and P. Becher, *Chem. Ber.* **90**, 2516 (1957).
281. J. L. Warnell and R. L. Shriner, *J. Am. Chem. Soc.* **79**, 3165 (1957).
282. R. H. Callighan, M. F. Tarker, Jr., and M. H. Wilt, *J. Org. Chem.* **26**, 1379 (1961).
283. H. Lettré and A. Jahn, *Justus Liebigs Ann. Chem.* **608**, 43 (1957).
284. H. Lettré and A. Jahn, *Angew. Chem.* **69**, 266 (1957).
285. H. Lettré and D. Hotz, *Justus Liebigs Ann. Chem.* **620**, 63 (1959).
286. H. Lettré and H. Schelling, *Justus Liebigs Ann. Chem.* **669**, 160 (1963).
287. L. Knof, *Justus Liebigs Ann. Chem.* **647**, 53 (1961).
288. G. Blust, Inaugural Dissertation, University of Karlsruhe, Germany, 1954.
289. R. W. Murray, R. D. Youssefyeh, and P. R. Story, *J. Am. Chem. Soc.* **88**, 3143 (1966).
290. S. Fliszár, J. Carles, and J. Renard, *J. Am. Chem. Soc.* **90**, 1364 (1968).
291. S. Fliszár and J. Carles, *J. Am. Chem. Soc.* **91**, 2637 (1969).
292. P. R. Story, C. E. Bishop, J. R. Burgess, R. W. Murray, and R. D. Youssefyeh, *J. Am. Chem. Soc.* **90**, 1907 (1968).
293. R. W. Murray and G. J. Williams, *Adv. Chem. Ser.* **77**, 32 (1968).
294. R. W. Murray and R. Hagen, *J. Org. Chem.* **36**, 1103 (1971).
295. C. W. Gillies and R. L. Kuczkowski, *J. Am. Chem. Soc.* **94**, 7609 (1972); C. W. Gillies, R. P. Lattimer, and R. L. Kuczkowskii, *ibid.* **96**, 1536 (1974).
296. R. P. Lattimer and R. L. Kuczkowski, *J. Am. Chem. Soc.* **96**, 6205 (1974).
297. K. L. Gallaher and R. L. Kuczkowski, *J. Org. Chem.* **41**, 892 (1976).
298. D. P. Higley and R. W. Murray, *J. Am. Chem. Soc.* **98**, 4526 (1976).
299. K. Griesbaum, *Chem. Ber.* **101**, 463 (1968).
300. M. Schulz, *Z. Chem.* **3**, 276 (1963).
301. M. Schulz and D. Becker, *Monatsber. Dtsch. Akad. Wiss. Berlin* **6**, 351 (1964).
302. A. Rieche and M. Schulz, *Chem. Ber.* **98**, 3623 (1965).
303. A. Rieche, M. Schulz, and D. Becker, *Chem. Ber.* **98**, 3627 (1965).
304. R. M. Carman and D. E. Cowley, *Tetrahedron Lett.* p. 2723 (1968).
305. R. W. Murray and A. Suzai, *J. Am. Chem. Soc.* **93**, 4963 (1971).
306 R. W. Murray and D. P. Higley, *J. Am. Chem. Soc.* **95**, 7886 (1973).
307. D. P. Higley and R. W. Murray, *J. Am. Chem. Soc.* **96**, 3330 (1974).
308. P. D. Bartlett and T. G. Traylor, *J. Am. Chem. Soc.* **84**, 3408 (1962).
309. F. L. Greenwood, *J. Am. Chem. Soc.* **88**, 3146 (1966).
310. R. W. Murray, *Trans. N.Y. Acad. Sci.* [2] **29**, 854 (1967).
311. O. S. Privett and E. C. Nickell, *J. Am. Oil. Chem. Soc.* **43**, 393 (1966).
312. M. Indo, *Nippon Kagaku Zasshi* **80**, 537 (1959); *Chem. Abstr.* **55**, 4404 (1961).
313. T. Vrbaski and R. J. Cvetanović, *Can. J. Chem.* **38**, 1063 (1960).
314. O. Lorenz and C. R. Parks, *Rubber Chem. Technol.* **36**, 201 (1963).
315. W. B. DeMore and C. L. Lin, *J. Org. Chem.* **38**, 985 (1973).
316. E. Briner, E. Dallwigk, and M. Ricca, *Helv. Chim. Acta* **41**, 1390 (1958).
317. E. Briner and M. Ricca, *Helv. Chim. Acta* **41**, 2178 (1958).
318. E. Briner and S. Fliszár, *Helv. Chim. Acta* **42**, 1310, 2063 (1959).
319. E. Briner, C. Christol, H. Christol, and S. Fliszár, *C. R. Hebd. Seances Acad. Sci.* **255**, 1170 (1962).
320. P. S. Bailey, A. Rustaiyan, and T. M. Ferrell, *J. Am. Chem. Soc.* **98**, 638 (1976).
321. R. W. Murray and R. Hagen, *J. Org. Chem.* **36**, 1098 (1971).

322. R. W. Murray, R. D. Youssefyeh, G. J. Williams, and P. R. Story, *Tetrahedron* **24**, 4347 (1968).
323. G. Schröder, Ph.D. Dissertation, University of Karlsruhe, Germany, 1959.
324. R. P. Lattimer, R. L. Kuczkowski, and C. W. Gillies, *J. Am. Chem. Soc.* **96**, 348 (1974).
325. H. Rubinstein, *J. Org. Chem.* **32**, 3236 (1967).
326. R. Criegee and H. Korber, *Chem. Ber.* **104**, 1807 (1971).
327. R. W. Murray, R. D. Youssefyeh, and P. R. Story, *J. Am. Chem. Soc.* **88**, 3655 (1966).
328. J. Renard and S. Fliszár, *J. Am. Chem. Soc.* **92**, 2628 (1970).
329. R. A. Rouse, *J. Am. Chem. Soc.* **95**, 3460 (1973).
330. R. A. Rouse, *J. Am. Chem. Soc.* **96**, 5095 (1974).
331. R. J. Blint and M. D. Newton, *J. Chem. Phys.* **59**, 6220 (1973).
332. A. Almenningen, P. Kolsaker, H. M. Seip, and T. Willadsen, *Acta Chem. Scand.* **23**, 3398 (1969).
333. C. W. Gillies and R. L. Kuczkowski, *J. Am. Chem. Soc.* **94**, 6337 (1972).
334. R. P. Lattimer, C. W. Gillies, and R. L. Kuczkowski, *J. Am. Chem. Soc.* **95**, 1348 (1973).
335. P. Groth, *Acta. Chem. Scand.* **24**, 2137 (1970).
336. R. A. Rouse, *Int. J. Quantum Chem., Symp.* **7**, 289 (1973).
337. S. Fliszár and J. Renard, *Can. J. Chem.* **48**, 3002 (1970).
338. S. Fliszár, J. Renard, and D. Z. Simon, *J. Am. Chem. Soc.* **93**, 6953 (1971).
339. N. C. Yang and R. V. Carr, *Tetrahedron Lett. p.* 5143 (1972).
339a. N. C. Yang and J. Libman, *J. Org. Chem.* **39**, 1782 (1974).
340. R. Criegee and A. Banciu, *Chem. Ztg.* **98**, 261 (1974).
341. C. E. Bishop and P. R. Story, *J. Am. Chem. Soc.* **90**, 1905 (1968).
342. J. Carles and S. Fliszár, *Can. J. Chem.* **47**, 1113 (1969).
343. J. Castonguay, M. Bertrand, J. Carles, S. Fliszár, and Y. Rousseau, *Can. J. Chem.* **47**, 919 (1969).
344. C. E. Bishop, D. D. Denson, and P. R. Story, *Tetrahedron Lett.* p. 5739 (1968).
345. G. Klutsch, J. Grignon, J. Renard, and S. Fliszár, *Can. J. Chem.* **48**, 1598 (1970).
346. G. Klopman and C. M. Joiner, *J. Am. Chem. Soc.* **97**, 5287 (1975).
347. R. Criegee, *Chem. Ber.* **108**, 743 (1975).
348. R. Criegee, A. Banciu, and H. Keul, *Chem. Ber.* **108**, 1642 (1975).
349. R. Criegee and A. Rustaiyan, *Chem. Ber.* **108**, 749 (1975).
350. H. H. Stechl, *Chem. Ber.* **97**, 2681 (1964).
351. S. Fliszár and J. B. Chylińska, *Can. J. Chem.* **45**, 29 (1967).
352. S. Fliszár and J. B. Chylińska, *Can. J. Chem.* **46**, 783 (1968).
353. S. Fliszár, *Can. J. Chem.* **44**, 2351 (1966).
354. E. Briner and S. de Nemitz, *Helv. Chim. Acta* **21**, 748 (1938).
355. A. L. Henne and W. L. Perilstein, *J. Am. Chem. Soc.* **65**, 2183 (1943).
356. F. G. Fischer, H. Düll, and L. Ertel, *Ber. Dtsch. Chem. Ges.* **65**, 1467 (1932).
357. R. W. Murray and G. J. Williams, *J. Org. Chem.* **34**, 1896 (1969).
358. S. D. Razumovskii, Y. N. Yur'ev, and V. K. Tsyskovskii, *Zh. Org. Khim.* **6**, 254 (1970); *J. Org. Chem. USSR (Engl. Transl.)* **6**, 246 (1970).
359. R. Criegee, G. Blust, and G. Lohaus, *Justus Liebigs Ann. Chem.* **583**, 2 (1953).
360. H. Kwart and D. M. Hoffman, *J. Org. Chem.* **31**, 419 (1966).
361. R. Criegee and P. Günther, *Chem. Ber.* **96**, 1564 (1963).
362. D. R. Kerur and D. G. M. Diaper, *Can J. Chem.* **47**, 43 (1969).
363. P. S. Bailey, T. M. Ferrell, A. Rustaiyan, S. Seyhan, and L. E. Unruh, *J. Am. Chem. Soc.*, **100**, 894 (1978).
364. K. Bühler, *Justus Liebigs Ann. Chem.* **721**, 225 (1969).
365. P. S. Bailey and T. P. Carter, Jr., unpublished data (1973).
366. P. S. Bailey and Y. G. Chang, *J. Org. Chem.* **27**, 1192 (1962).
367. S. Fliszár, C. Belżecki, and J. B. Chylińska, *Can J. Chem.* **45**, 221 (1967).

368. E. Bernatek, T. Ledaal, and T. Solbjør, *Acta Chem. Scand.* **19**, 1198 (1965).
369. E. Bernatek, P. Grønning, and T. Ledaal, *Acta Chem. Scand.* **18**, 1966 (1964).
370. M. I. Fremery and E. K. Fields, *J. Org. Chem.* **28**, 2537 (1963).
371. P. S. Bailey, *J. Am. Chem. Soc.* **78**, 3811 (1956).
372. P. S. Bailey and F. J. Garcia-Sharp, *J. Org. Chem.* **22**, 1008 (1957).
373. P. S. Bailey, S. S. Bath, F. Dobinson, F. J. Garcia-Sharp, and C. D. Johnson, *J. Org. Chem.* **29**, 697 (1964).
374. P. S. Bailey, *J. Org. Chem.* **22**, 1548 (1957).
375. P. S. Bailey, *Ind. Eng. Chem.* **50**, 993 (1958).
376. A. G. Davies, "Organic Peroxides," Chapter 6. Butterworth, London, 1961.
377. E. G. E. Hawkins, "Organic Peroxides," Chapters 5, 8, and 9. Van Nostrand-Reinhold, Princeton, New Jersey, 1961.
378. M. I. Fremery and E. K. Fields, *J. Org. Chem.* **29**, 2240 (1964).
379. E. K. Fields, *Adv. Chem. Ser.* **51**, 99 (1965).
380. J. C. Robertson and W. J. Verzino, Jr., *J. Org. Chem.* **35**, 545 (1970).
381. N. A. Milas, U. S. Patent 3,009,962 (1961).
382. M. Schulz, D. Becker, and A. Rieche, *Angew. Chem., Int. Ed. Engl.* **4**, 525 (1965); *Angew. Chem.* **77**, 548 (1965); M. Schulz, A. Rieche, and D. Becker, *Chem. Ber.* **99**, 3233 (1966).
383. R. Criegee, P. de Bruyn, and G. Lohaus, *Justus Liebigs Ann. Chem.* **583**, 19 (1953).
384. E. Berner and P. Kolsaker, *Acta Chem. Scand.* **23**, 597 (1969).
385. W. P. Keaveney and J. J. Pappas, *Tetrahedron Lett.* p. 841 (1969).
386. T. Ledaal, *Tetrahedron Lett.* p. 2425 (1966).
387. S. Fliszár, *Tetrahedron Lett.* p. 6083 (1966).
388. S. Fliszár and J. Renard, *Can. J. Chem.* **45**, 533 (1967).
389. S. Fliszár and M. Granger, *J. Am. Chem. Soc.* **91**, 3330 (1969).
390. S. D. Razumovskii and Y. N. Yur'ev, *Tetrahedron Lett.* p. 3939 (1967).
391. S. D. Razumovskii and Y. N. Yur'ev, *Zh. Org. Khim.* **4**, 1716 (1968); *J. Org. Chem. USSR (Engl. Transl.)* **4**, 1651 (1968).
392. S. Fliszár and M. Granger, *J. Am. Chem. Soc.* **92**, 3361 (1970).
393. R. H. Callighan and M. H. Wilt, *J. Org. Chem.* **26**, 4912 (1961).
394. J. Renard and S. Fliszár, *Can. J. Chem.* **47**, 3333 (1969).
395. P. Kolsaker and P. S. Bailey, *Acta Chem. Scand.* **21**, 537 (1967).
396. N. R. Raulins and L. A. Sibert, *J. Org. Chem.* **26**, 1382 (1961).
397. P. R. Story and J. R. Burgess, *Tetrahedron Lett.* p. 1287 (1968).
398. C. D. Johnson and P. S. Bailey, *J. Org. Chem.* **29**, 703 (1964).
399. A. G. Davies, "Organic Peroxides." Butterworth, London, 1961.
400. A. G. Davies, "Organic Peroxides," Chapter 9. Butterworth, London, 1961.
401. J. O. Edwards, in "Peroxide Reaction Mechanisms" (J. O. Edwards, ed.), p. 67. Wiley (Interscience), New York, 1962.
402. R. Curci and J. O. Edwards, *in* "Organic Peroxides" (D. Swern, ed.), Vol. I, Chapter 4, p. 218. Wiley (Interscience), New York, 1970.
402a. G. A. Razuvaev, V. A. Shushunov, V. A. Dodonov, and T. G. Brilkina, *in* "Organic Peroxides" (D. Swern, ed.), Vol. III, Chapter 3, p. 141. Wiley (Interscience), New York, 1972.
403. H. Boardman and G. E. Hulse, *J. Am. Chem. Soc.* **75**, 4272 (1953).
404. D. H. Fortnum, C. J. Battaglia, S. R. Cohen, and J. O. Edwards, *J. Am. Chem. Soc.* **82**, 778 (1960).
405. T. H. Fisher and J. C. Martin, *J. Am. Chem. Soc.* **88**, 3382 (1966).
406. J. Carles and S. Fliszár, *Can. J. Chem.* **50**, 2552 (1972).
407. P. R. Story, C. E. Bishop, J. R. Burgess, J. B. Olson, R. W. Murray, and R. D. Youssefyeh, *Adv. Chem. Ser.* **77**, 46 (1968).

408. A. V. Tobolsky and R. B. Mesrobian, "Organic Peroxides." Wiley (Interscience), New York, 1954.
409. R. D. Mair and A. J. Graupner, *Anal. Chem.* **36**, 194 (1964).
410. J. J. Pappas, W. P. Keaveney, E. Gancher, and M. Berger, *Tetrahedron Lett.* p. 4273 (1966).
411. P. L. Stotter and J. B. Eppner, *Tetrahedron Lett.* p. 2417 (1973).
412. O. Lorenz, *Anal. Chem.* **37**, 101 (1965).
413. J. J. Pappas, W. P. Keaveney, M. Berger, and R. V. Rush, *J. Org. Chem.* **33**, 787 (1968).
414. W. S. Knowles and Q. E. Thompson, *J. Org. Chem.* **25**, 1031 (1960).
415. F. L. Greenwood, *J. Org. Chem.* **20**, 803 (1955).
416. J. O. Sousa and A. L. Bluhm, *J. Org. Chem.* **25**, 108 (1960).
417. F. L. Greenwood and B. J. Haske, *J. Org. Chem.* **30**, 1276 (1965).
418. S. O. Lawesson and N. C. Yang, *J. Am. Chem. Soc.* **81**, 4230 (1959).
419. G. Slomp, Jr. and J. L. Johnson, *J. Am. Chem. Soc.* **80**, 915 (1958).
420. G. Slomp, Jr. and J. L. Johnson, *Adv. Chem. Ser.* **21**, 162 (1959).
421. K. Griesbaum, *Chem. Commun.* p. 920 (1966); *Brennst-Chem.* **50**, 212 (1969).
422. E. H. Pryde, D. E. Anders, H. M. Teeter, and J. C. Cowan, *J. Org. Chem.* **27**, 3055 (1962).
423. D. G. M. Diaper and W. M. J. Strachan, *Can. J. Chem.* **45**, 33 (1967).
424. P. Kolsaker, E. Bernatek, R. Johanson, and R. Hytta, *Acta Chem. Scand.* **27**, 1526 (1973).
425. H. Kwart, P. S. Starcher, and S. W. Tinsley, *Chem. Commun.* p. 335 (1967).
426. E. H. Pryde and J. C. Cowan, J. *Am. Oil Chem. Soc.* **39**, 496 (1962).
427. S. D. Razumovskii and Y. N. Yur'ev, *Zh. Org. Khim.* **3**, 251 (1967); *J. Org. Chem. USSR* (*Engl. Transl.*) **3**, 238 (1967).
428. E. Bernatek and T. Ledaal, *Tetrahedron Lett.* p. 30 (1960).
429. E. Bernatek, H. Hagen, and T. Ledaal, *Acta Chem. Scand.* **21**, 1555 (1967).
430. Y. N. Yur'ev, S. D. Razumovskii, L. V. Berezova, and E. S. Zelikman, *Zh. Org. Khim.* **11** 8 (1975); *J. Org. Chem. USSR* (*Engl. Transl.*) **11**, 5 (1975).
431. L. A. Subluskey, G. C. Harris, A. Maggiolo, and A. L. Tumolo, *Adv. Chem. Ser.* **21**, 149 (1959).
432. C. Nebel, *Chem. Commum.* p. 101 (1968).
433. E. H. Pryde, D. J. Moore, and J. C. Cowan, *J. Am. Oil Chem. Soc.* **45**, 888 (1968).
434. M. G. Sturrock, E. L. Cline, and K. R. Robinson, *J. Org. Chem.* **28**, 2340 (1963).
435. M. G. Sturrock, B. J. Cravy, and V. A. Wing, *Can. J. Chem.* **49**, 3047 (1971).
436. C. Harris, *Ber. Dtsch. Chem. Ges.* **39**, 3728 (1906).
437. R. Criegee and H. Korber, *Adv. Chem. Ser.* **112**, 22 (1972).
438. N. Kornblum and H. E. De La Mare, *J. Am. Chem. Soc.* **73**, 880 (1951).
439. R. M. Ellam and J. M. Padbury, *Chem. Commun.* p. 1094 (1971).
440. R. M. Ellam and J. M. Padbury, *Chem. Commun.* p. 1086 (1972).
441. P. S. Bailey and R. E. Erickson, *Org. Synth., Collect. Vol.* **5**, 493 (1973).
442. D. G. M. Diaper and D. L. Mitchell, *Can. J. Chem.* **43**, 319 (1965).
443. M. S. Kharasch, A. Fono, W. Nudenberg, and B. Bischof, *J. Org. Chem.* **17**, 207 (1952).
444. W. A. Bonner and C. J. Collins, *J. Am. Chem. Soc.* **75**, 3831 (1953).
445. Y. N. Yur'ev, S. D. Razumovskii, and V. K. Tsyskovskii, *Zh. Org. Khim.* **4**, 1720 (1968); *J. Org. Chem. USSR* (*Engl. Transl.*) **4**, 1656 (1968).
446. S. D. Razumovskii and Y. N. Yur'ev, *Zh. Org. Khim.* **4**, 1884 (1968); *J. Org. Chem. USSR* (*Engl. Transl.*) **4**, 1818 (1968).
447. S. D. Razumovskii, G. I. Bantsyrev, M. I. Cherkashin, G. Y. Zaikov, and A. A. Berlin, *Vysokomol Soedin., Ser. A* **13**, 206 (1971); *Polym. Sci. USSR* (*Engl. Transl.*) **13**, 236 (1971).
448. E. Bernatek and M. Hvatum, *Acta Chem. Scand.* **14**, 836 (1960).
449. E. C. Nickell and O. S. Privett, *Lipids* **1**, 166 (1966).

450. P. R. Story, W. H. Morrison, III, T. K. Hall, J. C. Farine, and C. E. Bishop, *Tetrahedron Lett.* p. 3291 (1968).
451. P. R. Story, T. K. Hall, W. H. Morrison, III, and J. C. Farine, *Tetrahedron Lett.* p. 5397 (1968).
452. P. R. Story, W. H. Morrison, III, and J. M. Butler, *J. Am. Chem. Soc.* **91**, 2398 (1969).
453. P. R. Story, D. D. Denson, C. E. Bishop, B. C. Clark, Jr., and J. C. Farine, *J. Am. Chem. Soc.* **90**, 817 (1968).
454. P. R. Story, B. Lee, C. E. Bishop, D. D. Denson, and P. Busch, *J. Org. Chem.* **35**, 3059 (1970). B. Lee, P. R. Story, and J. R. Sanderson, *ibid.* **41**, 2314 (1976).
455. L. A. Hull, I. C. Hisatsune, and J. Heicklen, *J. Phys. Chem.* **76**, 2659 (1972).
456. J. I. G. Cadogan, D. H. Hey, and W. A. Sanderson, *J. Chem. Soc.* p. 4897 (1960).
457. D. G. M. Diaper, *Can. J. Chem.* **46**, 3095 (1968).
458. M. Schulz and K. Kirschke, *Adv. Heterocycl. Chem.* **8**, 165 (1967).
459. M. Bertrand. S. Fliszár, and Y. Rousseau, *J. Org. Chem.* **33**, 1931 (1968).
460. J. Carles, Y. Rousseau, and S. Fliszár, *Can. J. Chem.* **48**, 2345 (1970).
461. J. Grignon and S. Fliszár, *Can. J. Chem.* **49**, 3127 (1971).
462. M. Bertrand, J. Carles, S. Fliszár, and Y. Rousseau, *Org. Mass Spectrom.* **9**, 297 (1974).
463. M. Schulz and K. Kirschke, *in* "Organic Peroxides" (D. Swern, ed.), Vol. III, Chapter 2, pp. 67–140. Wiley (Interscience), New York, 1972.
464. R. Criegee, *in* "Oxidation in Organic Chemistry" (K. B. Wiberg, ed.), Part A, Chapter 5, p. 302. Academic Press, New York, 1965.
465. C. Harries and R. Haarmann, *Ber, Dtsch. Chem. Ges.* **46**, 1737 (1913).
466. C. Harries and R. Haarmann, *Ber, Dtsch. Chem. Ges.* **48**, 32 (1915).
467. H. Depoorter, *Meded. Vlaam. Chem. Ver.* **14**, 105 (1952).
468. L. Ruzicka, P. A. Plattner, and A. Fürst, *Helv. Chim. Acta* **25**, 1364 (1942).
469. W. S. Johnson, B. Bannister, and R. Pappo, *J. Am. Chem. Soc.* **78**, 6331 (1956).
470. D. C. Aldridge, J. F. Grove, P. McCloskey, and W. Klyne, *J. Chem. Soc.* p. 2569 (1963).
471. E. J. Eisenbraun and S. M. McElvain, *J. Am. Chem. Soc.* **77**, 3383 (1955).
472. P. Vlad and M. Souček, *Collect. Czech. Chem. Commun.* **27**, 1726 (1962).
473. J. P. Wibaut and H. C. Beyerman, *Recl. Trav. Chim. Pays Bas* **70**, 977 (1951).
474. W. G. Young, A. C. McKinnis, I. D. Webb, and J. D. Roberts, *J. Am. Chem. Soc.* **68**, 293 (1946).
475. D. H. R. Barton, P. deMayo, and J. C. Orr, *J. Chem. Soc.* 2239 (1958).
476. T. L. Jacobs, and R. B. Brownfield, *J. Am. Chem. Soc.* **82**, 4033 (1960).
477. A. S. Dreiding and J. A. Hartman, *J. Am. Chem. Soc.* **75**, 939 (1953).
478. C. H. Hassal, *Org. React.* **9**, 73 (1957).
479. K. Aoki, M. Kato, M. Suzuki, Y. Hayakawa, H. Nakamura, K. Yamada, and Y. Hirata, *Chem. Commun.* p. 618 (1973).
480. G. W. Krakower, H. A. Van Dine, P. A. Diassi, and I. Bacso, *J. Org. Chem.* **32**, 184 (1967).
481. W. Oroshnik, G. Karmas, and A. D. Mebane, *J. Am. Chem. Soc.* **74**, 295 (1952).
482. W. Oroshnik, A. D. Mebane, and G. Karmas, *J. Am. Chem. Soc.* **75**, 1050 (1953).
483. J. Knights and E. S. Waight, *J. Chem. Soc.* p. 2830 (1955).
484. S. J. Rhoads, R. Raulins, and R. D. Reynolds, *J. Am. Chem. Soc.* **76**, 3456 (1954).
485. A. T. Nielsen, *J. Am. Chem. Soc.* **79**, 2518 (1957).
486. T. P. C. Mulholland, *J. Chem. Soc.* p. 2693 (1958).
487. Y. Hirata, H. Nakata, K. Yamada, K. Okuhara, and T. Naito, *Tetrahedron* **14**, 252 (1961).
488. P. deMayo, E. Y. Spencer, and R. W. White, *J. Am. Chem. Soc.* **84**, 494 (1962).
489. E. Mosettig, U. Beglinger, F. Dolder, H. Lichti, P. Quitt, and J. A. Waters, *J. Am. Chem. Soc.* **85**, 2305 (1963).
490. E. Demole, *Experientia* **20**, 609 (1964).
491. J. R. Hosking, *Ber. Dtsch. Chem. Ges.* **69**, 780 (1936).

492. G. R. Proctor, *J. Chem. Soc.* p. 3989 (1961).
493. D. Caunt, W. D. Crow, and R. D. Haworth, *J. Chem. Soc.* p. 1313 (1951).
494. E. Vogel, O. Roos, and K. H. Disch, *Justus Liebigs Ann. Chem.* **653**, 55 (1962).
495. D. H. R. Barton and P. deMayo, *J. Chem. Soc.* p. 142 (1956).
496. W. Herz, W. A. Rohde, K. Rabindran, P. Jayaraman, and N. Viswanathan, *J. Am. Chem. Soc.* **84**, 3857 (1962).
497. W. G. Dauben, H. G. Wight, and G. A. Boswell, *J. Org. Chem.* **23**, 1787 (1958).
498. G. Lefebvre, P. Germain, and R. Gay, *Bull. Soc. Chim. Fr.* p. 173 (1974).
499. J. J. Bonet, H. Wehrli, and K. Schaffner, *Helv. Chim. Acta* **46**, 1776 (1963).
500. D. Brookes, B. K. Tidd, and W. B. Turner, *J. Chem. Soc.* p. 5385 (1963).
501. D. H. R. Barton and G. S. Gupta, *J. Chem. Soc.* p. 1961 (1962).
502. E. Caspi, W. Schmid, and B. T. Khan, *Tetrahedron* **18**, 767 (1962).
503. E. Caspi, B. T. Khan, and S. N. Balasubrahmanyam, *Tetrahedron* **18**, 1013 (1962).
504. S. N. Balasubrahmanyam, E. Caspi, and B. T. Khan, *J. Chem. Soc.* p. 761 (1963).
505. O. R. Rodig and G. Zanati, *J. Org. Chem.* **33**, 914 (1968).
506. D. Yang and S. W. Pelletier, *Chem. Commun.* p. 1055 (1968).
507. H. J. Storesund and E. Bernatek, *Acta Chem. Scand.* **24**, 3237 (1970).
508. E. Bernatek and K. A. Straumsgard, *Acta Chem. Scand.* **13**, 178 (1959).
509. E. Bernatek and K. Valen, *Acta Chem. Scand.* **14**, 224 (1960).
510. E. Bernatek and A. N. Soteland, *Acta Chem. Scand.* **16**, 2054 (1962).
511. E. Bernatek, *Tetrahedron* **4**, 213 (1958).
512. E. Bernatek, "Ozonolyses in the Naphthoquinone and Benzofuran Series." Oslo Univ. Press, Oslo, Norway, 1960.
513. E. Bernatek, F. Christenssen, and T. Ledaal, *Acta Chem. Scand.* **21**, 822 (1967).
514. D. H. R. Barton and E. Seoane, *J. Chem. Soc.* p. 4150 (1956).
515. E. Späth and M. Pailer, *Ber. Dtsch. Chem. Ges.* **73**, 238 (1940).
516. E. Späth, M. Pailer, and G. Gergely, *Ber. Dtsch. Chem. Ges.* **73**, 795 (1940).
517. A. von Wacek and H. O. Eppinger, *Ber. Dtsch. Chem. Ges.* **73**, 664 (1940).
518. F. W. Semmler, *Ber. Dtsch. Chem. Ges.* **42**, 246 (1909).
519. M. de Botton, *C. R. Hebd. Seances Acad. Sci.* **260**, 4783 (1965).
520. K. Ebisu, L. B. Batty, J. M. Higaki, and H. O. Larson, *J. Am. Chem. Soc.* **88**, 1995 (1966).
521. J. R. Hanson, *J. Chem. Soc.* p. 5061 (1963).
522. R. T. Aplin, R. P. K. Chan, and T. G. Halsall, *J. Chem. Soc. C* p. 2322 (1969).
523. E. P. Blanchard, Jr., *Chem. Ind.* (*London*) p. 293 (1958).
524. F. Holloway, H. J. Anderson, and W. Rodin, *Ind. Eng. Chem.* **47**, 2111 (1955).
525. G. S. Fisher and J. S. Stinson, *Ind. Eng. Chem.* **47**, 1569 (1955).
526. G. Brus and G. Peyresblangues, *C. R. Hebd. Seances Acad. Sci.* **187**, 984 (1928).
527. K. H. Overton and P. Owen, *J. Chem. Soc., Perkin Trans. 1* p. 226 (1973).
528. W. G. Young and S. L. Linden, *J. Am. Chem. Soc.* **69**, 2042 (1947).
529. N. Müller and W. Hoffmann, *Synthesis* p. 781 (1975).
530. H. H. Inhoffen, F. Bohlmann, and G. Linhoff, *Justus Liebigs Ann. Chem.* **570**, 73 (1950).
531. J. Wolinsky and D. Chan, *J. Am. Chem. Soc.* **85**, 937 (1963).
532. A. Gagneux and C. A. Grob, *Helv. Chim. Acta* **42**, 2006 (1959).
533. F. Asinger, *Ber. Dtsch. Chem. Ges.* **75**, 656 (1942).
534. K. Ziegler, W. Hechelhammer, H. D. Wagner, and H. Wilms, *Justus Liebigs Ann. Chem.* **567**, 99 (1950).
535. C. S. Marvel, W. M. Schilling, D. J. Shields, C. Bluestein, O. R. Irwin, P. G. Sheth, and J. Honig, *J. Org. Chem.* **16**, 838 (1951).
536. A. I. Iakubchik, A. I. Spasskova, and G. I. Tikhomirov, *Zh. Obshch. Khim.* **28**, 916 (1958); *J. Gen. Chem. USSR* (*Engl. Transl.*) **28**, 890 (1958).
537. A. Greiner and G. Wagner, *Z. Phys. Chem.* (*Leipzig*) **215**, 109 (1960).

538. G. Büchi, R. E. Erickson, and N. Wakabayashi, *J. Am. Chem. Soc.* **83**, 927 (1961).
539. G. R. Clemo and J. M. Macdonald, *J. Chem. Soc.* p. 1294 (1935).
540. E. Briner and D. Frank, *Helv. Chim. Acta* **21**, 1297 (1938).
541. E. Briner and R. Meier, *Helv. Chim. Acta* **12**, 529 (1929).
542. E. Briner, *Helv. Chim. Acta* **22**, 591 (1939).
543. C. Bischoff and A. Rieche, *Z. Chem.* **5**, 97 (1905).
544. A. Greiner and U. Müller, *J. Prakt. Chem.* [4] **15**, 313 (1962).
545. A. Greiner, *J. Prakt. Chem.* [4] **27**, 69 (1965).
546. M. Stoll and A. Rouvé, *Helv. Chim. Acta* **27**, 950 (1944).
547. G. A. Howard and A. R. Tatchell, *J. Chem. Soc.* p. 2400 (1954).
548. M. Verzele and H. Depoorter, *Bull. Soc. Chim. Belg.* **68**, 85 (1959).
549. N. J. Leonard and J. A. Deyrup, *J. Am. Chem. Soc.* **84**, 2148 (1962).
550. M. S. Newman and G. Kaugars, *J. Org. Chem.* **30**, 3105 (1965).
551. F. Feist, *Ber. Dtsch. Chem. Ges.* **26**, 747 (1893).
552. F. Feist, *Justus Liebigs Ann. Chem.* **436**, 125 (1924).
553. F. R. Goss, C. K. Ingold, and J. F. Thorpe, *J. Chem. Soc.* **123**, 327 (1923).
554. M. G. Ettlinger, *J. Am. Chem. Soc.* **74**, 5805 (1952).
555. M. G. Ettlinger and F. Kennedy, *Chem. Ind.* (*London*) p. 166 (1956).
556. A. S. Kende, *Chem. Ind.* (*London*) pp. 437 and 544 (1956).
557. A. T. Bottini and J. D. Roberts, *J. Org. Chem.* **21**, 1169 (1956).
558. R. E. Erickson, *Tetrahedron Lett.* p. 1753 (1966).
559. R. E. Erickson and G. D. Mercer, *Adv. Chem. Ser.* **112**, 114 (1972).
560. J. Pasero and M. Naudet, *Rev. Fr. Corps Gras* **7**, 189 (1960).
561. J. Pasero, L. Comeau, and M. Naudet, *Bull. Soc. Chim. Fr.* p. 1794 (1963).
562. J. Pasero and M. Naudet, *Rev. Fr. Corps Gras* **10**, 453 (1963).
563. J. Pasero, L. Comeau, and M. Naudet, *Bull. Soc. Chim. Fr.* 493 (1965).
564. D. G. M. Diaper, *Can. J. Chem.* **44**, 2819 (1966).
565. D. G. M. Diaper, J. Pasero, and M. Naudet, *Can. J. Chem.* **46**, 2767 (1968).
566. H. E. Eschinazi, *J. Org. Chem.* **26**, 3072 (1961).
567. A. Verley, *Bull. Soc. Chim. Fr.* [4] **43**, 854 (1928).
568. V. Grignard and J. Doevre, *Bull. Soc. Chim. Fr.* [4] **45**, 809 (1929).
569. F. L. Benton, A. A. Kiess, and H. J. Harwood, *J. Am. Oil Chem. Soc.* **36**, 457 (1959).
570. O. S. Privett and C. Nickell, *J. Am. Oil Chem. Soc.* **39**, 414 (1962).
571. E. F. Ullman and W. A. Henderson, Jr., *J. Am. Chem. Soc.* **89**, 4390 (1967).
572. F. Straus, *Justus Liebigs Ann. Chem.* **393**, 235 (1912).
573. H. R. Henze and G. R. Allen, Jr., *J. Am. Chem. Soc.* **77**, 461 (1955).
574. A. N. Akopyan and V. S. Aslamazyan, *Zh. Obshch. Khim.* **33**, 1160 (1963); *J. Gen. Chem. USSR* (*Engl. Transl.*) **33**, 1141 (1963).
575. G. Cignarella, C. R. Pasqualucci, G. G. Gallo, and E. Testa, *Tetrahedron* **20**, 1057 (1964).
576. K. Griesbaum and H. Keul, *Angew. Chem., Int. Ed. Engl.* **14**, 716 (1975; *Angew. Chem.* **87**, 748 (1975).
577. K. Griesbaum and J. Brüggemann, *Chem. Ber.* **105**, 3638 (1972).
578. K. Griesbaum and P. Hofmann, *J. Am. Chem. Soc.* **98**, 2877 (1976).
579. F. Gozzo and G. Camaggi, *Chim. Ind.* (*Milan*) **50**, 197 (1968).
580. C. W. Gillies, *J. Am. Chem. Soc.* **97**, 1276 (1975).
581. F. Ramirez, R. B. Mitra, and N. B. Desai, *J. Am. Chem. Soc.* **82**, 5763 (1960).
582. F. Ramirez, N. B. Desai, and R. B. Mitra, *J. Am. Chem. Soc.* **83**, 492 (1961).
583. F. Ramirez, S. B. Bhatia, R. B. Mitra, Z. Hamlet, and N. B. Desai, *J. Am. Chem. Soc.* **86**, 4394 (1964).
584. K. Hunger, V. Hasserodt, and F. Korte, *Tetrahedron* **20**, 1593 (1964).

585. M. D. Bon, *Uch. Zap., Leningr Gos. Univ., Ser. Khim. Nauk* **3**, 3 (1938); *Chem. Abstr.* **33**, 3755 (1939); *Chem. Zentralbl.* **110**, II, 366 (1939).
586. H. D. Hartzler, *J. Am. Chem. Soc.* **83**, 4990 (1961).
587. W. J. Ball and S. R. Landor, *Proc. Chem. Soc., London* p. 143 (1961); *J. Chem. Soc.* p. 2298 (1962).
588. B. Smith, R. Ohlson, and A. M. Olson, *Acta Chem. Scand.* **16**, 1463 (1962).
589. W. R. Moore and H. R. Ward, *J. Org. Chem.* **27**, 4179 (1962).
590. W. R. Moore and R. C. Bertelson, *J. Org. Chem.* **27**, 4182 (1962).
591. L. Skattebøl, *Acta Chem. Scand.* **17**, 1683 (1963).
592. K. J. Crowley, *J. Am. Chem. Soc.* **85**, 1210 (1963).
593. D. R. Taylor, *Chem. Rev.* **67**, 317 (1967).
594. P. Kolsaker and B. Teige, *Acta Chem. Scand.* **24**, 2101 (1970).
595. J. K. Crandall and W. W. Conover, *J. Chem. Soc., Chem. Commun.* p. 340 (1973).
596. G. Maier, *Angew. Chem., Int. Ed. Engl.* **6**, 402 (1967); *Angew. Chem.* **79**, 446 (1967).
597. W. v. E. Doering, G. Laber, R. Vonderwahl, N. F. Chamberlain, and R. B. Williams, *J. Am. Chem. Soc.* **78**, 5448 (1956).
598. F. L. J. Sixma and E. Detilleux, *Recl. Trav. Chim. Pays Bas* **72**, 173 (1953).
599. E. J. Corey, H. J. Burke, and W. A. Remers, *J. Am. Chem. Soc.* **77**, 4941 (1955); E. J. Corey and H. J. Burke, *ibid.* **78**, 174 (1956).
600. K. Alder, K. Kaiser, and M. Schumacher, *Justus Liebigs Ann. Chem.* **602**, 80 (1957).
601. E. J. Corey, H. J. Burke, and W. A. Remers, *J. Am. Chem. Soc.* **78**, 180 (1956).
602. J. P. Wibaut and F. L. J. Sixma, *Recl. Trav. Chim. Pays Bas* **73**, 797 (1954).
603. S. L. Friess and V. Boekelheide, *J. Am. Chem. Soc.* **71**, 4145 (1949).
604. N. A. Milas, J. T. Nolan, Jr., and P. P. H. L. Otto, *J. Org. Chem.* **23**, 624 (1958).
605. M. Christl and G. Brüntrup, *Chem. Ber.* **107**, 3908 (1974).
606. G. Komppa and R. H. Roschier, *Chem. Zentralbl.* **89**, I, 623 (1918).
607. G. Komppa and R. H. Roschier, *Justus Liebigs Ann. Chem.* **470**, 129 (1929).
608. H. J. Backer, *Chem. Weekbl.* **36**, 214 (1939).
609. P. D. Bartlett and M. Stiles, *J. Am. Chem. Soc.* **77**, 2806 (1955); P. D. Bartlett and T. T. Tidwell, *ibid.* **90**, 4421 (1968).
610. S. Munavalli and G. Ourisson, *Bull. Soc. Chim. Fr.* p. 729 (1964).
611. C. A. Henrick and P. R. Jefferies, *Aust. J. Chem.* **17**, 915 (1964).
612. A. R. Hochstetler, *J. Org. Chem.* **40**, 1536 (1975).
613. C. R. Enzell and B. R. Thomas, *Tetrahedron Lett.* p. 225 (1965).
614. C. R. Enzell and B. R. Thomas, *Acta Chem. Scand.* **19**, 1875 (1965).
615. C. R. Enzell, *Acta Chem. Scand.* **15**, 1191 (1961); **16**, 1553 (1962).
616. R. Takasaki, *Chem. Pharm. Bull.* **10**, 445 (1962).
617. K. Tanabe, R. Takasaki, and R. Hayashi, *Chem. Pharm. Bull.* **9**, 7 (1961).
618. K. Tanabe, R. Hayashi, and R. Takasaki, *Chem. Pharm. Bull.* **9**, 1 (1961).
619. K. Alder and H. Söll, *Justus Liebigs Ann. Chem.* **565**, 57 (1949).
620. D. H. R. Barton and C. J. W. Brooks, *J. Chem. Soc.* p. 263 (1951).
621. G. P. Cotterell, T. G. Halsall, and M. J. Wriglesworth, *Chem. Commun.* p. 1121 (1967).
622. J. McLean, G. A. Silverstone, and F. S. Spring, *J. Chem. Soc.* p. 935 (1951).
623. I. A. Kaye, M. Fieser, and L. F. Fieser, *J. Am. Chem. Soc.* **77**, 5936 (1955).
624. G. M. Stanton, *Diss Abstr.* **23**, 2327 (1963).
625. R. E. Lutz, W. J. Welstead, Jr., R. G. Bass, and J. I. Dale, *J. Org. Chem.* **27**, 1111 (1962).
626. T. S. Work, *J. Chem. Soc.* p. 334 (1944).
627. S. H. Graham and A. J. S. Williams, *J. Chem. Soc.* p. 4066 (1959).
628. R. C. Fuson, M. D. Armstrong, W. E. Wallace, and J. W. Kneisley, *J. Am. Chem. Soc.* **66**, 1274 (1944).

629. R. C. Fuson, P. L. Southwick, and S. P. Rowland, *J. Am. Chem. Soc.* **66**, 1109 (1944).
630. P. S. Bailey, J. W. Ward, F. E. Potts, III, Y. G. Chang, and R. E. Hornish, *J. Am. Chem. Soc.* **96**, 7228 (1974).
631. R. C. Fuson, N. Rabjohn, and D. J. Byers, *J. Am. Chem. Soc.* **66**, 1272 (1944).
632. G. Baddeley, B. G. Heaton, and J. W. Rasburn, *J. Chem. Soc.* p. 4713 (1960).
633. P. Naffa and G. Ourisson, *Bull. Soc. Chim. Fr.* p. 1115 (1954).
634. G. G. Allan, J. M. Beaton, J. I. Shaw, F. S. Spring, R. Stevenson, J. L. Stewart, and W. S. Strachan, *Chem. Ind.* (*London*) p. 281 (1955).
635. M. Stoll, B. Willhalm, and G. Büchi, *Helv. Chim. Acta* **38**, 1573 (1955).
636. R. A. Eade, H. Page, A. Robertson, K. Turner, and W. B. Whalley, *J. Chem. Soc.* p. 4913 (1957).
637. R. C. Fuson, D. H. Chadwick, and M. L. Ward, *J. Am. Chem. Soc.* **68**, 389 (1946).
638. E. H. Farmer and R. C. Pitkethly, *J. Chem. Soc.* p. 287 (1938).
639. R. O. Uhler, H. Schechter, and G. V. D. Tiers, *J. Am. Chem. Soc.* **84**, 3397 (1962).
640. M. S. Ahamd and G. Baddeley, *J. Chem. Soc.* p. 4303 (1961).
641. F. M. Dean, T. Francis, and K. Manunapichu, *J. Chem. Soc.* p. 4551 (1958).
642. F. M. Dean, D. S. Deorha, J. C. Knight, and T. Francis, *J. Chem. Soc.* p. 327 (1961).
643. F. M. Dean and J. C. Knight, *J. Chem. Soc.* p. 4745 (1962).
644. L. H. Zalkow, M. V. Kulkarni, and N. N. Girotra, *J. Org. Chem.* **30**, 1679 (1965).
645. S. McLean and U. O. Trotz, *Can. J. Chem.* **49**, 863 (1971).
646. K. Alder and R. Rühmann, *Justus Liebigs Ann. Chem.* **566**, 1 (1950).
647. W. F. Johns, *J. Org. Chem.* **29**, 2545 (1964).
648. A. Bhati, *J. Org. Chem.* **27**, 4485 (1962).
649. P. S. Bailey and A. M. Reader, *Chem. Ind.* (*London*) p. 1063 (1961).
650. P. S. Bailey unpublished data, (1974).
651. R. W. Hoffmann and J. Schneider, *Chem. Ber.* **100**, 3698 (1967).
652. I. J. Borowitz and R. D. Rapp, *J. Org. Chem.* **34**, 1370 (1969).
653. P. G. Gassman and X. Creary, *Tetrahedron Lett.* pp. 4407 and 4411 (1972).
654. R. D. Clark and C. H. Heathcock, *J. Org. Chem.* **41**, 1396 (1976).
655. T. L. Jacobs and R. S. Bauer, *J. Am. Chem. Soc.* **81**, 606 (1959).
656. R. Wheland and P. D. Bartlett, *J. Am. Chem. Soc.* **92**, 6057 (1970).
657. J. K. Crandall and L. C. Crawley, *J. Org. Chem.* **39**, 489 (1974).
658. E. G. Janzen, I. G. Lopp, and J. W. Happ, *J. Chem. Soc. D* p. 1140 (1970).
659. P. S. Bailey, F. E. Potts, III, and J. W. Ward, *J. Am. Chem. Soc.* **92**, 230 (1970).
660. W. A. Mosher, *Adv. Chem. Ser.* **21**, 140 (1959).
661. R. Criegee and R. Askani, *Angew. Chem., Int. Ed. Engl.* **5**, 519 (1966); *Angew. Chem.* **78**, 494 (1966).
662. J. Drugman, *J. Chem. Soc.* **89**, 939 (1906).
663. E. W. Blair and T. S. Wheeler, *J. Soc. Chem. Ind., London, Trans. Commun.* **42**, 343T and 347T (1923).
664. E. Briner and P. Schnorf, *Helv. Chim. Acta* **12**, 154 (1929).
665. E. Briner and P. Schnorf, *Helv. Chim. Acta* **12**, 181 (1929).
666. A. J. Haagan-Smit, *Ind. Eng. Chem.* **44**, 1342 (1952); A. J. Haagan-Smit, E. F. Darley, M. Zaitlin, H. Hull, and W. Noble, *Plant Physiol.* **27**, 18 (1952).
667. H. Niki, E. E. Daby, and B. Weinstock, *Adv. Chem. Ser.* **113**, 16 (1972).
668. P. L. Hanst, E. R. Stephens, and W. E. Scott, *Proc., Am. Pet. Inst., Sect. 3* **35**, 175 (1955); *J. Air Pollut. Control Assoc.* **5**, 219 (1956).
669. W. E. Scott, E. R. Stephens, P. L. Hanst, and R. C. Doerr, *Proc., Am. Pet. Inst., Sect. 3* **37**, 171 (1957).
670. P. L. Hanst, E. R. Stephens, W. E. Scott, and R. C. Doerr, *Prepr., Div. Pet. Chem., Am. Chem. Soc.* **4**, A-7 (1959).

671. E. F. Darley, E. R. Stephens, J. T. Middleton, and P. L. Hanst, *Int. J. Air Pollut.* **1**, 155 (1959).
672. J. T. Middleton, *Annu. Rev. Plant Physiol.* **12**, 431 (1961).
673. P. A. Leighton, "Photochemistry of Air Pollution," Chapter 6. Academic Press, New York, 1961.
673a. A. P. Altshuller and J. J. Bufalini, *Photochem. Photobiol.* **4**, 97 (1965); *Environ. Sci. Technol.* **5**, 39 (1971).
674. B. E. Saltzman and N. Gilbert, *Ind. Eng. Chem.* **51**, 1415 (1959).
675. J. Heicklen, *J. Phys. Chem.* **70**, 477 (1966).
676. W. B. DeMore, *Int. J. Chem. Kinet.* **1**, 209 (1969).
677. L. A. Hull, I. C. Hisatsune, and J. Heicklen, *Can. J. Chem.* **51**, 1504 (1973).
678. D. H. Stedman, C. H. Wu, and H. Niki, *J. Phys. Chem.* **77**, 2511 (1973).
679. K. H. Becker, U. Schurath, and H. Seitz, *Int. J. Chem. Kinet.* **6**, 725 (1974).
680. J. T. Herron and R. E. Huie, *J. Phys. Chem.* **78**, 2085 (1974).
681. S. M. Japar, C. H. Wu, and H. Niki, *Environ. Lett.* **7**, 245 (1974).
682. S. M. Japar, C. H. Wu, and H. Niki, *J. Phys. Chem.* **78**, 2318 (1974).
683. F. S. Toby and S. Toby, *Int. J. Chem. Kinet.* **6**, 417 (1974).
684. F. S. Toby and S. Toby, *Int. J. Chem. Kinet., Symp.* No. 1, p. 197 (1975).
685. R. E. Huie and J. T. Herron, *Int. J. Chem. Kinet., Symp.* No. 1, p. 165 (1975).
686. E. P. Grimsrud, H. H. Westberg, and R. A. Rasmussen, *Int. J. Chem. Kinet., Symp.* No. 1, p. 183 (1975).
687. L. A. Ripperton, H. E. Jeffries, and O. White, *Adv. Chem. Ser.* **113**, 219 (1972).
688. R. A. Cox and S. A. Penkett, *J. Chem. Soc., Faraday Trans. 1* **68**, 1735 (1972).
689. E. A. Schuck, G. J. Doyle, and N. Endow, *Air Pollut. Found., Rep.* No. 29 (1959): No. 31 (1960).
690. B. E. Saltzman, *Ind. Eng. Chem.* **50**, 677 (1958).
691. B. J. Finlayson, J. N. Pitts, Jr., and H. Akimoto, *Chem. Phys. Lett.* **12**, 495 (1972).
692. B. J. Finlayson, J. N. Pitts, Jr., and R. Atkinson, *J. Am. Chem. Soc.* **96**, 5356 (1974).
693. R. H. Eastman and R. M. Silverstein, *J. Am. Chem. Soc.* **75**, 1493 (1953).
694. H. E. Smith and R. H. Eastman, *J. Am. Chem. Soc.* **83**, 4274 (1961).
695. J. N. Pitts, Jr., W. A. Kummer, R. P. Steer, and B. J. Finlayson, *Adv. Chem. Ser.* **113**, 246 (1972).
696. R. Atkinson, B. J. Finlayson, and J. N. Pitts, Jr., *J. Am. Chem. Soc.* **95**, 7592 (1973).
697. C. C. Spencer, W. I. Weaver, E. A. Oberright, H. J. Sykes, A. L. Barney, and A. L. Elder, *J. Org. Chem.* **5**, 610 (1940).
698. H. R. Schulten and U. Schurath, *J. Phys. Chem.* **79**, 51 (1975).
699. D. A. Hansen and J. N. Pitts, Jr., *Chem. Phys. Lett.* **35**, 569 (1975).
700. K. L. Cordes, *Chem. Ind.* (*London*) p. 340 (1966).
701. R. S. Sheinson, F. S. Toby, and S. Toby, *J. Am. Chem. Soc.* **97**, 6593 (1975).
702. E. Mathias, E. Sanhueza, I. C. Hisatsune, and J. Heicklen, *Can. J. Chem.* **52**, 3852 (1974).
703. H. E. O'Neal and C. Blumstein, *Int. J. Chem. Kinet.* **5**, 397 (1973).
704. U. Schurath, H. Güsten, and R. D. Penzhorn, *J. Photochem.* **5**, 33 (1976).
705. S. Jaffe and R. Loudon, *Adv. Chem. Ser.* **113**, 264 (1972).
706. S. Toby, *J. Lumin.* **8**, 94 (1973).
707. F. Gozzo, A. Bergomi, G. Camaggi, and V. Oprandi, *Eur. Polym. J.* **6**, 219 (1970).
708. B. W. Gay, Jr., P. L. Hanst, J. J. Bufalini, and R. C. Noonan, *Environ. Sci. Technol.* **10**, 58 (1976).
709. C. W. Blume, I. C. Hisatsune, and J. Heicklen, Report No. 385–75. Center for Air Environment Studies, Pennsylvania State University, University Park, 1975.
710. E. Sanhueza, I. C. Hisatsune, and J. Heicklen, Report No. 387–75, pp. 38–74. Center for Air Environment Studies, Pennsylvania State University, University Park, 1975.

711. J. P. Desvergne, J. Bouas-Laurent, E. V. Blackburn, and R. Lapouyade, *Tetrahedron Lett.* p. 947 (1974).

712. J. P. Desvergne and J. M. Thomas, *J. Chem. Soc.*, *Perkin Trans. 2* p. 594 (1975).

713. P. S. Bailey, *in* "Ozone in Water and Wastewater Treatment" (F. L. Evans, III, ed.), Chapter 3. Ann Arbor Sci. Publ., Ann Arbor, Michigan, 1972.

714. P. S. Bailey, *in* "Ozone for Water and Wastewater Treatment" (R. G. Rice and M. E. Browning, eds.), pp. 101–119. International Ozone Institute, Waterbury, Connecticut, 1975.

715. J. S. Murphy and J. R. Orr, eds., "Ozone Chemistry and Technology. A Review of the Literature: 1961–1974." Franklin Inst. Press, Philadelphia, Pennsylvania, 1975.

716. P. C. Hiberty, *J. Am. Chem. Soc.* **98**, 6088 (1976); Pap., *Joint Conf. Chem. Inst. Can. Am. Chem. Soc., 2nd, 1977* ORGN 089 (1977).

717. P. S. Bailey, Pap. *Joint Conf. Chem. Inst. Can. Am. Chem. Soc., 2nd, 1977* ORGN 087 (1977);

717a. P. S. Bailey and T. M. Ferrell, *J. Am. Chem. Soc.*, **100**, 699 (1978).

Index

D

P